Lecture Notes in Artificial Intelligence 9113

Subseries of Lecture Notes in Computer Science

LNAI Series Editors

Randy Goebel
University of Alberta, Edmonton, Canada
Yuzuru Tanaka
Hokkaido University, Sapporo, Japan
Wolfgang Wahlster
DFKI and Saarland University, Saarbrücken, Germany

LNAI Founding Series Editor

Joerg Siekmann
DFKI and Saarland University, Saarbrücken, Germany

More information about this series at http://www.springer.com/series/1244

Jaume Baixeries · Christian Sacarea
Manuel Ojeda-Aciego (Eds.)

Formal
Concept Analysis

13th International Conference, ICFCA 2015
Nerja, Spain, June 23–26, 2015
Proceedings

 Springer

Editors
Jaume Baixeries
Universitat Politècnica de Catalunya
Barcelona
Spain

Christian Sacarea
Babeș-Bolyai University
Cluj-Napoca
Romania

Manuel Ojeda-Aciego
Departamento de Matemática Aplicada
Universidad de Málaga
Malaga
Spain

ISSN 0302-9743   ISSN 1611-3349   (electronic)
Lecture Notes in Artificial Intelligence
ISBN 978-3-319-19544-5   ISBN 978-3-319-19545-2   (eBook)
DOI 10.1007/978-3-319-19545-2

Library of Congress Control Number: 2015939667

LNCS Sublibrary: SL7 – Artificial Intelligence

Springer Cham Heidelberg New York Dordrecht London

Printed on acid-free paper

Springer International Publishing AG Switzerland is part of Springer Science+Business Media
(www.springer.com)

Preface

Formal concept analysis (FCA) is a mathematical field rooted in lattice and order theory, which, although being of such a theoretical nature, has proved to be of interest to various applied fields such as knowledge discovery and data mining, database theory, data visualization, and many others.

The goal of the International Conference on Formal Concept Analysis (ICFCA) is to offer researchers from different backgrounds the possibility to present and discuss their research related to FCA. Since the first ICFCA conference in 2003 in Darmstadt, Germany, it has been held annually in different countries in Europe, Africa, America, and Australia.

The 13th ICFCA took place during June 23–26, 2015 in Nerja (Málaga), Spain, organized by members of the Universidad de Málaga. There were 38 submissions by authors from 16 different countries. Each paper was reviewed by three members of the Program Committee (sometimes four). Sixteen significant papers were chosen for publication in this volume, amounting to an acceptance rate of 42 %. Seven other works in progress were considered for presentation during the conference and included in a supplementary volume.

This volume presents a set of articles that cover a wide range of fields about or related to FCA. The papers included in this volume are divided into different sections: **Theory**, with papers that discuss various theoretical aspects on FCA, **Methods and Applications**, with papers that show the application of FCA to different fields, and **Enhanced FCA**, with papers that present new trends in FCA, for instance, pattern structures of fuzzy FCA.

As an addition to previous conferences, the new section **Graphs and FCA** includes a selection of papers that relate graph theory or graph applications to FCA, from various points of view. These papers were presented in the special session **Practical Graph Applications in FCA**.

Apart from the regular sections, the conference included a presentation, for the first time, of a survey paper on FCA (*Formal Concept Analysis and Information Retrieval: A Survey*), which shows the maturity of this discipline.

We were also delighted that four prestigious researchers accepted the invitation to give a talk, and we also included their corresponding papers:

- *Measuring the Implications of the D-basis in Analysis of Data in Biomedical Studies* by Prof. Dr. Kira Adaricheva, Nazarbayev University (Kazakhstan);
- *Quantitative Redundancy in Partial Implications* by Prof. Dr. José Luís Balcázar, Universitat Politècnica de Catalunya (Spain);
- *Close encounters with transitivity in various relational frameworks* by Prof. Dr. Bernard De Baets, Ghent University (Belgium);
- *Formal Concept Analysis from the Standpoint of Possibility Theory* by Prof. Dr. Henri Prade, IRIT (France).

Our deepest gratitude goes to all the authors of submitted papers. Choosing ICFCA 2015 as a forum to publish their research was key to the success of the conference. Besides the submitted papers, the high quality of this published volume would not have been possible without the strong commitment of the authors, the Program Committee and editorial board members, and the external reviewers. Working with the efficient and capable team of local organizers was a constant pleasure. We are deeply indebted to all of them for making this conference a successful forum on FCA.

Last, but not least, we are most grateful to Springer for showing, for the 13th consecutive year, their trust in the International Conference on Formal Concept Analysis, as well as to the organizations that sponsored this event: the Universidad de Málaga, Andalucía Tech (International Campus of Excellence), the Patronato de Turismo de la Costa del Sol and the Área de Turismo del Ayuntamiento de Nerja, all in Spain. Finally, we would like to emphasize the great help of EasyChair in making the technical duties easier.

June 2015

Jaume Baixeries
Christian Sacarea
Manuel Ojeda-Aciego

Organization

Executive Committee

Conference Chair

Manuel Ojeda-Aciego   Universidad de Málaga, Spain

Conference Organizing Committee

Manuel Ojeda-Aciego	Universidad de Málaga, Spain
Inma P. Cabrera	Universidad de Málaga, Spain
Pablo Cordero	Universidad de Málaga, Spain
Estrella Rodríguez-Lorenzo	Universidad de Málaga, Spain
José Manuel Rodríguez	Universidad de Málaga, Spain

Program and Conference Proceedings

Program Chairs

Jaume Baixeries	Universitat Politècnica de Catalunya, Catalunya, Spain
Christian Sacarea	Babeş-Bolyai University, Cluj-Napoca, Romania

Editorial Board

Peggy Cellier	IRISA, INSA Rennes, France
Felix Distel	Technische Universität Dresden, Germany
Florent Domenach	University of Nicosia, Cyprus
Peter Eklund	University of Wollongong, Australia
Sebastien Ferré	Université de Rennes 1, France
Bernhard Ganter	Technische Universität Dresden, Germany
Cynthia-Vera Glodeanu	Technische Universität Dresden, Germany
Robert Godin	Université du Québec à Montréal, Canada
Robert Jäschke	Leibniz Universität Hannover, Germany
Mehdi Kaytoue	Université de Lyon, France
Sergei O. Kuznetsov	Higher School of Economics, Russia
Leonard Kwuida	Bern University of Applied Sciences, Switzerland
Rokia Missaoui	Université du Québec en Outaouais, Canada
Sergei Obiedkov	Higher School of Economics, Russia
Uta Priss	Ostfalia University of Applied Sciences, Germany
Sebastian Rudolph	Technische Universität Dresden, Germany
Stefan E. Schmidt	Technische Universität Dresden, Germany
Petko Valtchev	Universit du Québec à Montréal, Canada
Karl Erich Wolff	University of Applied Sciences, Germany

Honorary Member

Rudolf Wille   Technische Universität Darmstadt, Germany

Program Committee

Simon Andrews   University of Sheffield, UK
Karell Bertet   L3I Université de La Rochelle, France
François Brucker   Centrale Marseille, France
Claudio Carpineto   Fondazione Ugo Bordoni, Italy
Pablo Cordero   Universidad de Málaga, Spain
Stephan Doerfel   University of Kassel, Germany
Vincent Duquenne   ECP6-CNRS, Université Paris 6, France
Alain Gély   Université Paul Verlaine, France
Marianne Huchard   LIRMM, Université Montpellier, France
Dmitry Ignatov   Higher School of Economics, Russia
Michal Krupka   Palacký University, Czech Republic
Marzena Kryszkiewicz   Warsaw University of Technology, Poland
Wilfried Lex   Universität Clausthal, Germany
Jesús Medina   Universidad de Cádiz, Spain
Engelbert Mephu Nguifo   LIMOS, Université de Clermont Ferrand 2, France
Amedeo Napoli   LORIA, Nancy, France
Lhouari Nourine   Université Blaise Pascal, France
Jan Outrata   Palacký University, Czech Republic
Jean-Marc Petit   LIRIS, INSA de Lyon, France
Sandor Radeleczki   University of Miskolc, Hungary
Laszlo Szathmary   University of Debrecen, Hungary
Andreja Tepavčević   University of Novi Sad, Serbia

External Reviewers

María José Benítez   Universidad de Cádiz, Spain
Aleksey Buzmakov   LORIA, Nancy, France
Maria Eugenia Cornejo   Universidad de Cádiz, Spain
Madalina Croitoru   LIRMM, Université Montpellier, France
Ingrid Falk   Université de Strasbourg, France
Philippe Fournier-Viger   University of Moncton, Canada
Mathieu Lafourcade   LIRMM, Université Montpellier, France
Valentín Liñeiro-Barea   Universidad de Cádiz, Spain
Mohamed Nader Jelassi   ISIMA, France
Julien Rabatel   LIRMM, Université Montpellier, France
Eloisa Ramírez Poussa   Universidad de Cádiz, Spain
Ali Tayari   University of Nicosia, Cyprus
Norbert Tsopze   University of Yaounde, Cameroun
Jens Zumbraegel   Technische Universität Dresden, Germany

Sponsoring Institutions

Universidad de Málaga, Spain
Andalucía Tech (International Campus of Excellence), Spain
Patronato de Turismo de la Costa del Sol, Spain
Área de Turismo del Ayuntamiento de Nerja, Spain

Contents

Invited Talks

Quantitative Redundancy in Partial Implications 3
José L. Balcázar

Formal Concept Analysis from the Standpoint of Possibility Theory 21
Didier Dubois and Henri Prade

Measuring the Implications of the D-Basis in Analysis of Data in
Biomedical Studies 39
Kira Adaricheva, J.B. Nation, Gordon Okimoto, Vyacheslav Adarichev,
Adina Amanbekkyzy, Shuchismita Sarkar, Alibek Sailanbayev,
Nazar Seidalin, and Kenneth Alibek

Survey

Formal Concept Analysis and Information Retrieval – A Survey 61
Victor Codocedo and Amedeo Napoli

Theory

Bonds Between L-Fuzzy Contexts Over Different Structures of
Truth-Degrees 81
Jan Konecny

The Linear Algebra in Formal Concept Analysis over Idempotent
Semifields 97
Francisco J. Valverde-Albacete and Carmen Peláez-Moreno

On Closure Systems and Adjunctions Between Fuzzy Preordered Sets 114
F. García-Pardo, I.P. Cabrera, P. Cordero, and M. Ojeda-Aciego

Extensional Confluences and Local Closure Operators 128
Henry Soldano

A Note on Pattern Structures and Their Projections 145
Lars Lumpe and Stefan E. Schmidt

Enhanced FCA

Exploring Pattern Structures of Syntactic Trees for Relation Extraction   153
Artuur Leeuwenberg, Aleksey Buzmakov, Yannick Toussaint,
and Amedeo Napoli

Totally Balanced Formal Context Representation.   169
François Brucker and Pascal Préa

Randomized Fuzzy Formal Contexts and Relevance of One-Sided
Concepts   183
Lubomir Antoni, Stanislav Krajči, and Ondrej Krídlo

Revisiting Pattern Structure Projections   200
Aleksey Buzmakov, Sergei O. Kuznetsov, and Amedeo Napoli

Methods and Applications

Exploring Faulty Data   219
Daniel Borchmann

Automatic Validation of Terminology by Means of Formal Concept
Analysis.   236
Luis Felipe Melo Mora and Yannick Toussaint

Towards a Navigation Paradigm for Triadic Concepts   252
Sebastian Rudolph, Christian Săcărea, and Diana Troancă

Graphs and FCA

A Proposal for Extending Formal Concept Analysis to Knowledge
Graphs.   271
Sébastien Ferré

Simple Undirected Graphs as Formal Contexts   287
Giampiero Chiaselotti, Davide Ciucci, and Tommaso Gentile

A Conceptual Approach for Relational IR: Application to Legal
Collections.   303
Nada Mimouni, Adeline Nazarenko, and Sylvie Salotti

Author Index   319

Invited Talks

Quantitative Redundancy in Partial Implications

José L. Balcázar[(✉)]

LARCA Research Group, Department of Computer Science,
Universitat Politècnica de Catalunya, Barcelona, Spain
jose.luis.balcazar@upc.edu

Abstract. We survey the different properties of an intuitive notion of redundancy, as a function of the precise semantics given to the notion of partial implication.

1 Introduction

The discovery of regularities in large scale data is a multifaceted current challenge. Each syntactic mechanism proposed to represent such regularities opens the door to wide research questions. We focus on a specific sort of regularities sometimes found in transactional data, that is, data where each observation is a set of items, and defined in terms of pairs of sets of items.

Syntactically, the fact that this sort of regularity holds for a given pair (X, Y) of sets of items is often denoted as an implication: $X \rightarrow Y$. However, whereas in Logic an implication like this is true if and only if Y holds whenever X does, in our context, namely, partial implications and association rules, it is enough if Y holds "most of the times" X does. Thus, in association mining, the aim is to find out which expressions of that sort are valid for a given transactional dataset: for what X and what Y, the transactions that contain X "tend to contain" Y as well.

In many current works, that syntax is defined as if its meaning was sufficiently clear. Then, any of a number of "measures of interestingness" is chosen to apply to them, in order to select some to be output by a data analysis process on a particular dataset. Actually, the mere notation $X \rightarrow Y$ is utterly insufficient: any useful perspective requires to endow these expressions with a definite semantics that makes precise how that naïve intuition of "most of the times" is formalized; only then can we study and clarify the algorithmic properties of these syntactical expressions. Thus, we are not really to "choose a measure of interestingness" but plainly to *define* what $X \rightarrow Y$ means, and there are many acceptable ways of doing this.

J.L. Balcázar—Partially supported by project BASMATI (TIN2011-27479-C04-04) of Programa Nacional de Investigación (Ministerio de Ciencia e Innovación, Spain) and grant 2014SGR 890 (MACDA) from AGAUR, Generalitat de Catalunya.

J. Baixeries et al. (Eds.): ICFCA 2015, LNAI 9113, pp. 3–20, 2015.
DOI: 10.1007/978-3-319-19545-2_1

This idea of a relaxed implication connective is a relatively natural concept, and versions sensibly defined by resorting to conditional probability have been proposed in different research communities: a common semantics of $X \to Y$ is through a lower bound on its "confidence", the conditional probability of Y given X. This meaning appears already in the "partial implications" of [27] (actually, "implications partielles", with confidence christened there "prècision"). Some contributions based on Mathematical Logic develop notions related to these partial implications defined in terms of conditional probability: see [18]. However, it must be acknowledged that the contribution that turned on the spotlights on partial implications was [3] and the improved algorithm in [4]: the proposal of exploring large datasets in search for association rules of high support and confidence has led to huge amounts of research since. Association rules are partial implications that impose the additional condition that the consequent is a single item.

Three of the major foci of research in association rules and partial implications are as follows. First, the quantity of candidate itemsets for both the antecedent X and, sometimes, the consequent Y grows exponentially with the number of items. Hence, the space to explore is potentially enormous: on real world data, very soon we run already into billions of candidate antecedents. Most existing solutions are based on the acceptance that, as not all of them can be considered within reasonable running times, we make do with those that obey the support constraint ("frequent itemsets"). The support constraint combines well with confidence in order to avoid reporting mere statistical artifacts [28] but its major role is to reduce the search space. A wide repertory of algorithms for frequent sets and association rule mining exists by now [1].

Second, many variations have been explored: for instance, cases of more complicated structures in the data and, also, combinations with other machine-learning models or tasks like in [20,41].

This paper surveys part of a research line that belongs to a third focus: in a vast majority of practical applications, if any partial implication is found at all, it often happens that the search returns hundreds of thousands of them. It is far from trivial to design an associator able to choose well, among them, a handful to show to an impatient user. This is tantamount to modifying the semantics of the partial implication connective, by adding or changing the conditions under which one such expression is deemed valid and is to be reported. Most often, but not always (as we report in Sects. 3 and 4) this approach takes the form of "quality evaluations" performed to select which partial implications are to be highlighted for the user. We do not consider this problem solved yet, but deep progresses have been achieved so far; we survey a humble handful of those, where the present author was actively involved. For a wider perspective of all these three aspects of association rule mining, see Part II of [43].

The main link along this paper can be described informally as follows: human intuition, maybe on the basis of our experience with full, standard implications, tends to expect that smaller antecedents are better than larger ones, and larger consequents are better than smaller ones. We call this statement here the *central intuition* of this paper; many references express, in various variants, this intuition (e.g. [12, 22, 26, 30, 33, 36] just to name a few). This intuition is only partially true in implications, where the GD basis gets to be minimal through the use of subtly enlarged antecedents [17]. This survey paper discusses, essentially, the particular fact that, on partial implications, this intuition is both true and false... as a function, of course, of the actual semantics given to the partial implication connective.

2 Notation and Preliminary Definitions

Our datasets are transactional. This means that they are composed of transactions, each of which consists of an itemset with a unique transaction identifier. Itemsets are simply subsets of some fixed set $\mathcal{U}$ of items. We will denote itemsets by capital letters from the end of the alphabet, and use juxtaposition to denote union, as in XY. The inclusion sign as in $X \subset Y$ denotes proper subset, whereas improper inclusion is denoted $X \subseteq Y$. The cardinality of a set X (either an itemset or a set of transactions) is denoted $|X|$.

2.1 Partial Implications

As indicated in the Introduction, the most common semantics of partial implication is its *confidence*: the conditional empirical probability of the consequent given the antecedent, that is, the ratio between the number of transactions in which X and Y are seen together and the number of transactions that contain X. We will see below that this semantics may be somewhat misleading. In most application cases, the search space is additionally restricted by a minimal *support* criterion, thus avoiding itemsets that appear very seldom in the dataset.

More precisely, for a given dataset $\mathcal{D}$, consisting of n transactions, the *supporting set* $\mathcal{D}_X \subseteq \mathcal{D}$ of an itemset X is the subset of transactions that include X. (For the reader familiar with the FP-growth frequent set miner [19], these are the same as their "projected databases", except for the minor detail that, here, we do not remove X from the transactions.)

The *support* $s_\mathcal{D}(X) = |\mathcal{D}_X|/n \in [0,1]$ of an itemset X is the cardinality of the set of transactions that contain X divided by n; it corresponds to the relative frequency or empirical probability of X. An alternative rendering of support is its unnormalized version, but some of the notions that will play a major role later on are simpler to handle with normalized supports. Now, the *confidence* of a partial implication $X \to Y$ is $c_\mathcal{D}(X \to Y) = s_\mathcal{D}(XY)/s_\mathcal{D}(X)$: that is, the empirical approximation to the corresponding conditional probability. The *support* of a partial implication $X \to Y$ is $s_\mathcal{D}(X \to Y) = s_\mathcal{D}(XY)$. In both expressions, we will omit the subscript $\mathcal{D}$ whenever the dataset is clear from the context. Clearly, $s_{\mathcal{D}_Z}(X) = \frac{|\mathcal{D}_{XZ}|}{|\mathcal{D}_Z|} = c(Z \to X)$.

Often, we will assume that $X \cap Y = \emptyset$ in partial implications $X \to Y$. Some works impose this condition globally; we will mention it explicitly whenever it is relevant, but, generally speaking, we allow X and Y to intersect or, even, to fulfill $X \subseteq Y$. Note that, if only support and confidence are at play, then

$c_D(X \to XY) = c_D(X \to Y)$ and $s_D(X \to XY) = s_D(X \to Y)$. Of course, in practical terms, after a partial implication mining process, only the part of Y that does not appear in X would be shown to the user.

We do allow $X = \emptyset$ as antecedent of a partial implication: then, its confidence coincides with the support, $c_D(\emptyset \to Y) = s_D(Y)$, since $s_D(\emptyset) = 1$. Allowing $Y = \emptyset$ as consequent as well is possible but turns out not to be very useful; therefore, empty-consequent partial implications are always omitted from consideration. All along the paper, there are occassional glitches where the empty set needs to require separate consideration. Being interested in the general picture, here we will mostly ignore these issues, but the reader can check that these cases are given careful treatment in the original references provided for each part of our discussion.

By $X \Rightarrow Y$ we denote full, standard logical implication; this expression will be called the *full counterpart* of the partial implication $X \to Y$.

2.2   Partial Implications Versus Association Rules

Association rules were defined originally as partial implications $X \to Y$ with singleton consequents: $|Y| = 1$; we abbreviate $X \to \{A\}$ as $X \to A$. This decision allows one to reduce association mining to a simple postprocessing after finding frequent sets. Due to the illusion of augmentation, many users are satisfied with this syntax, but, however, more items in the consequent provide more information.

Indeed, in full implications, the expression $(A \Rightarrow B) \wedge (A \Rightarrow C)$ is fully equivalent to $A \Rightarrow BC$, and we lose little by enforcing singleton consequents (equivalently, definite Horn clauses); an exception is the discussion of minimal bases, where nonsingleton consequents allow for canonical bases that are unreachable in the Horn clause syntax [17]. But, in partial implications, $A \to BC$ says more than the conjunction of $A \to B$ and $A \to C$, namely, B and C abound *jointly* in D_A. Whenever possible, $A \to BC$ is better, being both more economical and more informative. This can be ilustrated by the following example from [8], to which we will return later on.

Example 1. Consider a dataset on $\mathcal{U} = \{A, B, C, D, E\}$ consisting of 12 transactions: 6 of them include all of $\mathcal{U}$, 2 consist of ABC, 2 more are AB, and then one each of CDE and BC. It can be seen that the confidence of both $B \to A$ and $B \to C$ is 9/11, whereas the confidence of $B \to AC$ is 8/11.

Actually, even restricted to association rules, the output of confidence-based associators is often still too large: the rest of this paper discusses how to reduce the output with no loss of information, first, and, then, as the outcome is often still too large in practice, we will need to allow for a carefully tuned loss of information.

3   Redundancy in Confidence-Based Partial Implications

We start our discussion by "proving correct" our *central intuition*, that is, providing a natural semantics under which that intuition is correct. For this section, we work under confidence and support thresholds, and it turns out to be convenient to explicitly assume that the left-hand side of each partial implication is included in the right-hand side. We force that inclusion using notations in the style of $X \to XY$.

Several references ([2] for one) have considered the following argument: assume that we could know beforehand that, in all datasets, the confidence and support of $X_0 \to X_0Y_0$ are always larger than or equal to those of $X_1 \to X_1Y_1$. Then, whenever we are mining some dataset under confidence and support thresholds, assume that we find $X_1 \to X_1Y_1$: we should not bother to report as well $X_0 \to X_0Y_0$, since it must be there anyhow, and its presence in the output is uninformative. In a very strong sense, $X_0 \to X_0Y_0$ is redundant with respect to $X_1 \to X_1Y_1$. Irredundant partial implications according to this criterion are called "essential rules" in [2] and *representative rules* in [21]; we will follow this last term.

Lemma 1. *Consider two partial implications, $X_0 \to X_0Y_0$ and $X_1 \to X_1Y_1$. The following are equivalent:*

1. *The confidence and support of $X_0 \to X_0Y_0$ are larger than or equal to those of $X_1 \to X_1Y_1$, in all datasets: for every $\mathcal{D}$, $c_{\mathcal{D}}(X_0 \to X_0Y_0) \geq c_{\mathcal{D}}(X_1 \to X_1Y_1)$ and $s_{\mathcal{D}}(X_0 \to X_0Y_0) \geq s_{\mathcal{D}}(X_1 \to X_1Y_1)$.*
2. *The confidence of $X_0 \to X_0Y_0$ is larger than or equal to that of $X_1 \to X_1Y_1$, in all datasets: for every $\mathcal{D}$, $c_{\mathcal{D}}(X_0 \to X_0Y_0) \geq c_{\mathcal{D}}(X_1 \to X_1Y_1)$.*
3. $X_1 \subseteq X_0 \subseteq X_0Y_0 \subseteq X_1Y_1$.

When these cases hold, we say that $X_1 \to X_1Y_1$ makes $X_0 \to X_0Y_0$ *redundant*. The fact that the inequality on support follows from the inequality on confidence is particularly striking. This lemma can be interpreted as proving correct the *central intuition* that smaller antecedents and larger consequents are better, by indentifying a semantics of the partial implication connective that makes this true and by pointing out that it is not just the consequent that is to be maximized, but the union of antecedent and consequent. If only consequents are maximized separately, and are kept disjoint from the antecedents, then one gets to a quite more complicated situation discussed below.

Definition 1. *Fix a dataset and confidence and support thresholds. The* representative rule basis *for that dataset at these support and confidence thresholds consists of those partial implications that pass both thresholds in the dataset, and are not made redundant, in the sense of the previous paragraph, by other partial implications also above the thresholds.*

Hence, a redundant partial implication is so because we can know beforehand, from the information in the basis, that its confidence is above the threshold. We have:

Proposition 1. *(Essentially, from [21].) For a fixed dataset $\mathcal{D}$ and a fixed confidence threshold γ:*

1. *Every partial implication of confidence at least γ is made redundant by some representative rule.*
2. *Partial implication $X \to Y$ with $X \subseteq Y$ is a representative rule if and only if $c_D(X \to Y) \geq \gamma$ but there is no X' and Y' with $X' \subseteq X$ and $XY \subseteq X'Y'$ such that $c_D(X' \to Y') \geq \gamma$, except $X = X'$ and $Y = Y'$.*

According to statement *(3)* in Lemma 1, that last point means that a representative rule is not redundant with respect to any partial implication (different from itself) that has confidence at least γ in the dataset. It is interesting to note that one does not need to mention support in this last proposition, the reason being, of course, statement *(2)* in Lemma 1. The fact that statement *(3)* implies statement *(1)* was already pointed out in [2,21,31] (in somewhat different terms). The remaining implications are from [7]; see this reference as well for proofs of additional properties, including the fact the representative basis has the minimum possible size among all bases for this notion of redundancy, and for discussions of other related redundancy notions. In particular, several other natural proposals are shown there to be equivalent to this redundancy. Also [8] provides further properties of the representative rules. These references discuss as well the connection with a similar notion in [42].

In Example 1, at confidence threshold 0.8, the representative rule basis consists of seven partial implications: $\emptyset \to C$, $B \to C$, $\emptyset \to AB$, $C \to AB$, $A \to BC$, $D \to ABCE$, and $E \to ABCD$.

3.1 Quantitative Evaluation of Non-redundancy: Confidence Width

Redundancy is a qualitative property; still, it allows for a quantitative discussion. Consider a representative rule $X \to XY$: at confidence $c(X \to XY)$, no partial implication makes it redundant. But we could consider now to what extent we need to reduce the confidence threshold in order to find a partial implication that would make this one redundant. If a partial implication of almost the same confidence can be found to make $X \to XY$ redundant, then our partial implication is not so interesting. According to this idea, one can define a parameter, the *confidence width* [6], that, in a sense, evaluates how different is our partial implication from other similar ones. We do not discuss this parameter further, but a related quantity is treated below in Sect. 6.2.

3.2 Closure-Aware Redundancy Notions

Redundancy of one partial implication with respect to another can be redefined as well in a similar but slightly more sophisticate form by taking into account the closure operator obtained from the data (see [15]). Often, this variant yields a more economical basis because the full implications are described by their often very short Guigues-Duquenne basis [17]; see again [7] for the details.

4 Redundancy with Multiple Premises

The previous section indicates precisely when "one partial implication follows logically from another". It is natural to ask whether a stronger, more useful notion to reduce the size of a set of partial implications could be based on partial implications following logically from several others together, beyond the single-premise case.

Simply considering standard examples with full implications like Augmentation (from $X \Rightarrow Y$ and $X' \Rightarrow Y'$ it follows $XX' \Rightarrow YY'$) or Transitivity (from $X \Rightarrow Y$ and $Y \Rightarrow Z$ it follows $X \Rightarrow Z$), it is easy to see that these cases fail badly for partial implications. Indeed, one might suspect, as this author did for quite some time, that one partial implication would not follow logically from several premises unless it follows from one of them.

Generally speaking, however, this suspicion is wrong. It is indeed true for confidence thresholds $\gamma \in (0, 0.5)$, but these are not very useful in practice, as an association rule $X \to A$ of confidence less than 0.5 means that, in $\mathcal{D}_X$, the absence of A is more frequent than its presence.

And, for $\gamma \in [0.5, 1)$, it turns out that, for instance, from $A \to BC$ and $A \to BD$ it follows $ACD \to B$, in the sense that if both premises have confidence at least γ in any dataset, then the conclusion also does. The general case for two premises was fully characterized in [7], but the case of arbitrary premise sets has remained elusive for some years. Eventually, a very recent result from [5] proved that redundancy with respect to a set of premises that are partial implications hinges on a complicated combinatorial property of the premises themselves. We give that property a short (if admittedly uninformative) name here:

Definition 2. *Let* $X_1 \to Y_1, \ldots, X_k \to Y_k$ *be a set of partial implications. We say that it is nice if* $X_1 \Rightarrow Y_1, \ldots, X_k \Rightarrow Y_k \models X_i \Rightarrow U$, *for all* $i \in 1 \ldots k$, *where* $U = X_1 Y_1 \cdots X_k Y_k$.

Here we use the standard symbol $\models$ for logical entailment; that is, whenever the implications at the left-hand side are true, the one at the right-hand side must be as well.

Note that the definition of nicety of a set of partial implications states a property, not of the partial implications themselves, but of their full counterparts. Then, we can characterize entailment among partial implications for high enough thresholds of confidence, as follows:

Theorem 1 *[5]. Let* $X_1 \to Y_1, \ldots, X_k \to Y_k$ *be a set of partial implications with* $k \geq 1$, *candidates to premises, and a candidate conclusion* $X_0 \to Y_0$. *If* $\gamma \geq (k-1)/k$, *then the following are equivalent:*

1. *in any dataset where the confidence of the premises* $X_1 \to Y_1, \ldots, X_k \to Y_k$ *is at least* γ, $c(X_0 \to Y_0) \geq \gamma$ *as well;*
2. *either* $Y_0 \subseteq X_0$, *or there is a non-empty* $L \subseteq \{1 \ldots k\}$ *such that the following conditions hold:*

(a) $\{X_i \rightarrow Y_i : i \in L\}$ *is nice,*
(b) $\bigcup_{i \in L} X_i \subseteq X_0 \subseteq \bigcup_{i \in L} X_i Y_i,$
(c) $Y_0 \subseteq X_0 \cup \bigcap_{i \in L} Y_i.$

Interestingly, the last couple of conditions are reasonably correlated, for the case of several premises, with the *central intuition* that smaller antecedents are better than larger ones, and larger consequents are better than smaller ones. The premises actually necessary must all include the consequent of the conclusion, and their antecedents are to be included in the antecedent of the conclusion. Even the additional fact that the antecedent of the conclusion does not have "extra items" not present in the premises also makes sense.

However, there is the additional condition that only nice sets of partial implications may have a nontrivial logical consequence, and all this just for high enough confidence thresholds. The proof is complex and we refrain from discussing it here; see [5], where, additionally, the case of $\gamma < 1/k$ is also characterized and the pretty complicated picture for intermediate values of γ is discussed.

We do indicate, though, that the notion of "nicety", in practice, turns out to be so restrictive that we have not found any case of nontrivial entailment from more than one premise in a number of tests with stardard benchmark datasets. Therefore, this approach is not particularly useful in practice to reduce the size of the outcome of an associator.

4.1   Ongoing Developments

As for representative rules (Subsect. 3.2), there exists a natural variant of the question of redundancy, whereby full implications are handled separately; essentially, the redundancy notion becomes "closure-based". This extension was fully characterized as well for the case of two premises in [7], but it is current work in progress how to extend the scheme to the case of arbitrary quantities of premises.

5   Alternative Evaluation Measures

We move on to discuss how to reinterpret the *central intuition* as we change the semantics of the partial implication connective. Confidence is widely used as a definition of partial implication but, in practice, presents two drawbacks. First, it does not detect negative correlations; and, second, as already indicated, often lets pass far too many rules and, moreover, fiddling with the confidence threshold turns out to be a mediocre or just useless solution. Examples of both disadvantages are both easy to construct and easy to find on popular benchmark datasets. Both objections can be addressed by changing the semantics of the expression $X \rightarrow Y$, by either replacing the confidence measure or by strengthening it with extra conditions. The literature on this topic is huge and cannot be reviewed here: see [16, 25, 35] and their references for information about the relevant developments published along these issues. We focus here on just a tiny subset of all these studies.

The first objection alluded to in the previous paragraph can be naturally solved via an extra normalization (more precisely, dividing the confidence by the support of the consequent). The outcome is *lift*, a well-known expression in basic probability; a closely related parameter is *leverage*:

Definition 3. *Assume* $X \cap Y = \emptyset$. *The* lift *of partial implication* $X \to Y$ *is* $\ell_{\mathcal{D}}(X \to Y) = \frac{c_{\mathcal{D}}(X \to Y)}{s_{\mathcal{D}}(Y)} = \frac{s_{\mathcal{D}}(XY)}{s_{\mathcal{D}}(X) \times s_{\mathcal{D}}(Y)}$. *The* leverage *of partial implication* $X \to Y$ *is* $\lambda_{\mathcal{D}}(X \to Y) = s_{\mathcal{D}}(XY) - s_{\mathcal{D}}(X) \times s_{\mathcal{D}}(Y)$.

If supports are unnormalized, extra factors n are necessary. In case of independence of both sides of a partial implication $X \to Y$, we would have $s(XY) = s(X)s(Y)$; therefore, both lift and leverage are measuring deviation from independence: lift is the multiplicative deviation, whereas leverage measures it rather as an additive distance instead. Leverage was introduced in [32] and, under the name "Novelty", in [24], and received much attention via the Magnum Opus associator [38]. We find lift in the references going by several different names: it has been called *interest* [34] or, in a slightly different but fully equivalent form, *strength* [33]; *lift* seems to be catching up as a short name, possibly aided by the fact that the Intelligent Miner system from IBM employed that name. These notions allow us to exemplify that we are modifying the semantics of our expressions: if we define the meaning of $X \to Y$ through confidence, then partial implications of the form $X \to Y$ and $X \to XY$ are always equivalent, whereas, if we use lift, then they may not be. Note that, in case $X = \emptyset$, the lift trivializes to 1. Also, if we are to use lift, then we must be careful to keep the right-hand side Y disjoint from the left-hand side: $X \cap Y = \emptyset$.

A related notion is:

Definition 4 *[24]. The* relative confidence *of partial implication* $X \to Y$, *also called* centered confidence *or* relative accuracy, *is* $r_{\mathcal{D}}(X \to Y) = c_{\mathcal{D}}(X \to Y) - c_{\mathcal{D}}(\emptyset \to Y)$.

Therefore, the relative confidence is measuring additively the effect, on the support of the consequent Y, of "adding the condition" or antecedent X. Since $c_{\mathcal{D}}(\emptyset \to Y) = s_{\mathcal{D}}(Y)$, lift can be seen as comparing $c_{\mathcal{D}}(X \to Y)$ with $c_{\mathcal{D}}(\emptyset \to Y)$, that is, effecting the same comparison but multiplicatively this time: $\ell(X \to Y) = \frac{s(XY)}{s(X) \times s(Y)} = \frac{c(X \to Y)}{s(Y)} = \frac{c(X \to Y)}{c(\emptyset \to Y)}$. Also, it is easy to check that leverage can be rewritten as $\lambda_{\mathcal{D}}(X \to Y) = s_{\mathcal{D}}(X) \times r_{\mathcal{D}}(X \to Y)$ and is therefore called also *weighted relative accuracy* [24]. Relative confidence has the potential to solve the "negative correlation" objection to confidence, and all subsequent measures to be described here inherit this property as well.

An objection of a different sort is that lift and leverage are symmetric. As the implicational syntax is asymmetric, they do not fit very well the directional intuition of an expression like $X \to Y$; that is one of the reasons behind the exploration of many other options. However, to date, none of the more sophisticate attempts seems to have gained a really noticeable "market share". Most common implementations either offer a long list of options of measures for the user to choose from (like [13] for one), or employ the simpler notions of confidence, support, lift, or leverage (for instance, Magnum Opus [38]). We believe that one must keep close to confidence and to deviation from independence. Confidence is the most natural option for many educated domain experts not specialized in data mining, and it provides actually a directionality to our partial implications.

The vast majority of these alternatives attempt at defining the quality of partial implication $X \to Y$ relying only on the supports of X, Y, XY, or their complements. One major exception is *improvement* [12], which is the added confidence obtained by using the given antecedent as opposed to any properly smaller one. We discuss it and two other related quantities next. They are motivated again by our *central intuition*: if the confidence of a partial implication with a smaller antecedent and the same consequent is sufficiently high, the larger partial implication should not be provided in the output. They have in common that their computation requires exploration of a larger space, however; we return to this point in the next section.

5.1 Improvement: Additive and Multiplicative

The key observation for this section is that $X \to Y$ and $Z \to Y$, for $Z \subset X$, provide different, independent information. From the perspective of confidence, either may have it arbitrarily higher than the other. For inequality in one direction, suppose that almost all transactions with X have Y, but they are just a small fraction of those supporting Z, which mostly lack Y; conversely, Y might hold for most transactions having Z, but the only transactions having all of X can be those without Y. In Example 1, one can see that $c(\emptyset \to BC) < c(A \to BC)$ whereas $c(\emptyset \to C) > c(B \to C)$.

This fact underlies the difficulty in choosing a proper confidence bound. Assume that there exists a mild correlation giving, say, $c(Z \to A) = 2/3$. If the threshold is set higher, of course this rule is not found; but an undesirable side effect may appear: there may be many ways of choosing subsets of the support of Z, by enlarging it a bit, where Y is frequent enough to pass the threshold. Thus, often, in practice, the algorithms enlarge Z into various supersets X_i so that all the confidences $c(X_i \to A)$ do pass, and then $Z \to A$ is not seen, but generates dozens of very similar "noisy" rules, to be manually explored and filtered. Finding the appropriate threshold becomes difficult, also because, for different partial implications, this sort of phenomenon may appear at several threshold values simultaneously.

Relative confidence tests confidence by a comparison to what happens if the antecedent is replaced by one of its subsets in particular, namely $\emptyset$. Improvement generalizes it by considering not only the alternative partial implication $\emptyset \to Y$ but all proper subsets of the antecedent, as alternative antecedents, and in the same additive form:

Definition 5. *The* improvement $X \to Y$, *where* $X \neq \emptyset$, *is* $i(X \to Y) = \min\{c(X \to Y) - c(Z \to Y) \mid Z \subset X\}$.

The definition is due to [12], where only association rules are considered, that is, cases where $|Y| = 1$. The work on productive rules [39] is related: these coincide with the rules of *positive improvement*. In [26], improvement is combined with further pruning on the basis of the χ^2 value. We literally quote from [12]: "A rule with negative improvement is typically undesirable because the rule can be simplified to yield a proper sub-rule that is more predictive, and applies to an equal or larger population due to the antecedent containment relationship. An improvement greater than 0 is thus a desirable constraint in almost any application of association rule mining. A larger minimum on improvement is also often justified because most rules in dense data-sets are not useful due to conditions or combinations of conditions that add only a marginal increase in confidence."

The same process, and with the same intuitive justification, can be applied to lift, which is, actually, a multiplicative, instead of additive, version of relative confidence as indicated above: $\ell(X \to Y) = c(X \to Y)/c(\emptyset \to Y)$. Taking inspiration in this correspondence, we studied in [9] a multiplicative variant of improvement that generalizes lift, exactly in the same way as improvement generalizes relative confidence:

Definition 6. *The* multiplicative improvement *of* $X \to Y$, *where* $X \neq \emptyset$, *is* $m(X \to Y) = \min\{c(X \to Y)/c(Z \to Y)|Z \subset X\}$.

In Example 1, the facts that $c(A \to BC) = 4/5$ and $c(\emptyset \to BC) = 3/4$ lead to $i(A \to BC) = 4/5 - 3/4 = 0.05$ and $m(A \to BC) = (4/5)/(3/4) \approx 1.066$. Here, as the size of the antecedent is 1, there is one single candidate $Z = \emptyset$ to proper subset of the antecedent and, therefore, improvement coincides with relative confidence, and multiplicative improvement coincides with lift. For larger left-hand sides, the values will be different in general.

5.2   Rule Blocking

Attempting at formalizing the same part of the *central intuition*, we proposed in [6] a notion of "rule blocking", where a smaller antecedent $Z \subset X$ would "block" (that is, suggest to omit) a given partial implication $X \to Y$. We will compare the number of tuples having XY (that is, having Y within the supporting set of X) with the quantity that would be predicted from the confidence of the partial implication $Z \to Y$, that applies to a larger supporting set: we are going to bound the relative error incurred if the support $s(X)$ and the confidence of $Z \to Y$ are employed to approximate the confidence of $X \to Y$.

More precisely, let $c(Z \to ZY) = c$. If Y is distributed along the support of X at the same ratio as along the larger support of Z, we would expect $s(XY) \approx c \times s(X)$: we consider the relative error committed by $c \times s(X)$ used as an approximation to $s(XY)$ and, if the error is low, we consider that $Z \to Y$ is sufficient information about $X \to Y$ and dispose of this last one.

Definition 7 *[6]. $Z \subset X$ blocks $X \to Y$ at blocking threshold ϵ when*

$$\frac{s(XY) - c(Z \to Y)s(X)}{c(Z \to Y)s(X)} \leq \epsilon.$$

In case the difference in the numerator is negative, it would mean that $s(XY)$ is even lower than what $Z \to Y$ would suggest. If it is positive but the quotient is low, $c(Z \to Y) \times s(X)$ still suggests a good approximation to $c(X \to Y)$, and the larger partial implication $X \to Y$ does not bring high enough confidence to be considered besides $Z \to Y$, a simpler one: it remains blocked. But, if the quotient is larger, and this happens for all Z, then $X \to Y$ becomes interesting since its confidence is higher enough than suggested by other partial implications of the form $Z \to Y$ for smaller antecedents Z. Of course, the higher the block threshold, the more demanding the constraint is. Note that, in the presence of a support threshold τ, $s(ZY) \geq s(XY) > \tau$ or a similar inequality would be additionally required. The value ϵ is intended to take positive but small values, say around 0.2 or lower. In Example 1, $\emptyset$ blocks $A \to BC$ at blocking threshold $1/15 \approx 0.066$.

Rule blocking relates to multiplicative improvement as follows:

Proposition 2. *The smallest blocking threshold at which $X \to Y$ is blocked is $m(X \to Y) - 1$.*

Proof. As everything around is finite, this is equivalent to proving that $Z \subset X$ blocks $X \to Y$ at block threshold ϵ if and only if $\frac{c(X \to Y)}{c(Z \to Y)} - 1 \leq \epsilon$, for all such Z. Starting from the definition of blocking, multiplying both sides of the inequality by $c(Z \to Y)$, separating the two terms of the left-hand side, replacing $s(XY)/s(X)$ by its meaning, $c(X \to Y)$, and then solving first for $c(Z \to Y)$ and finally for ϵ, we find the stated equivalence. All the algebraic manipulations are reversible.

5.3   Ongoing: Conditional Weighted Versions of Lift and Leverage

We propose here one additional step to enhance the flexibility of both lift and leverage by considering their action, on the same partial implication, but with respect to many different subsets of the dataset, and under a weighting scheme that leads to different existing measures according to the weights chosen.

For a given partial implication $X \to Y$, we consider many limited views of the dataset, namely, all its projections into subsets of the antecedent. We propose to measure a weighted variant of the lift and/or the leverage of the same partial implication in all these projections, and evaluate as the quality of the partial implication the minimum value thus obtained. That is, we want our high-quality partial implications not only to have high lift or leverage, but also to maintain it when we consider projections of the dataset on the subsets of the antecedent. We call the measures obtained *conditional weighted* lift and leverage.

Definition 8. *Assume $X \cap Y = \emptyset$. Let w be a weighting function associating a weight (either a positive real number or ∞) to each proper subset of X. The* conditional weighted lift *of partial implication $X \to Y$ is $\ell'_{\mathcal{D},w}(X \to Y) = \min\{w(Z)\ell_{\mathcal{D}_Z}(X \to Y) \mid Z \subseteq X\}$. The* conditional weighted leverage *of partial implication $X \to Y$ is $\lambda'_{\mathcal{D},w}(X \to Y) = \min\{w(Z)g_{\mathcal{D}_Z}(X \to Y) \mid Z \subseteq X\}$.*

These notions can be connected to other existing notions with unificatory effects. We only state here one such connection. Further development will be provided in a future paper in preparation.

Proposition 3. *For inverse confidence weights, conditional weighted leverage is* improvement: *for all $X \to Y$, $\lambda'_{\mathcal{D},w}(X \to Y) = i(X \to Y)$ holds for the weighting function $w_r(Z) = c_{\mathcal{D}}(Z \to X)^{-1}$.*

6   Support Ratio and Confidence Boost

From the perspective of our *central intuition*, the previous section has developed, essentially issues related to smallish antecedents. This is fully appropriate for the discussion of association rules, which were defined originally as partial implications with singleton consequents. We now briefly concentrate on largish consequents, and then join both perspectives.

6.1   Support Ratio

The support ratio was employed first, to our knowledge, in [23], where no particular name was assigned to it. Together with other similar quotients, it was introduced in order to help obtaining faster algorithmics.

Definition 9. *In the presence of a support threshold τ, the* support ratio *of a partial implication $X \to Y$ is*

$$\sigma(X \to Y) = \frac{s(XY)}{\max\{s(Z) \mid XY \subset Z, \, s(Z) > \tau\}}.$$

We see that this quantity depends on XY but not on the antecedent X itself. In Example 1, we find that $\sigma(A \to BC) = 4/3$.

6.2   Confidence Boost

Definition 10. *The* confidence boost *of a partial implication $X \to Y$ (always with $X \cap Y = \emptyset$) is $\beta(X \to Y) =$*

$$\frac{c(X \to XY)}{\max\{c(X' \to X'Y') \mid (X \to XY) \not\equiv (X' \to X'Y'), \, X' \subseteq X, \, Y \subseteq Y'\}}.$$

where the partial implications in the denominator are implicitly required to clear the support threshold, in case one is enforced: $s(X' \to X'Y') > \tau$.

Let us explain the interpretation of this parameter. Suppose that $\beta(X \rightarrow Y)$ is low, say $\beta(X \rightarrow Y) \leq b$, where b is just slightly larger than 1. Then, according to the definition, there must exist some *different* partial implication $X' \rightarrow X'Y'$, with $X' \subseteq X$ and $Y \subseteq X'Y'$, such that $\frac{c(X \rightarrow Y)}{c(X' \rightarrow Y')} \leq b$, or $c(X' \rightarrow Y') \geq c(X \rightarrow Y)/b$. This inequality says that the partial implication $X' \rightarrow Y'$, stating that transactions with X' tend to have $X'Y'$, has a confidence relatively high, not much lower than that of $X \rightarrow Y$; equivalently, the confidence of $X \rightarrow Y$ is not much higher (it could be lower) than that of $X' \rightarrow Y'$. But all transactions having X do have X', and all transactions having Y' have Y, so that the confidence found for $X \rightarrow Y$ is not really that novel, given that it does not give so much additional confidence over a partial implication that states such a similarly confident, and intuitively stronger, fact, namely $X' \rightarrow Y'$.

This author has developed a quite successful open-source partial implication miner based on confidence boost (`yacaree.sf.net`); all readers are welcome to experiment with it and provide feedback. We note also that the confidence width alluded to in Sect. 3.1, while having different theoretical and practical properties, is surprisingly close in definition to confidence boost. See [8] for further discussion of all these issues. Confidence boost fits the general picture as follows:

Proposition 4. $\beta(X \rightarrow Y) = \min\{\sigma(X \rightarrow Y), m(X \rightarrow Y)\}$.

The inequalities $\beta(X \rightarrow Y) \leq \sigma(X \rightarrow Y)$ (due to [11]) and $\beta(X \rightarrow Y) \leq m(X \rightarrow Y)$ are simple to argue: the consequent leading to the support ratio, or the antecedent leading to the multiplicative improvement, take a role in the denominator of confidence boost. Conversely, taking the maximizing partial implication in the denominator, if it has the same antecedent X then one obtains a bound on the support ratio whereas, if the antecedent is properly smaller, a bound on the multiplicative improvement follows.

In Example 1, since $\sigma(A \rightarrow BC) = 4/3$ and $m(A \rightarrow BC) = (4/5)/(3/4)$, which is smaller, we obtain $\beta(A \rightarrow BC) = (4/5)/(3/4) \approx 1.066$.

A related proposal in [22] suggests to minimize directly the antecedents and maximizing the consequents, within the confidence bound, and in a context where antecedents and consequents are kept disjoint. This is similar to statement (3) in Lemma 1, except that, there, one maximizes jointly consequent and antecedent. If consequents are maximized separately, then the *central intuition* fails, but there is an interesting connection with confidence boost; see [8].

The measures in this family of improvement, including conditional weighted variants and also confidence boost, tend to require exploration of larger spaces of antecedents compared to simpler rule quality measures. This objection turns out not to be too relevant because human-readable partial implications have often just a few items in the antecedent. Nontrivial algorithmic proposals for handling this issue appear as well in [8].

6.3   Ongoing Developments

We briefly mention here the following observations. First, like in Sect. 3.2, a variant of confidence boost appropriate for closure-based analysis exists [8].

Second, both variants trivialize if they are applied directly, in their literal terms, to full implications. However, the intuitions leading to confidence boost can be applied as well to full implications. In future work, currently in preparation, we will discuss proposals for formalizing the same intuition in the context of full implications.

7   Evaluation of Evaluation Measures

We have covered just a small fraction of the evaluation measures proposed to endow with useful semantics the partial implication connective. All of those attempt, actually, at capturing a potential (but maybe nonexisting) "naïve concept" of interesting partial implication from the perspective of an end user. Eventually, we would like to find one such semantics that fits as best as possible that hypothetical naïve concept.

We can see no choice but to embark, at some point, in the creation of resources where, for specific datasets, the interest of particular implications is recorded as per the assessment of individual humans. Some approximations to this plan are Sect. 5.2 of [8], where the author, as a scientific expert, subjectively evaluates partial implications obtained from abstracts or scientific papers; a similar approach in [14] using PKDD abstracts; and the work in [10,44] where partial implications found on educational datasets from university course logs are evaluated by the teachers of the corresponding courses. These preliminary experiments are positive and we hope that a more ambitious attempt could be made in the future along these lines.

The idea of evaluating associators through the predictive capabilities of the rules found has been put forward in several sources, e.g. [29]. The usage of association rules for direct prediction (where the "class" attribute is forced to occur in the consequent) has been widely studied (e.g. [41]). In [29], two different associators are employed to find rules with the "class" as consequent, and they are compared in terms of predictive accuracy. This scheme is inappropriate to evaluate our proposals for the semantics of partial implications, because, first, we must focus on single pairs of attribute and value as right-hand side, thus making it useless to consider larger right-hand sides; and, also, the classification will only be sensible to minimal left-hand sides independently of their confidences.

In [9], we have deployed an alternative framework that allows us to evaluate the diverse options of semantics for association rules, in terms of their usefulnes for subsequent predictive tasks. By means of a mechanism akin to the AUC measure for predictor evaluation, we have focused on potential accuracy improvements of predictors on given, public, standard benchmark datasets, if one more Boolean column is added, namely, one that is true exactly for those observations that are exceptions to one association rule: the antecedent holds but the consequent does not. In a sense, we use the association rule as a "hint of outliers", but, instead of removing them, we simply offer direct access to this label to the predictor, through the extra column. Of course, in general this may lead astray the predictor instead of helping it. Our experiments suggest that leverage, support, and multiplicative improvement tend to be better than the other measures with respect to this evaluation score.

7.1 Ongoing Developments

We are currently developing yet new frameworks that, hopefully, might be helpful in assessing the relative merits of the different candidates for semantics of partial implications, put forward often as rule quality measures. One of them resorts to an empirical application of approximations to the MDL principle along the lines of Krimp [37]. A second idea is to make explicit the dependence on alternative partial implications, in the sense that $X \rightarrow Y$ would mean, intuitively, that Y appears often on the support of X and that, barring the presence of some other partial implication to the contrary, it is approximately uniformly distributed there. These avenues will be hopefully explored along the coming months or years. A common thread is that additional statistical knowledge, along the lines of the self-sufficient itemsets of Webb [40], for instance, is expected to be at play in the future developments of the issue of endowing the partial implication connective with the right intuitive semantics.

References

1. Aggarwal, C.C., Han, J. (eds.): Frequent Pattern Mining. Springer, Switzerland (2014)
2. Aggarwal, C.C., Yu, P.S.: A new approach to online generation of association rules. IEEE Trans. Knowl. Data Eng. **13**(4), 527–540 (2001)
3. Agrawal, R., Imielinski, T., Swami, A.N.: Mining association rules between sets of items in large databases. In: Buneman, P., Jajodia, S. (eds.) SIGMOD Conference, pp. 207–216. ACM Press (1993)
4. Agrawal, R., Mannila, H., Srikant, R., Toivonen, H., Verkamo, A.I.: Fast discovery of association rules. In: Piatetsky-Shapiro, G., Frawley, W. (eds.) Advances in Knowledge Discovery and Data Mining, pp. 307–328. AAAI/MIT Press, Menlo Park (1996)
5. Atserias, A., Balcázar, J.L.: Entailment among probabilistic implications. Accepted for presentation at LICS (2015)
6. Balcázar, J.L.: Two measures of objective novelty in association rule mining. In: Theeramunkong, T., Nattee, C., Adeodato, P.J.L., Chawla, N., Christen, P., Lenca, P., Poon, J., Williams, G. (eds.) New Frontiers in Applied Data Mining. LNCS, vol. 5669, pp. 76–98. Springer, Heidelberg (2010)
7. Balcázar, J.L.: Redundancy, deduction schemes, and minimum-size bases for association rules. Logical Methods Comput. Sci. **6**(2), 1–33 (2010)
8. Balcázar, J.L.: Formal and computational properties of the confidence boost of association rules. TKDD **7**(4), 19 (2013)
9. Balcázar, J.L., Dogbey, F.: Evaluation of association rule quality measures through feature extraction. In: Tucker, A., Höppner, F., Siebes, A., Swift, S. (eds.) IDA 2013. LNCS, vol. 8207, pp. 68–79. Springer, Heidelberg (2013)
10. Balcázar, J.L., Tîrnăucă, C., Zorrilla, M.: Mining educational data for patterns with negations and high confidence boost. Taller de Minería de Datos TAMIDA (2010). http://personales.unican.es/tirnaucac

11. Balcázar, J. L., Tîrnăucă, C., Zorrilla, M.E.: Filtering association rules with negations on the basis of their confidence boost. KDIR (2010). http://personales.unican. es/tirnaucac
12. Bayardo, R., Agrawal, R., Gunopulos, D.: Constraint-based rule mining in large, dense databases. In: ICDE, pp. 188–197 (1999)
13. Borgelt, C.: Efficient implementations of Apriori and Eclat. In: Goethals, B., Zaki, M.J.(eds.) FIMI, CEUR Workshop Proceedings, vol. 90 (2003). CEUR-WS.org
14. Gallo, A., De Bie, T., Cristianini, N.: MINI: mining informative non-redundant itemsets. In: Kok, J.N., Koronacki, J., Lopez de Mantaras, R., Matwin, S., Mladenič, D., Skowron, A. (eds.) PKDD 2007. LNCS (LNAI), vol. 4702, pp. 438–445. Springer, Heidelberg (2007)
15. Ganter, B., Wille, R.: Formal Concept Analysis: Mathematical Foundations. Springer, Heidelberg (1999)
16. Geng, L., Hamilton, H.J.: Interestingness measures for data mining: a survey. ACM Comput. Surv. **38**(3), 1–32 (2006)
17. Guigues, J., Duquenne, V.: Familles minimales d'implications informatives resultants d'un tableau de données binaires. Math. Sci. Hum. **95**, 5–18 (1986)
18. Hájek, P., Holeňa, M., Rauch, J.: The GUHA method and its meaning for data mining. J. Comput. Syst. Sci. **76**(1), 34–48 (2010)
19. Han, J., Pei, J., Yin, Y., Mao, R.: Mining frequent patterns without candidate generation:a frequent-pattern tree approach. Data Min. Knowl. Discov. **8**(1), 53–87 (2004)
20. Jaroszewicz, S., Scheffer, T., Simovici, D.A.: Scalable pattern mining with bayesian networks as background knowledge. Data Min. Knowl. Discov. **18**(1), 56–100 (2009)
21. Kryszkiewicz, M.: Representative association rules. In: Wu, X., Kotagiri, R., Korb, K.B. (eds.) PAKDD 1998. LNCS, vol. 1394, pp. 198–209. Springer, Heidelberg (1998)
22. Kryszkiewicz, M.: Representative association rules and minimum condition maximum consequence association rules. In: Żytkow, J.M., Quafafou, M. (eds.) PKDD 1998. LNCS, vol. 1510, pp. 361–369. Springer, Heidelberg (1998)
23. Kryszkiewicz, M.: Closed set based discovery of representative association rules. In: Hoffmann, F., Adams, N., Fisher, D., Guimarães, G., Hand, D.J. (eds.) IDA 2001. LNCS, vol. 2189, pp. 350–359. Springer, Heidelberg (2001)
24. Lavrač, N., Flach, P.A., Zupan, B.: Rule evaluation measures: a unifying view. In: Džeroski, S., Flach, P.A. (eds.) ILP 1999. LNCS (LNAI), vol. 1634, pp. 174–185. Springer, Heidelberg (1999)
25. Lenca, P., Meyer, P., Vaillant, B., Lallich, S.: On selecting interestingness measures for association rules: user oriented description and multiple criteria decision aid. Eur. J. Oper. Res. **184**(2), 610–626 (2008)
26. Liu, B., Hsu, W., Ma, Y.: Pruning and summarizing the discovered associations. In: Proceeding of the Knowledge Discovery in Databases, pp. 125–134 (1999)
27. Luxenburger, M.: Implications partielles dans un contexte. Mathématiques et Sciences Humaines **29**, 35–55 (1991)
28. Megiddo, N., Srikant, R.: Discovering predictive association rules. In: Proceeding of the Knowledge Discovery in Databases, pp. 274–278 (1998)
29. Mutter, S., Hall, M., Frank, E.: Using classification to evaluate the output of confidence-based association rule mining. In: Webb, G.I., Yu, X. (eds.) AI 2004. LNCS (LNAI), vol. 3339, pp. 538–549. Springer, Heidelberg (2004)

30. Padmanabhan, B., Tuzhilin, A.: Small is beautiful: discovering the minimal set of unexpected patterns. In: Proceeding of the Knowledge Discovery in Databases, pp. 54–63 (2000)
31. Phan-Luong, V.: The representative basis for association rules. In: Cercone, V., Lin, T.Y., Wu, X. (eds.) Proceeding of the 2001 IEEE International Conference on Data Mining (ICDM), pp. 639–640. IEEE Computer Society (2001)
32. Piatetsky-Shapiro, G.: Discovery, analysis, and presentation of strong rules. In: Proceeding of the Knowledge Discovery in Databases, pp. 229–248 (1991)
33. Shah, D., Lakshmanan, L., Ramamritham, K., Sudarshan, S.: Interestingness and pruning of mined patterns. In: ACM SIGMOD Workshop on Research Issues in Data Mining and Knowledge Discovery (1999)
34. Silverstein, C., Brin, S., Motwani, R.: Beyond market baskets: generalizing association rules to dependence rules. Data Min. Knowl. Discov. **2**(1), 39–68 (1998)
35. Tan, P.-N., Kumar, V., Srivastava, J.: Selecting the right objective measure for association analysis. Inf. Syst. **29**(4), 293–313 (2004)
36. Toivonen, H., Klemettinen, M., Ronkainen, P., Hätönen, K., Mannila, H.: Pruning and grouping discovered association rules. In: ECML1995 Workshop on Statistics, Machine Learning, and Knowledge Discovery in Databases, pp. 47–52 (1995)
37. Vreeken, J., van Leeuwen, M., Siebes, A.: Krimp: mining itemsets that compress. Data Min. Knowl. Discov. **23**(1), 169–214 (2011)
38. Webb, G.I.: Efficient search for association rules. In: Ramakrishnan, R., Stolfo, S.J., Bayardo, R.J., Parsa, I. (eds.) Proceedings of the Sixth ACM SIGKDD International Conference on Knowledge Discovery and Data Mining, Boston, MA, USA, 20–23 August 2000, pp. 99–107. ACM (2000)
39. Webb, G.I.: Discovering significant patterns. Mach. Learn. **68**(1), 1–33 (2007)
40. Webb, G.I.: Self-sufficient itemsets: an approach to screening potentially interesting associations between items. TKDD **4**(1), 1–20 (2010)
41. Yin, X., Han, J.: CPAR: classification based on predictive association rules. In: Barbará, D., Kamath, C. (eds.) Proceedings of the Third SIAM International Conference on Data Mining, San Francisco, CA, USA, 1–3 May 2003, pp. 331–335. SIAM (2003)
42. Zaki, M.J.: Mining non-redundant association rules. Data Min. Knowl. Discov. **9**(3), 223–248 (2004)
43. Zaki, M.J., Wagner Meira, J.: Data Mining and Analysis: Fundamental Concepts and Algorithms. Cambridge University Press, New York (2014)
44. Zorrilla, M.E., García-Sáiz, D., Balcázar, J.L.: Towards parameter-free data mining: mining educational data with yacaree. In: Pechenizkiy, M., Calders, T., Conati, C., Ventura, S., Romero, C., Stamper, J.C. (eds.) EDM, pp. 363–364 (2011). www.educationaldatamining.org

Formal Concept Analysis from the Standpoint of Possibility Theory

Didier Dubois and Henri Prade[✉]

IRIT, Université Paul Sabatier, 118 Route de Narbonne,
31062 Toulouse Cedex 09, France
{dubois,prade}@irit.fr

Abstract. Formal concept analysis (FCA) and possibility theory (PoTh) have been developed independently. They address different concerns in information processing: while FCA exploits relations linking objects and properties, and has applications in data mining and clustering, PoTh deals with the modeling of (graded) epistemic uncertainty. However, making a formal parallel between FCA and PoTh is fruitful. The four set-functions at work in PoTh have meaningful counterparts in FCA; this leads to consider operators neglected in FCA, and thus new fixed point equations. One of these pairs of equations, paralleling the one defining formal concepts in FCA, defines independent sub-contexts of objects and properties that have nothing in common. The similarity of the structures underlying FCA and PoTh is still more striking, using a cube of opposition (a device extending the traditional square of opposition in logic). Beyond the parallel between FCA and PoTh, this invited contribution, which largely relies on several past publications by the authors, also addresses issues pertaining to the possible meanings, degree of satisfaction vs. degree of certainty, of graded object-property links, which calls for distinct manners of handling the degrees. Other lines of interest for further research are briefly mentioned.

1 Introduction

Formal concept analysis (FCA) and possibility theory (PoTh) are two theoretical frameworks that are addressing different concerns in the processing of information. Namely FCA builds concepts from a relation linking objects to the properties they satisfy, which has applications in data mining, clustering and related fields, while PoTh deals with the modeling of (graded) epistemic uncertainty. This difference of focus explains why the two settings have been developed completely independently for a very long time. However, it is possible to build a formal analogy between FCA and PoTh. Both theories heavily rely on the comparison of sets, in terms of containment or overlap. The four set-functions at work in PoTh actually determine all possible relative positions of two sets. Then the FCA operator defining the set of objects sharing a set of properties, which is at the basis of the definition of formal concepts, appears to be the counterpart of the set function expressing strong (or guaranteed) possibility in PoTh. Then, it

© Springer International Publishing Switzerland 2015
J. Baixeries et al. (Eds.): ICFCA 2015, LNAI 9113, pp. 21–38, 2015.
DOI: 10.1007/978-3-319-19545-2_2

suggests that the three other set functions existing in PoTh should also make sense in FCA, which leads to consider their FCA counterparts and new fixed point equations in terms of the new operators. One of these pairs of equations, paralleling the one defining formal concepts, define independent sub-contexts of objects and properties that have nothing in common.

The parallel of FCA with PoTh can still be made more striking using a cube of opposition (a device extending the traditional square of opposition existing in logic, and exhibiting a structure at work in many theories aiming at representing some aspects of the handling of information).

In this survey we shall indicate various issues pertaining to FCA that could be worth studying in the future. For instance, the object-property links in formal contexts of FCA may be a matter of degree. These degrees may refer to very different notions, such as the degree of satisfaction of a gradual property, the degree of certainty that an object has, or not, a property, or still the typicality of an object with respect to a set of properties. These different intended semantics call for distinct manners of handling the degrees, as advocated in the presentation.

Lastly, other examples of lines of interest for further research, such as the extension of the parallel of FCA with PoTh to conceptual pattern structures, or the applications to the fusion of conflicting pieces of information, to the clustering of sets of objects on the basis of approximate concepts, or to the building of conceptual analogical proportions, are briefly mentioned.

2   Possibility Theory and Formal Concept Analysis - A Parallel

Formal concept analysis [5, 30, 43] associates objects with the set of their properties, through a formal context which is a binary relation R on the Cartesian product of the set of objects $\mathcal{O}$ and the set of properties $\mathcal{P}$. Thus, knowing only that an object x has some property y, the set $R^t(y) = \{x \in \mathcal{O} | (x, y) \in R\}$ is the set of the *possible* objects corresponding to the elementary piece of knowledge "the object has property y" (in the context R). This suggests a possibilistic reading of formal concept analysis and leads to considering the formal counterpart to possibility theory set-functions in this framework. After introducing some notations, we first provide a short refresher on possibility theory [18, 21, 47].

2.1   Describing Objects

An object, or item, is denoted by x, or x_i in case we consider several ones at the same time. A subset of objects is denoted by a capital letter X, and we write $X = \{x_1, \ldots, x_i, \ldots, x_m\}$. A set of objects associated with their respective sets of properties defines a formal context $R \subseteq \mathcal{O} \times \mathcal{P}$ [30]. An object x is associated with its description, denoted $\partial(x)$. In the following, we only consider simple descriptions, expressible in terms of a subset Y of properties y_j, namely, $Y = \{y_1, \ldots, y_j, \ldots, y_n\}$. In such a case, we write $\partial(x) = Y$.

Besides, a useful kind of structured description is in terms of attributes. Let a, and $A = \{a_1, \ldots, a_k, \ldots, a_r\}$, respectively denote an attribute, and a set of attributes. The value of attribute a for x is denoted $a(x) = u$, where u belongs to the attribute domain U_a. In this case, we shall write $\partial(x) = (a_1(x), \ldots, a_k(x), \ldots, a_r(x)) = (u_1, \ldots, u_k, \ldots, u_r)$. This corresponds to a completely informed situation where all the considered attribute values are known for x. When it is not the case, the precise value $a_k(x)$ will be replaced by the possibility distribution $\pi_{a_k(x)}$. Such a possibility distribution [47] is a mapping from U_{a_k} to $[0,1]$, or more generally any linearly ordered scale. Then $\pi_{a_k(x)}(u) \in [0,1]$ estimates to what extent it is possible that the value of a_k for x is u. 0 means impossibility; several distinct values may be fully possible (i.e. at degree 1). The characteristic function of an ordinary subset is a particular case of a possibility distribution. Precise information corresponds to the characteristic function of singletons. An elementary property y can be viewed as a subset of a single attribute domain, i.e. $y \subseteq U$. Note that while a set of properties Y is *conjunctive* (in the sense that an object possesses *all* properties in Y), each property y corresponds to a subset of some attribute domain U that is *disjunctive* [23]: it is a set of mutually exclusive values, since object x having property y possesses a single attribute value $a(x) = u$ in U.

Taking inspiration from the existence of four set functions in possibility theory [20], new operators have been suggested in the setting of formal concept analysis [16]. These set functions are now recalled, emphasizing the symmetrical roles played by the object x and the attribute value u, a point of view unusual in possibility theory, but echoing the symmetrical role played by objects and properties in formal concept analysis. See [20,21] for more complete introductions and surveys on possibility theory.

2.2   Possibility Theory

Let $\pi_{a(x)}(u)$ denote the possibility that object x has value $u \in U$ (for attribute a). For simplicity, we only consider the single-attribute case here. We assume that π_a is bi-normalized: $\forall x \exists u \, \pi_{a(x)}(u) = 1$ and $\forall u \exists x \, \pi_{a(x)}(u) = 1$. This means that for any object x, there is some fully possible value for attribute a, and that for any value u there is an object x that takes this value. Let X be a set of objects, and $y \subseteq U$ be a property. Then, one can define

(i) the possibility measures [47], denoted by Π:

$$\Pi(X) = \max_{x \in X} \pi_{a(x)}(u) \text{ and } \Pi(y) = \max_{u \in y} \pi_{a(x)}(u).$$

$\Pi(X)$ estimates to what extent it is possible that there is an object in X having value u, while $\Pi(y)$ is the possibility that object x has property y. Π is an indicator of non-empty intersection of the fuzzy set induced by the possibility distribution with an ordinary subset. They are measures of "*weak*, or potential *possibility*". Clearly, Π is max-decomposable with respect to set union.

(ii) the dual measures of necessity N (or "strong or actual necessity") [17]:

$$N(X) = \min_{x \notin X} 1 - \pi_{a(x)}(u) \text{ and } N(y) = \min_{u \notin y} 1 - \pi_{a(x)}(u)$$

$N(X)$ estimates to what extent it is certain (necessarily true) that an object has value u is in X, while $N(y)$ is the certainty that object x has property y. Note that $N(y) = 1 - \Pi(\bar{y})$ where $\bar{y} = U \setminus y$. N may be viewed as a degree of inclusion of the fuzzy set induced by the possibility distribution into an ordinary subset. N is min-decomposable with respect to set intersection.

(iii) the measures of "strong (or actual, or guaranteed) possibility" [19]

$$\Delta(X) = \min_{x \in X} \pi_{a(x)}(u) \text{ and } \Delta(y) = \min_{u \in y} \pi_{a(x)}(u)$$

$\Delta(X)$ estimates to what extent it is possible that *all* objects in X have value u, while $\Delta(y)$ estimates the possibility that object x takes any value in y. Δ may be viewed as a degree of inclusion of an ordinary subset into the fuzzy set induced by the possibility distribution. Δ is min-decomposable with respect to set union.

(iv) the dual measures of "weak (or potential) necessity or certainty" [19]

$$\nabla(X) = 1 - \min_{x \notin X} \pi_{a(x)}(u) \text{ and } \nabla(y) = 1 - \min_{u \notin y} \pi_{a(x)}(u)$$

$\nabla(X)$ estimates to what extent there exists at least one object outside X that has a low degree of possibility of having value u, while $\nabla(y)$ measures to what extent x has a low possibility value outside y. Note that $\nabla(y) = 1 - \Delta(\bar{y})$. ∇ is an indicator of non-full coverage of the considered universe by the fuzzy set induced by the possibility distribution together with an ordinary subset. ∇ is max-decomposable with respect to set intersection.

2.3   Formal Context Setting

The classical setting of formal concept analysis defined from a formal context relies on a single operator that associates a subset of objects with the set of properties shared by them (and the dual operator). In [16], this framework has been enlarged with the introduction of three other operators. We now recall the four operators which are counterparts to the possibility theory set functions in the setting of a formal context.

Namely, let R be the formal context. Then $R(x) = \{y \in \mathcal{P}|(x,y) \in R\}$ is the set of properties of object x, and $R^t(y) = \{x \in \mathcal{O}|(x,y) \in R\}$ is the set of objects having properties y. Then, four remarkable sets can be associated with a subset X of objects (the notations have been chosen here in order to emphasize the parallel with possibility theory):

– the set $R^{\Pi}(X)$ of properties that are possessed by *at least one* object in X:

$$R^{\Pi}(X) = \{y \in \mathcal{P}|R^t(y) \cap X \neq \emptyset\} = \bigcup_{x \in X} R(x).$$

Clearly, we have $R^{\Pi}(X_1 \cup X_2) = R^{\Pi}(X_1) \cup R^{\Pi}(X_2)$.

– the set $R^N(X)$ of properties s. t. any object that satisfies *one* of them is necessarily in X:

$$R^N(X) = \{y \in \mathcal{P}|R^t(y) \subseteq X\} = \bigcap_{x \notin X} \overline{R(x)}.$$

In other words, having any property in $R^N(X)$ is a sufficient condition for belonging to X. Moreover, we have $R^N(X) = \overline{R^{\Pi}(\overline{X})} = \mathcal{P} \setminus R^{\Pi}(\overline{X})$, and $R^N(X_1 \cap X_2) = R^N(X_1) \cap R^N(X_2)$.

– the set $R^{\triangle}(X)$ of properties shared by *all* objects in X:

$$R^{\triangle}(X) = \{y \in \mathcal{P}|R^t(y) \supseteq X\} = \bigcap_{x \in X} R(x).$$

In other words, satisfying all properties in $R^{\triangle}(X)$ is a necessary condition for an object for belonging to X. $R^{\triangle}(X)$ is a partial conceptual characterization of objects in X: objects in X have all the properties of $R^{\triangle}(X)$ and may have some others (that are not shared by all objects in X). It is worth noticing that $\overline{R^{\Pi}(X)}$ provides a negative conceptual characterization of objects in X since it gathers all the properties that are never satisfied by any object in X. Moreover, we have $R^{\triangle}(X_1 \cup X_2) = R^{\triangle}(X_1) \cap R^{\triangle}(X_2)$. Besides, as can be seen, $R^N(X) \cap R^{\triangle}(X)$ is the set of properties possessed by all objects in X and only by them.

– the set $R^{\triangledown}(X)$ of properties that are not satisfied by at least one object in $\overline{X}$.

$$R^{\triangledown}(X) = \{y \in \mathcal{P}|R^t(y) \cup X \neq \mathcal{O}\} = \bigcup_{x \notin X} \overline{R(x)}.$$

Note that $R^{\triangledown}(X) = \overline{R^{\triangle}(\overline{X})} = \mathcal{P} \setminus R^{\triangle}(\overline{X})$. In other words, in context R, for any property in $R^{\triangledown}(X)$, there exists at least one object outside X that misses it. Moreover, we have $R^{\triangledown}(X_1 \cap X_2) = R^{\triangledown}(X_1) \cup R^{\triangledown}(X_2)$.

Note that $R^{\Pi}(X)$ and $R^N(X)$ become larger when X increases, while $R^{\triangle}(X)$ and $R^{\triangledown}(X)$ get smaller. The four subsets $R^{\Pi}(X)$, $R^N(X)$, $R^{\triangle}(X)$, and $R^{\triangledown}(X)$ have been considered by different authors (with different notations) without any reference to possibility theory. Düntsch et al. [26,27] calls $R^{\triangle}$ a *sufficiency* operator, and its representation capabilities are studied in the theory of Boolean algebras. Taking inspiration as the previous authors from rough sets [40], Yao [45,46] also considers these four subsets. In both cases, the four operators were introduced. See also [33,41].

2.4 The Cube of Opposition in FCA

Before being able to present the structures of opposition relating the four operators introduced in the previous section, we need to start with a refresher on the Aristotelian square of opposition [39]. The traditional square involves four logically related statements exhibiting universal or existential quantifications: it has been noticed that a statement **A** of the form "every x is p" is negated by the statement **O** "some x is not p", while a statement like **E** "no x is p" is clearly in even stronger opposition to the first statement **A**. These three statements, together with the negation of the last one, namely **I** "some x is p", give birth to the Aristotelian square of opposition in terms of quantifiers **A**: $\forall x\ p(x)$, **E**: $\forall x\ \neg p(x)$, **I**: $\exists x\ p(x)$, **O**: $\exists x\ \neg p(x)$, pictured in Fig. 1. Such a square is usually denoted by the letters **A**, **I** (affirmative half) and **E**, **O** (negative half). The names of the vertices come from a traditional Latin reading: **Aff**I**rmo**, n**E**g**O**).

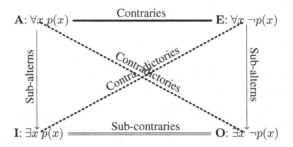

Fig. 1. Square of opposition

As can be seen, different relations hold between the vertices. Namely,

(a) **A** and **O** are the negation of each other, as well as **E** and **I**;
(b) **A** entails **I**, and **E** entails **O** (we assume that there are some x for avoiding existential import problems);
(c) **A** and **E** cannot be true together, but may be false together;
(d) **I** and **O** cannot be false together, but may be true together.

Recently, it has been noticed that such a square can be generated by a binary relation and a subset that can be composed together [12]. Indeed, let R be a binary relation on a Cartesian product $\mathcal{X} \times \mathcal{Y}$ (nothing forbids $\mathcal{Y} = \mathcal{X}$ in the construction we are going to describe). We assume $R \neq \emptyset$. Let R^t denote the transposed relation $((y, x) \in R^t$ iff $(x, y) \in R)$. Moreover, we assume that $\forall x$, $R(x) \neq \emptyset$, which means that the relation R is *serial*, namely $\forall x, \exists y$ such that $(x, y) \in R$; this is also referred to in the following as the $\mathcal{X}$-*normalization condition*. In the same way R^t is also supposed to be serial, i.e., $\forall y$, $R^t(y) \neq \emptyset$ ($\mathcal{Y}$-normalization). We further assume that the complementary relation $\overline{R}$ $((x, y) \in \overline{R}$ iff $(x, y) \notin R)$, and its transpose are also serial, i.e. $\forall x$, $R(x) \neq \mathcal{Y}$ and $\forall y$, $R^t(y) \neq \mathcal{X}$. These conditions enforce a non trivial relation between $\mathcal{X}$ and $\mathcal{Y}$. In the following, set complementations will be denoted by means of overbars.

Let S be a subset of $\mathcal{Y}$. We assume $S \neq \emptyset$ and $S \neq \mathcal{Y}$. The relation R and the subset S give birth to the following subset of X, namely the (left) image of S by R

$$R(S) = \{x \in X \mid \exists s \in S, (x, s) \in R\} = \{x \in X \mid S \cap R(x) \neq \emptyset\}.$$

Similarly, we consider $R(\overline{S})$, $\overline{R(S)}$, and $\overline{R(\overline{S})} = \{x \in X \mid \forall s \in \overline{S}, (x, s) \notin R\} = \{x \in X \mid R(x) \subseteq S\}$. The four subsets thus defined can be nicely organized into a square of opposition. See Fig. 2. Indeed, it can be checked that the set counterparts of the relations existing between the logical statements of the traditional square of oppositions still hold here. Namely, $R(\overline{S})$ and $\overline{R(\overline{S})}$ are complements of each other, as $\overline{R(S)}$ and $R(S)$; we have $\overline{R(\overline{S})} \subseteq R(S)$ and $\overline{R(S)} \subseteq R(\overline{S})$, thanks to X-normalization condition; $\overline{R(\overline{S})} \cap \overline{R(S)} = \emptyset$; $R(S) \cup R(\overline{S}) = X$.

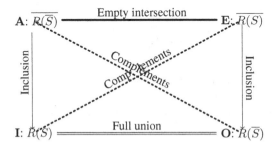

Fig. 2. Square of oppositions induced by a relation R and a subset S

Let us now consider the complementary relation $\overline{R}$. We further assume that $\overline{R} \neq \emptyset$ (i.e., $R \neq X \times Y$). Moreover we have also assumed the X-normalization of $\overline{R}$, i.e. $\forall x, \exists y\ (x, y) \notin R$. In the same way as previously, we get four other subsets of X from $\overline{R}$. Namely, $\overline{R}(\overline{S}) = \{x \in X \mid \exists s \in \overline{S}, (x, s) \notin R\} = \{x \in X \mid S \cup R(x) \neq X\}$; $\overline{R}(S)$; $\overline{\overline{R}(\overline{S})}$; $\overline{\overline{R}(S)} = \{x \in X \mid \forall s \in S, (x, s) \in R\} = \{x \in X \mid S \subseteq R(x)\}$. This generates a second square of opposition denoted by **aeoi**.

As can be seen, when R is a formal context (i.e., $X = O$, $Y = P$), we have $R^{\Pi}(S) = R(S)$, $R^N(S) = \overline{R(\overline{S})}$, $R^{\vartriangle}(S) = \overline{\overline{R}(S)}$, $R^{\triangledown}(S) = \overline{R}(\overline{S})$. The eight subsets involving R and its complement can be organized into a cube of opposition as in Fig. 3. The four formal concept analysis operators correspond to the left side facet of the cube of oppositions. The full cube is then obtained by introducing their complements, giving birth to the right side facet. Since $\overline{R^{\Pi}(S)} = R^N(\overline{S})$, and $\overline{R^{\vartriangle}(S)} = R^{\triangledown}(\overline{S})$, the classical square of oppositions **AEOI** is given by the four corners $R^N(S)$, $R^N(\overline{S})$, $R^{\Pi}(\overline{S})$, and $R^{\Pi}(S)$, and the second square **aeoi** on the back of the cube is given by $R^{\vartriangle}(S)$, $R^{\vartriangle}(\overline{S})$, $R^{\triangledown}(\overline{S})$, and $R^{\triangledown}(S)$.

Moreover, in the side facets, all edges are uni-directed, including the diagonal ones, and express inclusions. Indeed, as already established in [16], under the X- and Y-normalization hypotheses, the following inclusion relation holds:

$$R^N(S) \cup R^{\vartriangle}(S) \subseteq R^{\Pi}(S) \cap R^{\triangledown}(S).$$

$R^N(S)$, $R^A(S)$, $R^\Pi(S)$, and $R^\nabla(S)$ constitute, four distinct pieces of information [16], which are only (weakly) related by the above relation.

Lastly, it can be checked that we also have $R^A(S) \cap R^N(\overline{S}) = \emptyset$ and $R^A(\overline{S}) \cap R^N(S) = \emptyset$ on the one hand, and $R^\nabla(S) \cup R^\Pi(\overline{S}) = \mathcal{X}$ and $R^\nabla(\overline{S}) \cup R^\Pi(S) = \mathcal{X}$ on the other hand. These are the relations that holds on the top and on the bottom facets of the cube respectively.

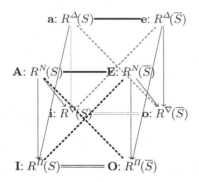

Fig. 3. Cube of opposition in formal concept analysis

The cube of oppositions not only underlie FCA (and PoTh) [25], but also is a setting of interest for building bridges with rough set theory [40] (see [12]), or even formal argumentation [1]!

3   Formal Context Decomposition

In FCA, a formal concept [30] is defined as a pair $(X, Y) \in \mathcal{O} \times \mathcal{P}$ such that

$$R^\triangle(X) = Y \text{ and } R^{t\triangle}(Y) = X,$$

where $R^{t\triangle}(Y) = \{x \in \mathcal{O} | R(x) \supseteq Y\} = \bigcap_{y \in Y} R^t(y)$ is the set X of objects having all properties in Y, and in this case Y is also the maximal set of properties shared by all objects in X. A formal concept (X, Y) is a maximal sub-rectangle in the formal context, i.e. is such that $X \times Y \subseteq R$. It can be checked that R^∇ gives back the same Galois connection as the one defined from $R^\triangle$, while R^N (or R^Π) induces another connection, which is now described.

Consider the connection defined from R^N in a similar formal way as when defining formal concepts. It was proposed by Popescu [41] and studied in a general setting of residuated algebras, but not in the usual Boolean setting. Namely, let us consider pairs (X, Y) s.t. $R^N(X) = Y$ and $R^{tN}(Y) = X$. As suggested in [22], the pairs (X, Y) s.t. $R^N(X) = Y$ and $R^{tN}(Y) = X$ allow us to characterize independent sub-contexts (i.e. that have no common objects and no common properties). They are thus of interest for the decomposition of a formal context into smaller independent ones. This is expressed through the following property, proved in [13,24]:[1]

Proposition 1. *The following properties of pairs* (X, Y) *are equivalent*

1. $R^N(X) = Y$ *and* $R^{tN}(Y) = X$
2. $R^N(\overline{X}) = \overline{Y}$ *and* $R^{tN}(\overline{Y}) = \overline{X}$
3. $R^{\Pi}(X) = Y$ *and* $R^{t\Pi}(Y) = X$
4. $R \subseteq (X \times Y) \cup (\overline{X} \times \overline{Y})$

Proof. Let us first show that Property 1 implies Property 4. First it is clear that:
$R^N(X) = Y \Leftrightarrow \bigcap_{x \in \overline{X}} \overline{R(x)} = Y \Leftrightarrow \bigcup_{x \in \overline{X}} R(x) = \overline{Y}.$

Denoting $X + Y = \overline{X} \times \overline{Y}$, it implies $R \subseteq X + \overline{Y}$.

Likewise due to $R^{tN}(Y) = X, \quad R^t \subseteq Y + \overline{X}$ holds.

Finally: $R \subseteq (X + \overline{Y}) \cap (Y + \overline{X})$, which equivalently writes: $R \subseteq (X \times Y) \cup (\overline{X} \times \overline{Y})$.

Conversely assume Property 4. Then it is clear that $R^N(X) \subseteq Y$ and $R^{tN}(Y) \subseteq X$ hold since there is no property possessed by any object in X outside Y, and no object outside X that possesses a property outside Y. Suppose $R^N(X) \subset Y$, i.e. $\exists y^* \in Y$ such that property y^* is possessed by objects outside X. But then $R(x, y^*) = 1$ for some $x \in \overline{X}, y \in \overline{Y}$. So Property 4 does not hold. Contradiction.

The invariance of Property 4 with respect to complementation proves that the choice of (X, Y) versus $(\overline{X}, \overline{Y})$ in Property 1 is immaterial. Hence the equivalence with Property 2. For Property 3, note that $R^N(X) = Y$ is equivalent to $R^{\Pi}(\overline{X}) = \overline{Y}.$ □

Thus, (X, Y) and $(\overline{X}, \overline{Y})$ are two independent sub-contexts in R, in the sense that there is no object / property pair (x, y) of the context R either in $X \times \overline{Y}$ or in $\overline{X} \times Y$. The above proposition does not involve any minimality in the inclusion Property 4 of the above proposition. In particular, the pair $(\mathcal{O}, \mathcal{P})$ trivially satisfies it. However, this result leads to a decomposition of R into a disjoint union of *minimal* independent sub-contexts. Indeed, suppose two pairs (X_1, Y_1), (X_2, Y_2) satisfy Proposition 1. It implies that for instance, the pair $(X_1 \cap X_2, Y_1 \cap Y_2)$ satisfies it (it can be checked that $R^N(X_1 \cap X_2) = Y_1 \cap Y_2$), and likewise with any element of the partition refining both partitions $(X_1, \overline{X_1})$ and $(X_2, \overline{X_2})$. Due to point 4 of Proposition 1, it yields

$$R \subseteq ((X_1 \times Y_1) \cup (\overline{X_1} \times \overline{Y_1})) \cap ((X_2 \times Y_2) \cup (\overline{X_2} \times \overline{Y_2})),$$

where the intersection on the right-hand side comes down to the union of subcontexts $(X_1 \cap X_2) \times (Y_1 \cap Y_2)$, $(X_1 \cap \overline{X_2}) \times (Y_1 \cap \overline{Y_2})$, $(\overline{X_1} \cap X_2) \times (\overline{Y_1} \cap Y_2)$, $(\overline{X_1} \cap \overline{X_2}) \times (\overline{Y_1} \cap \overline{Y_2})$. The decomposition of R into minimal subcontexts is achieved by taking the following intersection

$$\bigcap_{(X,Y):R^N(X)=Y, R^{tN}(Y)=X} (X \times Y) \cup (\overline{X} \times \overline{Y}).$$

---

[1] We again provide the proof for the sake of self-containedness.

Example 1. The table below presents a formal context. Pairs $(\{6,7,8\}, \{c,d,e\})$, or $(\{5,6,7,8\}, \{d,e\})$, or $(\{2,3,4\}, \{g,h\})$ are examples of formal concepts, while $(\{5,6,7,8\}, \{a,b,c,d,e\})$, $(\{2,3,4\}, (\{f,g,h\}))$, $(\{1\}, \{i\})$ are minimal subcontexts.

		objects							
p		1	2	3	4	5	6	7	8
r	a						×		
o	b				×	×			
p	c						×	×	×
e	d					×	×	×	×
r	e					×	×	×	×
t	f	×		×					
i	g	×	×	×					
e	h	×	×	×					
s	i	×							

Thus, through the notions of formal sub-contexts and of formal concepts, one sees two key aspects of granulation at work. Namely, on the one hand independent sub-contexts are separated, while *inside* each sub-context, formal concepts (X, Y) are identified where each object in X is associated with each property in Y. However, objects in the extension of a formal concept may not be fully similar since they may also possess properties outside the intension of the concept. They are only similar with respect to the properties associated to the formal concept.

Thus, the classical Galois connection founding formal concept analysis (associated with the actual possibility operator), and the other connection induced by the actual necessity operator, respectively embed two basic ideas associated with the idea of a cluster (see, e.g., [35]), namely

1. any pair of elements in a cluster should be closely related in some sense, and
2. any element of a cluster should be sufficiently separated from any element outside it.

Moreover, formal concept analysis is also useful for conceptual clustering, where clusters should be associated with labels, obtained in this case as a conjunction of the properties shared by the objects in the cluster [11].

Such an idea can be also stated in terms of graph clustering, taking advantage of an exact parallel between formal concept analysis and bipartite graph analysis [31], as viewing an (ideal) cluster as a group of vertices

1. either with no missing link inside the group,
2. or with no link with vertices outside the group.

These two complementary views are also clearly at the basis of cluster analysis for unipartite graphs [42].

In practice, it is important to introduce some tolerance in the evaluation of the similarity between the members of a cluster and in the separatedness of the clusters, leading to a more permissive and approximate view of granules or clusters; see, e.g., [32].

The other (mixed) connections $R^\Theta(S) = T$ and $R^{t\Lambda}(T) = S$ where $\Theta, \Lambda \in \{\Pi, N, \Delta, \nabla\}$ with $\Theta \neq \Lambda$ are also worth studying. They have still to be better understood and to be investigated systematically. See [15] for a preliminary discussion, and [7,9] for results in the graded case.

4   Graded Links Between a Property and an Object

Fuzzy extensions of FCA where R is a fuzzy relation in $L^{\mathcal{O} \times \mathcal{P}}$ with L often taken as the unit interval have been proposed early [6,10]. However, the development of a fuzzy formal concept analysis theory requires an appropriate algebra of fuzzy sets [6,8]. While many theoretical studies have been developed, the different gradual interpretations of a fuzzy formal context have not been much discussed. Following [14], this section highlights some basic issues regarding the fact that a "fuzzy" or graded extension of binary formal contexts may convey different semantics: graded satisfaction of properties vs. uncertainty.

4.1   Gradual Properties: Unipolar Vs. Bipolar Scale Interpretation

In this first interpretation, the values in the table (which are scalars in L) may be understood as providing a refinement of the cross marks. Namely, they represent to what extent an object has a property, while in the classical model, this relationship was not a matter of degree. It is important to remark that in this view, we do not refine the absence of a property for an object (the blank mark is always replaced by the bottom element 0 of L). This view will be referred to as the *positive unipolar interpretation*. In this interpretation, $R^t(y)$ (resp. $R(x)$) is considered as the support of the fuzzy set of objects (resp. properties) satisfying the property y (resp. the object x). One could also consider the opposite convention namely the *negative unipolar interpretation* where degrees would represent to which extent an object does not have a property and equivalently provide a refinement of the blank marks.

The most commonly used interpretations, through existing FCA proposals, are implicitly based on the positive unipolar interpretation that allows to map a formal context with quantitative attributes into a fuzzy formal context. In this spirit, conceptual scale theory [44] may be used to achieve a suitable (Boolean) representation by successive subsumptions.

Example 2. For instance, the formal context illustrated in Table 2 is obtained from Table 1 by a conceptual scaling of both many-valued attributes "Age" and "Salary". As can be seen, we have two sets of properties with obvious subsumption relations between them. Pairs ($\{Peter, Sophie, Mike, Joe\}, \{age \geq 20, salary \geq 1000\}$), ($\{Sophie, Mike\}, \{age \geq 20, age \geq 25, salary \geq 1000, salary \geq 1200\}$), or ($\{Mike\}, \{age \geq 20, age \geq 25, age \geq 30, salary \geq 1000, salary \geq 1200, salary \geq 1400\}$) are formal concepts.

Table 1. Many-valued relation

R_1	Pierre	Sophie	Mike	Nahla
Age	22	28	30	22
Salary	1100	1300	1500	1500

Table 2. Context subsumption

R_2	Pierre	Sophie	Mike	Nahla
age $\geq$ 20	×	×	×	×
age $\geq$ 25		×	×	
age $\geq$ 30			×	
salaire $\geq$ 1000	×	×	×	×
salaire $\geq$ 1200		×	×	×
salaire $\geq$ 1400			×	×

Table 3. Context summarization

R_3	Pierre	Sophie	Mike	Nahla
age 'young'	1	0.7	0.6	1
salary 'low'	1	0.8	0.6	0.6

Knowing the ages and the salaries, the formal context R_2 can be re-encoded in a more compact way, using two fuzzy sets 'young' and 'small' with decreasing membership functions, as illustrated in Table 3.

Observe also that R_3 offers a more precise representation of initial data than Table 2. The context in Table 3, event though more compact than Table 2 highlights the fact that *Mike*, and to a lesser extent *Sophie* are not very young and have a salary that is not really low. It constitutes in some sense the negative of the picture shown on Table 1. Note that the type of representation on Table 3 can be obtained even without providing interpretable fuzzy sets and thus, by normalizing in L the domain of attribute values. This approach is used in [36].

Another interpretation of the degrees, maybe more in the standard spirit of fuzzy logic would be to replace both the *cross* marks and the *blank* marks by values in the scale L ($L = [0, 1]$). Then L possesses a mid-point acting as a pivoting value between the situations where the object possesses the property to some extent and the converse situation where the object possesses the opposite property to some extent. Under this view, a fuzzy formal concept should be learnt together with its negation. This view corresponds to a *bipolar scale interpretation*.

4.2   Uncertainty

Neither the standard FCA approach nor its fuzzy extension are equipped for representing situations of partial or complete ignorance. To this end, in the

Boolean case, we need to introduce a proper representation of partial uncertainty including ignorance in the relational table of the formal context. One may think of introducing gradations of uncertainty by changing crosses and blanks in the table into probability degrees, or by possibility or necessity degrees. In the probabilistic case, one number shall assess the probability that a considered property holds for a given object (its complement to 1 corresponding to the probability it does not hold). However, this is assuming precise knowledge on the probability values, which is not really appropriate if we have to model the state of complete ignorance. It is why we investigate the use of the possibilistic setting in the following.

In the possibilistic setting, crosses may be replaced by positive degrees of necessity for expressing some certainty that an object satisfies a property. The blanks could be refined by possibility degrees less than 1, expressing that it is little possible that an object satisfies a property. However, this convention using a single number in the unit interval for each entry in the context may be misleading as when the number replaces a blank or a cross, the meaning of the number is not the same.

In the possibilistic setting, possibility and necessity functions are related by the duality relation $N(A) = 1 - \Pi(\overline{A})$, that holds for any event A, where $\overline{A}$ denotes the opposite event [18]. Then, for entries (x, y) in the table, we use a representation as *a pair* of necessity degrees $(\alpha, 1 - \beta)$ where $\alpha = N((x, y) \in R)$ (resp. $1 - \beta = N((x, y) \notin R)$) corresponds to the necessity (certainty) that object x has (resp. does not have) property y. Moreover, we should respect the property $\min(\alpha, 1 - \beta) = 0$, since $\min(N(A), N(\overline{A})) = 0$ in agreement with complete ignorance, in which case nothing (i.e., neither A nor $\overline{A}$) is even somewhat certain. Pairs (1,0) and (0,1) correspond to completely informed situations where it is known that object x has, respectively does not have, property y. The pair (0,0) reflects total ignorance, whereas pairs $(\alpha, 1 - \beta)$ s.t. $1 > \max(\alpha, 1 - \beta) > 0$ correspond to partial ignorance.

An uncertain formal context is thus represented by

$$R^U = \{(\alpha(x, y), 1 - \beta(x, y)) \mid x \in \mathcal{O}, y \in \mathcal{P}\}$$

where $\alpha(x, y) \in [0, 1]$, $\beta(x, y) \in [0, 1]$. A relational database with fuzzily-known attribute values is *theoretically* equivalent to the fuzzy set of all ordinary databases corresponding to the different possible ways of completing the information consistently with the fuzzy restrictions on the attribute values. So, an uncertain formal context may be viewed as a weighted family of all standard formal contexts obtained by changing uncertain entries into sure ones. More precisely, one may consider all the completions of an uncertain formal context. This is done by substituting entries (x, y) that are uncertain, i.e., such that $1 > \max(\alpha(x, y), 1 - \beta(x, y))$ by a pair (1,0), or a pair (0,1). Replacing $(\alpha(x, y), 1 - \beta(x, y))$ by (1, 0) is possible at degree $\beta(x, y)$, the possibility that x has property y. Similarly, replacing $(\alpha(x, y), 1 - \beta(x, y))$ by (0, 1) is possible at degree $1 - \alpha(x, y)$, the possibility that x does *not* have the property y. In this way, one may determine to what extent a particular completion (a context C) is possible, by aggregating the possibility degrees associated with each completed entry (using min operator). Formally, one can write

$$\pi(C) = \min(\min_{(x,y):(x,y)\in C}\beta(x,y), \min_{(x,y):(x,y)\notin C}1 - \alpha(x,y)).$$

Likewise the degree of possibility that (X,Y) is a formal concept of R^U is

$$\pi(X,Y) = \sup\{\pi(C) : C \text{ such that } (X,Y) \text{ is a formal concept of } C\}.$$

Useful completions are those where partial certainty becomes full certainty. Indeed, given an uncertain formal context and a threshold pair (u, v), let us replace all entries of the form $(\alpha, 0)$ such that $\alpha \geqslant u$ with $(1, 0)$ and entries of the form $(0, 1 - \beta)$ such that $1 - \beta \geqslant v$ with $(0, 1)$. All such replacements have possibility 1 according to the above formula. Remaining entries, which are more uncertain, can be systematically substituted either by $(1,0)$, or by $(0,1)$. Considering, the two extreme cases where *all* such entries are changed into $(1,0)$ and the case when where *all* such entries are changed into $(0,1)$ gives birth to upper and lower completions, respectively. In this way, two classical (Boolean) formal contexts, denoted $R^*_{(u,v)}$ and $R_{*(u,v)}$ are obtained as respective results of the two completions. They allow to determine, for a given threshold (u, v), maximal extensions (resp. minimal intensions) and minimal extensions (resp. maximal intensions) of uncertain formal concepts. It is clear that $R_{*(u,v)} \subseteq R^*_{(u,v)}$. Let us illustrate the idea with an example.

Example 3. Table 4 exhibits a formal context where some entries are pervaded with uncertainty. Let us examine the situation regarding formal concepts. Take $u = 0.7, v = 0.5$ for instance. In context $R_{*(0.7,0.5)}$, examples of formal concepts are pairs $(\{6,7,8\}, \{c,d,e\})$, or $(\{5,6,7,8\}, \{d,e\})$, or $(\{2,3,4\}, \{g,h\})$, although with $u = 0.9$, the last formal concept would reduce to $(\{2,3\}, \{g,h\})$, i.e., the extent of the concept is smaller.

Now consider $R^*_{(0.7,0.5)}$, where the entries with low certainty levels (either in favor or against the existence of the link between x and y) are turned into

Table 4. Uncertain formal concepts

	1	2	3	4	5	6	7	8
a						×		
b					×	×		
c					(0.5,0)	×	×	×
d					×	×	×	×
e					×	×	×	×
f		(0, 0.8)		×	(0, 0.3)			
g		×	×	(0.8, 0)				
h		×	×	(0.8, 0)				
i	×							

positive links. Then, $(\{2,3,4\}, \{g,h\})$ remains unchanged as a formal concept, while a larger concept now emerges, namely $(\{5,6,7,8\}, \{c,d,e\})$. However, one may prefer to consider the results obtained from $R_{*(0.7,0.5)}$, where only the almost certain information is changed into positive links. In the example, if we move down u to 0.5, and use $R_{*(0.5,0.5)}$ we still validate the larger former concept $(\{5,6,7,8\}, \{c,d,e\})$. This illustrates the fact that becoming less and less demanding on the level of certainty, may enable the fusion of close concepts (here $(\{6,7,8\}, \{c,d,e\})$, and $(\{5,6,7,8\}, \{d,e\})$), providing a more synthetic view of the formal context.

This small example is intended to illustrate several points. First of all, it should be clear that being uncertain about the existence of a link between an object and a property is not the same as being certain about a gradual link. Second, under uncertainty, there are formal concepts whose boundaries are not affected by uncertainty, while others are. Lastly, regarding certain enough pieces of information as fully certain may help simplifying the analysis of the formal context. Besides, the proposed setting may also handle inconsistent information by relaxing the constraint $\min(\alpha, 1 - \beta) = 0$. This would amount to introducing paraconsistent links between objects and properties.

5   More Lines for Further Research

Let us briefly conclude this survey of works in FCA inspired by PoTh by mentioning other examples of lines of interest for further research:

- The parallel of FCA with PoTh leading to the introduction of new operators extends to conceptual pattern structures [28,29], where the description $\partial(x)$ of an object x, may, e.g., be a possibilistic knowledge base [2];
- Applications of FCA to the fusion of conflicting pieces of information issued from multiple sources using pattern structures for labeling sets of possible values in terms of sources supporting them [3];
- The clustering of sets of objects on the basis of approximate concepts [24,32], with labeling of the clusters [38];
- The building of conceptual analogical proportions [37] on the basis of the formal definition of analogical proportions in non-distributive lattices [34], conceptualization and analogical reasoning being two basic cognitive activities [4].

References

1. Amgoud, L., Prade, H.: A formal concept view of abstract argumentation. In: van der Gaag, L.C. (ed.) ECSQARU 2013. LNCS, vol. 7958, pp. 1–12. Springer, Heidelberg (2013)
2. Assaghir, Z., Kaytoue, M., Prade, H.: A possibility theory-oriented discussion of conceptual pattern structures. In: Deshpande, A., Hunter, A. (eds.) SUM 2010. LNCS, vol. 6379, pp. 70–83. Springer, Heidelberg (2010)

3. Assaghir, Z., Napoli, A., Kaytoue, M., Dubois, D., Prade, H.: Numerical information fusion: lattice of answers with supporting arguments. In: Proceedings of 23rd IEEE International Conference on Tools with Artificial Intelligence (ICTAI'11), pp. 621–628, Boca Raton, 7–9 November 2011
4. Barbot, N., Miclet, L., Prade, H.: Analogical Proportions and the Factorization of Information in Distributive Lattices. In: CEUR Workshop Proceedings of the Tenth International Conference on Concept Lattices and Their Applications CLA 2013, pp. 175-186, La Rochelle, 15-18 October 2013
5. Barbut, M., Monjardet, B.: Ordre et Classification. Algèbre et Combinatoire. Tome 2. Hachette, Paris (1970)
6. Bělohlávek, R.: Fuzzy Galois connections. Math. Logic Q. **45**, 497–504 (1999)
7. Bělohlávek, R.: Optimal triangular decompositions of matrices with entries from residuated lattices. Int. J. Approximate Reasoning **50**(8), 1250–1258 (2009)
8. Bělohlávek, R., Vychodil, V.: What is a fuzzy concept lattice. In: Proceedings CLA2005, pp. 34–45, Olomuc, Czech Republic (2005)
9. Bělohlávek, R., Vychodil, V.: Discovery of optimal factors in binary data via a novel method of matrix decomposition. J. Comput. Syst. Sci. **76**(1), 3–20 (2010)
10. Burusco, A., Fuentes-Gonzalez, R.: The study of the L-fuzzy concept lattice. Mathw. Soft Comput. **3**, 209–218 (1994)
11. Carpineto, C., Romano, G.: GALOIS: an order-theoretic approach to conceptual clustering. In: Proceedings of 10th International Conference on Machine Learning, pp. 33–40. Morgan Kaufmann, Amherst, 27–29 June 1993
12. Ciucci, D., Dubois, D., Prade, H.: The structure of oppositions in rough set theory and formal concept analysis - toward a new bridge between the two settings. In: Beierle, C., Meghini, C. (eds.) FoIKS 2014. LNCS, vol. 8367, pp. 154–173. Springer, Heidelberg (2014)
13. Djouadi, Y., Dubois, D., Prade, H.: Possibility theory and formal concept analysis: context decomposition and uncertainty handling. In: Hüllermeier, E., Kruse, R., Hoffmann, F. (eds.) IPMU 2010. LNCS, vol. 6178, pp. 260–269. Springer, Heidelberg (2010)
14. Djouadi, Y., Dubois, D., Prade, H.: Graduality, uncertainty and typicality in formal concept analysis. In: Cornelis, C., Deschrijver, G., Nachtegael, M., Schockaert, S., Shi, Y. (eds.) 35 years of Fuzzy Sets Theory - Celebratory Volume Dedicated to the Retirement of Etienne E. Kerre. SFSC 261, pp. 127–147. Springer, Heidelberg (2011)
15. Djouadi, Y., Prade, H.: Possibility-theoretic extension of derivation operators in formal concept analysis over fuzzy lattices. Fuzzy Optim. Decis. Making **10**(4), 287–309 (2011)
16. Dubois, D., Dupin de Saint Cyr, F., Prade, H.: A possibility-theoretic view of formal concept analysis. Fundamenta Informaticae **75**(1–4), 195–213 (2007)
17. Dubois, D., Prade, H.: Fuzzy Sets and Systems: Theory and Applications. Academic Press, New York (1980)
18. Dubois, D., Prade, H.: Possibility Theory: An Approach to Computerized Processing of Uncertainty. Plenum Press, New York (1988)
19. Dubois, D., Prade, H.: Possibility theory as a basis for preference propagation in automated reasoning. In: Proceedings of 1st IEEE International Conference on Fuzzy Systems (FUZZ-IEEE 1992), San Diego, pp. 821–832, 8–12 March 1992
20. Dubois, D., Prade, H.: Possibility theory: qualitative and quantitative aspects. In: Gabbay, D., Smets, P. (eds.) Quantified Representation of Uncertainty and Imprecision of Handbook of Defeasible Reasoning and Uncertainty Management Systems, vol. 1, pp. 169–226. Kluwer Academic Publisher, Dordercht (1998)

21. Dubois, D., Prade, H.: Possibility theory. In: Meyers, R. (ed.) Encyclopedia of Complexity and Systems Science, pp. 6927–6939. Springer, New York (2009)
22. Dubois, D., Prade, H.: Possibility theory and formal concept analysis in information systems. In: Proceedings of International Fuzzy Systems Association World Congress and Conference of the Europe Society for Fuzzy Logic and Technology (IFSA-EUSFLAT 2009), pp. 1021–1026, Lisbon, 20–24 July 2009
23. Dubois, D., Prade, H.: Gradualness, uncertainty and bipolarity: making sense of fuzzy sets. Fuzzy Sets Syst. **192**, 3–24 (2012)
24. Dubois, D., Prade, H.: Possibility theory and formal concept analysis: characterizing independent sub-contexts. Fuzzy Sets Syst. **196**, 4–16 (2012)
25. Dubois, D., Prade, H.: From Blanché's hexagonal organization of concepts to formal concept analysis and possibility theory. Logica Universalis **6**(1–2), 149–169 (2012)
26. Düntsch, I., Gediga, G.: Approximation operators in qualitative data analysis. In: de Swart, H., Orlowska, E., Schmidt, G., Roubens, M. (eds.) Theory and Application of Relational Structures as Knowledge Instruments, pp. 216–233. Springer, Heidelberg (2003)
27. Düntsch, I., Orlowska, E.: Mixing modal and sufficiency operators. Bull. Sect. Log. Pol. Acad. Sci. **28**(2), 99–106 (1999)
28. Ferré, S., Ridoux, O.: A logical generalization of formal concept analysis. In: Ganter, B., Mineau, G.W. (eds.) ICCS 2000. LNCS, vol. 1867. Springer, Heidelberg (2000)
29. Ganter, B., Kuznetsov, S.O.: Pattern structures and their projections. In: Delugach, H.S., Stumme, G. (eds.) ICCS 2001. LNCS (LNAI), vol. 2120, p. 129. Springer, Heidelberg (2001)
30. Ganter, B., Wille, R.: Formal Concept Analysis. Springer-Verlag, Heidelberg (1999)
31. Gaume, B., Navarro, E., Prade, H.: A parallel between extended formal concept analysis and bipartite graphs analysis. In: Hüllermeier, E., Kruse, R., Hoffmann, F. (eds.) IPMU 2010. LNCS, vol. 6178, pp. 270–280. Springer, Heidelberg (2010)
32. Gaume, B., Navarro, E., Prade, H.: Clustering bipartite graphs in terms of approximate formal concepts and sub-contexts. Int. J. Comput. Intell. Syst. **6**(6), 1125–1142 (2013)
33. Georgescu, G., Popescu, A.: Non-dual fuzzy connections. Arch. Math. Log. **43**(8), 1009–1039 (2004)
34. Miclet, L., Prade, H.,Guennec, D.: Looking for analogical properties in a formal concept analysis setting. In: Napoli, A., Vychodil, V. (eds.) proceedings of 8th International Conference on Concept Lattices and Their Application (CLA 2011), pp. 295–307. INRIA, Nancy, 17–20 October 2011
35. Höppner, F., Klawonn, F., Kruse, R., Runkler, T.: Fuzzy Cluster Analysis: Methods for Classification. Data Analysis and Image Recognition. Wiley, New York (1999)
36. Messai, N., Devignes, M., Napoli, A., Tabbone, M.: Many-valued concept lattices for conceptual clustering and information retrieval. In: Proceedings of 18th European Conference on Artificial Intelligence, pp. 722–727, Patras (2008)
37. Miclet, L., Barbot, N., Prade, H.: From analogical proportions in lattices to proportional analogies in formal concepts. In: Schaub, T., Friedrich, G., O'Sullivan, B. (eds.) Proceedings of 21st European Conference on Artificial Intelligence (ECAI14), pp. 627–632. IOS Press, Prague, 18–22 August 2014
38. Navarro, E., Prade, H., Gaume, B.: Clustering sets of objects using concepts-objects bipartite graphs. In: Hüllermeier, E., Link, S., Fober, T., Seeger, B. (eds.) SUM 2012. LNCS, vol. 7520, pp. 420–432. Springer, Heidelberg (2012)
39. Parsons, T.: The traditional square of opposition. In: Zalta, E.N. (ed.) The Stanford Encyclopedia of Philosophy, Fall 2008 edn (2008)

40. Pawlak, Z.: Rough Sets. Theoretical Aspects of. Reasoning about Data. Kluwer Academic Publisher, Norwell (1991)
41. Popescu, A.: A general approach to fuzzy concepts. Math. Log. Q. **50**, 265–280 (2004)
42. Schaeffer, E.: Graph clustering. Comput. Sci. Rev. **1**, 27–64 (2007)
43. Wille, R.: Restructuring lattice theory: an approach based on hierarchies of concepts. In: Rival, I. (ed.) Ordered Sets, pp. 445–470. Reidel, Dordrecht (1982)
44. Wolff, K.E.: Concepts in fuzzy scaling theory: order and granularity. Fuzzy Sets Syst. **132**, 63–75 (2002)
45. Yao, Y., Chen, Y.: Rough set approximations in formal concept analysis. In: Peters, J.F., Skowron, A. (eds.) Transactions on Rough Sets V. LNCS, vol. 4100, pp. 285–305. Springer, Heidelberg (2006)
46. Yao, Y.: A comparative study of formal concept analysis and rough set theory in data analysis. In: Tsumoto, S., Słowiński, R., Komorowski, J., Grzymała-Busse, J.W. (eds.) RSCTC 2004. LNCS (LNAI), vol. 3066, pp. 59–68. Springer, Heidelberg (2004)
47. Zadeh, L.A.: Fuzzy sets as a basis for a theory of possibility. Fuzzy Sets Syst. **1**, 3–28 (1978)

Measuring the Implications of the D-Basis in Analysis of Data in Biomedical Studies

Kira Adaricheva[1,2]([✉]), J.B. Nation[3], Gordon Okimoto[4],
Vyacheslav Adarichev[1], Adina Amanbekkyzy[1], Shuchismita Sarkar[1],
Alibek Sailanbayev[1], Nazar Seidalin[5], and Kenneth Alibek[6]

[1] School of Science and Technology, Nazarbayev University, Astana, Kazakhstan
kira.adaricheva@nu.edu.kz
[2] Yeshiva University, New York, USA
[3] University of Hawaii, Honolulu, USA
[4] University of Hawaii Cancer Center, Honolulu, USA
[5] Medical Holding, Astana, Kazakhstan
[6] Graduate School of Medicine, Nazarbayev University, Astana, Kazakhstan

Abstract. We introduce the parameter of relevance of an attribute of a binary table to another attribute of the same table, computed with respect to an implicational basis of a closure system associated with the table. This enables a ranking of all attributes, by relevance parameter to the same fixed attribute, and, as a consequence, reveals the implications of the basis most relevant to this attribute. As an application of this new metric, we test the algorithm for D-basis extraction presented in Adaricheva and Nation [1] on biomedical data related to the survival groups of patients with particular types of cancer. Each test case requires a specialized approach in converting the real-valued data into binary data and careful analysis of the transformed data in a multi-disciplinary environment of cross-field collaboration.

Keywords: Binary table · Galois lattice · Implicational basis · D-basis · Support · Relevance · Gene expression · Survival · Response to treatment · Immune markers · Blood biochemistry · Infection

Knowledge retrieval from large data sets is an essential problem in economy, biology and medical sciences. The data is often recorded in tables with rows consisting of the objects and columns of the attributes. The dependencies existing between subsets of the attributes in the form of *association rules* can uncover the laws, causalities and trends hidden in the data.

In data mining, the retrieval and sorting of association rules is a research problem of considerable interest. The benchmark algorithms, such as *Apriori* in

K. Adaricheva and J.B. Natio—Partial support provided by grant N 13/42 (2013–2015) of Nazarbayev University.

J.B. Nation, V. Adarichev, N. Seidalin and K. Alibek—Partial support provided by grant N 0112PK02175 (2012–2014) of Ministry of Healthcare and Social Development of RK.

© Springer International Publishing Switzerland 2015
J. Baixeries et al. (Eds.): ICFCA 2015, LNAI 9113, pp. 39–57, 2015.
DOI: 10.1007/978-3-319-19545-2_3

Agrawal et al. [5], have the complexity that is exponential in the size of a table. Moreover, the number of association rules is staggering, and thus it requires further tools for filtering to obtain a short subset of rules that are significant. There are no strong mathematical results confirming a particular choice of such short subsets, and numerous approaches to the filtering process are described in various publications devoted to the topic. See, for example, Kryszkiewicz [14] and Balcázar [7].

One particular subset of association rules, the *implications*, or rules of full confidence, merit particular attention in data mining, as well as being the center of on-going theoretical study, supported by a number of strong mathematical statements. This could be explained by the fact that implications constitute one of the facets of closure systems. In particular, they closely relate to the structure of finite lattices.

Representation of a binary table and its concept (Galois) lattice *via sets of implications* continues to be a primary research goal of concept analysis (FCA). The target for many years was the retrieval of the *canonical basis*, or Guigues-Duquenne basis, of implications for closure systems defined by a Galois connection on the binary table. Nevertheless, recent results confirm that algorithmic solutions to such a task have complexity that is at least exponential in the size of the table; see, for example, Distel and Sertkaya [10] and Babin and Kuznetsov [6].

In Adaricheva and Nation [1], the authors suggest using a new type of basis, called the *D-basis*, which was introduced in Adaricheva, Nation, and Rand [2]. This basis has a lattice-theoretical flavor, for its generating notion is that of a minimal cover of a join irreducible element in a finite lattice. The *D*-basis is usually a proper subset of the *canonical direct unit basis* (this latter is different from the Guigues-Duquenne basis, see Bertet and Monjardet [8]), while it enjoys the property of being *ordered direct*.

The advantage of this basis in relation to the canonical basis, for the representation of a binary table, is in the possibility of reducing the task to dualization of an associated hypergraph. It is known that the hypergraph dualization problem has a sub-exponential algorithmic solution, see Fredman and Khachiyan [11]. The algorithm in [1] avoids generating the Galois lattice from the table, and only uses the arrow relations, which can be computed in polynomial time, to produce a hypergraph for each requested attribute. In that way, the existing code for hypergraph dualization, such as in Murakami and Uno [15], can be borrowed for execution.

In the current paper, we employ the code implementation of this algorithm as a working approach for data analysis in biomedical studies.

Since 2013 we have been working with two data sources, both connected with cancer research. One of them is the data sets provided by the bio-informatics group at the University of Hawaii Cancer Center, which relate the gene expression of patients with various types of cancer with their survival parameters. Another source is provided by medical research group at Medical Holding in Astana, Kazakhstan. The data relates the immune, viral and blood parameters of patients with brain tumors with their response to a new regimen of treatment.

While both data sets are essentially different sorts of real-valued medical data, we developed customized approaches to convert them into binary tables. In the case of the Astana research group, we dealt with temporal data, which included several measurements of parameters during the time of treatment of patients. The target of our tests was to reveal possible connections between the dynamics of sets of several parameters with the response to treatment. In analyzing the data from Honolulu, we worked in close collaboration with the Hawaii bio-informatics group, applying the implicational algorithms after the genetic data had been reduced to manageable size by other methods.

As in data mining, the main obstacle to analysis of binary tables *via* their representation by implications is the impressive number of implications in the basis. The algorithm in [1] allows us to retrieve only those implications $X \rightarrow b$ in the D-basis that have a fixed attribute b as a conclusion. In our case, b would represent a particular indicator, for example that a patient belongs to the group of long-survivors, say those who lived for longer than 1300 days beyond the day of diagnosis. Such subsets of the basis may contain close to 1500 implications when the table has just twenty attributes (columns). This number can easily increase to 1,000,000 when the number of columns is in the range of 250.

Our goal was to identify small groups of parameters whose appearance in the requested sector of the basis indicated the influence of such groups to the target parameter b. In order to rank the implications from the subset of the basis having b as a conclusion, we introduced new metric for the attributes of the table. We call the new metric the *relevance* of attribute a with respect to b, and it is computed based on frequency of a appearing in the antecedents of implications related to b in two bases: one for original table, and the other for the table, where attribute b is replaced with its complement $\neg b$. The computation of this parameter also takes into account the support of each individual implication in the basis where a appears.

After computing the relevance parameter for all attributes of the table, one can rank the implications $X \rightarrow b$ by taking the average of relevance parameters for all $x \in X$. For each individual data set, it is up to a specialist in the data to establish the lower threshold for the relevance parameter of implications, to separate the small portion of them which might have impact for further study.

We believe that our testing provides some first insights into the possibilities for using implication bases in biomedical studies that involve relatively large data sets.

One of the main achievements of our tests was to demonstrate that the algorithm can handle tables with the number of columns/attributes exceeding those reported in the literature, with respect to canonical or canonical direct unit bases; see for example Ryssel, Distel and Borchmann [16]. We had successful runs of the algorithm on a table with 287 columns, for medical data with relatively high density (proportion of ones in the table), while for less dense data sets, such as transaction tables, there were successful runs for tables with more than 500 attributes.

The paper is organized as follows. We provide the background information on closure operators and associated implicational bases, as well as their connection to binary tables, in the first section. Discussion of the D-basis and related information on other bases used in applications is given in Sect. 2. In the third section we introduce the definition of the parameter of the relevance of an attribute with respect to the fixed target attribute of the binary table. The computation of this parameter is illustrated in Sect. 4, which uses as input the gene expression data related to ovarian cancer provided by the University of Hawaii Cancer Center. A larger data set of the same type is discussed in the next section, where we also discuss some variations in computation of relevance metric. In Sect. 6 we discuss the test results on the temporal medical data provided by Medical Holding in Astana, related to a group of patients with brain tumors. The final section gives an overview of future testing and collaboration.

1   Short Introduction to Implications

A *closure operator* ϕ on a set A is an increasing, monotone and idempotent function $\phi : 2^A \rightarrow 2^A$. It is well known that any closure operator ϕ defined on finite set A can be fully represented by a *set of implications* $X \rightarrow y$ with $X \subseteq A$, $y \in A$. Any individual implication $X \rightarrow y$ can be considered as partial information about ϕ, saying that the ϕ-closure of X, i.e., the set $\phi(X) \subseteq A$, contains y. Implications $X \rightarrow y$ are also called *unit* implications, indicating that a single element y is on the right side of the arrow symbol. The unit implications can be *aggregated*: if there are several unit implications with the same left side X, then one can take the union of the right sides into a subset Y and represent these unit implications *via* $X \rightarrow Y$. However, for the algorithms used in this paper, it is better *not* to aggregate the unit implications.

The standard approach for the study and storage of the data related to a closure operator is to record an essential subset of the set of all implications of ϕ, called a *basis*, from which all valid implications (and thus the closure operator itself) can be recovered. There are many types of bases which have been targets for theoretical research, such as the canonical basis of Guigues-Duquenne [12] or the canonical direct unit basis; see the survey article [8].

Another type of basis, called the D-*basis*, was introduced in [2]. This basis is a subset of the canonical direct unit basis, and tends to be noticeably shorter. In our tests, the size of the D-basis (the number of implications) was on the average about 30 % shorter than the size of the canonical direct unit basis. On the other hand, the canonical direct unit basis is *direct*, meaning that closures of sets can be computed in one pass. The D-basis retains this property in a slightly modified form, called *ordered directness*.

One special case of the closure operator exists in any data presented by a binary table. By a binary table we understand a triple (U, A, R), where $R \subseteq U \times A$ is a relation between sets U and A, where U is the set of objects (corresponding to rows of the table) and A is the set of attributes (corresponding to columns). If $r = (u, a) \in R$, then the position in row u and column a is marked by 1.

This can be interpreted as object u possesses attribute a. Otherwise, the position is marked with a 0.

In order to recover the closure operator on the set of attributes, defined by a given binary table, we introduce two functions between subsets of attributes and objects.

The *support function* $S_A : 2^A \to 2^U$ is defined, for every $X \subseteq A$, by $S_A(X) = \{u \in U : (u, x) \in R \text{ for all } x \in X\}$. Thus, row u is in the support of set of columns X, if all intersections with columns from X, along this row, are marked by 1, or equivalently, if the object u possesses all the attributes from X.

Similarly, the support function $S_U : 2^U \to 2^A$ is defined for all $Y \subseteq U$ as $S_U(Y) = \{a \in A : (y, a) \in R \text{ for all } y \in Y\}$.

It is straightforward to show that the operator $\phi_A : 2^A \to 2^A$ defined as $\phi_A(X) = S_U(S_A(X))$ for $X \in 2^A$ is, in fact, a closure operator on A. Any implication $X \to y$ which holds ϕ_A can be directly interpreted from the table as follows: for each row of the matrix, whenever all intersections of this row with columns from set X are all marked by 1, the position at column y is also marked by 1. Note that the actual number of rows where intersections with X are marked by 1 is usually just a portion of total number of rows, and the set of such rows will be denoted $\sup(X)$, instead of $S_A(X)$, to match the notation used in data mining literature.

Let us illustrate these concepts in the following example. Consider the table with a set U of 6 objects and a set A of 7 attributes (Table 1).

Table 1.

	b	a_1	a_2	c_1	c_2	y	z
1	0	1	0	1	0	0	0
2	1	0	0	0	0	1	0
3	0	0	1	0	1	0	0
4	0	0	0	1	1	0	0
5	0	0	0	1	1	0	0
6	1	1	0	0	1	1	0

Consider $X = \{a_1, c_2\} \subseteq A$. Then $S_A(X) = \sup(X) = \{6\}$ and $\phi_A(X) = S_U(S_A(X)) = \{a_1, c_2, b, y\}$. Hence, we will have implications $X \to b$ and $X \to y$ in the set of all implications describing the operator ϕ_A. The logical statement *"if X then y"* holds in all rows of the matrix, while assumption X holds only in row 6.

From the point of view of data mining, the implication $X \to y$ is an *association rule* between columns of the given matrix with the support parameter $\frac{\sup(X \cup y)}{|U|} = 0.17$, which is just a normalized version of the support, showing the relative frequency of the rows where all attributes from X are marked.

The second essential parameter used for measuring the association rules is the *confidence*:

$$c(X \to y) = \frac{\sup(X \cup y)}{\sup(X)}.$$

If $X \to y$ is an implication, then the confidence is always 1, i.e., the highest among all possible values of this parameter. In general, an association rule may have confidence strictly lower than 1, and a lower bound threshold is used to filter the association rules of importance. For example, we may consider the association rule $c_1 \to c_2$, for which the normalized support is $\frac{1}{3}$, and the confidence is $\frac{2}{3} = 0.66$. This association rule might be discarded from consideration, assuming that the lower bound threshold is established, say, at $c = 0.75$. Among all association rules which can be considered for the attributes of the tabled data, the implications can be characterized as those with the confidence of 1.

Having established the connection with the field of data mining, we will deal in the sequel only with the implications describing the closure operator ϕ_A on the set of attributes of a binary table.

2   Comparison of the *D*-Basis with Other Bases for the Purposes of Table Description

The algorithm in [1] enables us to obtain the D-basis for the set of implications defining the operator ϕ_A on the set of attributes of a binary table. A critical difference with the other existing algorithms is that, instead of creating an intermediate algebraic structure known as a concept (Galois) lattice, or equivalently, finding all ϕ_A-closed sets, this algorithm retrieves only partial information about the structure in the form of *up-arrows, down-arrows* and *up-down-arrows*, which replace some of the 0-entries of the table. This additional information is enough to form an instance of the well-known *hypergraph dualization* problem, for which algorithmic solutions already exist and are realized in fast-executed computer programs; see, for example, Boros et al. [9] and Murakami and Uno [15].

Another essential difference between the structure of the D-basis and, say, canonical basis of Guigues-Duquenne, is that the D-basis is oriented toward finding, for any fixed $b \in B$, all subsets $X \subseteq A$ such that $X \to b$ is an implication for the operator ϕ_A, and X satisfies some irreducibility property with respect to b. In contrast, finding the canonical basis requires finding all *pseudo-closed* sets of operator ϕ_A, which will serve as antecedents of implications $X \to y$, and this search is irrelevant of what we want to find as the right side of the implications.

One consequence of the irreducibility property for $X \to y$ in the D-basis is that $X' \to b$ is not longer an implication for ϕ_A, for any proper subset $X' \subset X$. The latter property also holds for implications included into the canonical direct unit basis mentioned earlier. At the same time, the irreducibility property required for implications of the D-basis is stronger, which explains why the D-basis is normally a proper subset of the canonical direct unit basis.

Recently, U. Russel et al. [16] proposed a method of retrieval of the canonical direct unit basis that would employ the hypergraph dualization algorithm. It does not employ the D-relation, which we use for the purposes of obtaining the D-basis. The D-relation is a binary relation that can be computed in polynomial time in the size of the table, using the information about the up- and down-arrows mentioned above. This allows us to reduce the size of the hypergraphs for which the dualization should be computed, compared to the algorithm in [16].

3   Measurement of Relevance of Implications with Respect to a Fixed Attribute

In this section we describe our new approach to the measurement of the implications in a particular implicational basis, with the goal of distinguishing a small subset of implications relevant to a sector of the basis targeting a particular fixed parameter b in the set of attributes.

Given any closure system (A, ϕ) on the set A, and any *unit* implicational basis β defining this closure system, we can define a subset $\beta(b) \subseteq \beta$ with respect to any element $b \in A$ as follows:

$$\beta(b) = \{(X \to y) \in \beta : \phi(y) = \phi(b)\}.$$

For example, when β is the D-basis of the operator ϕ_A for the table given in Sect. 1, we have $\beta(b) = \{b \to y, y \to b\} \cup \{\{a_1, c_2\} \to t, \{a_2, c_1\} \to t : t = b, y\}$. Formally, column y can be stripped from the table, since it is identical to column b, which implies $\phi_A(y) = \phi_A(b)$. One can find the D-basis on the table without y, then extend the information we know for column b to its twin column y.

The algorithm presented in [1] is based on the retrieval of $\beta(b)$, for each $b \in A$, where the closure system is defined on the set of attributes A of a given binary table, and where β is a D-basis of this closure system. It is critical that the retrieval of the basis is done separately for each attribute $b \in A$, so that parallel processing could be done to optimize the required time to obtain the whole basis.

More often, though, the whole basis is not what is needed in a particular study, and the implications of the form $X \to b$, for some particular fixed $b \in A$, are of higher importance than others. Then the choice of the basis will be based on the possibility to compute $\beta(b)$ much faster than the whole β.

When $\beta(b)$ is available, the main task is to rank the implications thus obtained with respect to relevance of antecedent X to attribute b.

From the extensive list of parameters known for the filtering the association rules in data mining, the only parameter that can be applied for ranking of implications is the parameter of *support*. Indeed, while many parameters in data mining make extensive use of the parameter of confidence, that does not apply in the case of implications, as observed in Sect. 1.

We believe that, for each attribute $a \in A \setminus b$, the important parameter of relevance of this attribute to $b \in A$ is a parameter of *total support*, computed with respect to basis β:

$$\text{tsup}_b(a) = \Sigma\{\frac{|sup(X)|}{|X|} : a \in X, (X \to b) \in \beta\}.$$

Thus $\text{tsup}_b(a)$ shows the frequency of parameter a appearing together with some other attributes in implications $X \to b$ of the basis β. The contribution of each implication $X \to b$, where $a \in X$, into the computation of total support of a is higher when the support of X is higher, i.e., column a is marked by 1 in more rows of the table, together with other attributes from X, but also when X has fewer other attributes besides a.

While the frequent appearance of a particular attribute a in implications $X \to b$ might indicate the relevance of a to b, the same attribute may appear in implications $X \to \neg b$. The attribute $\neg b$ may not be present in the table and can be obtained by converting the column of attribute b into its complement.

Let $\beta(\neg b)$ be the basis of closure system obtained after replacing the original column of attribute b by its complement column $\neg b$. Then the *total support* of $\neg b$ can be computed, for each $a \in A \setminus b$, as before:

$$\text{tsup}_{\neg b}(a) = \Sigma\{\frac{|sup(X)|}{|X|} : a \in X, (X \to \neg b) \in \beta(\neg b)\}.$$

Define now the parameter of relevance of parameter $a \in A \setminus b$ to parameter b, with respect to basis β:

$$\text{rel}_b(a) = \frac{\text{tsup}_b(a)}{\text{tsup}_{\neg b}(a) + 1}.$$

The highest relevance of a is achieved by a combination of high total support of a in implications $X \to b$ and low total support in implications $X \to \neg b$. This parameter provides the ranking of all parameters $a \in A \setminus b$, but also allows us to rank implications $X \to b$ in the basis, by computing the average of $\text{rel}_b(x)$ for $x \in X$:

$$\text{rel}_b(X \to b) = \frac{\Sigma\{\text{rel}_b(x) : x \in X\}}{|X|}.$$

We emphasize that while the measurement of relevance of an attribute a with respect to b can be done for any basis, there should be some assumption about irreducibility of the antecedents in implications. Indeed, for each implication $X \to b$ one may add to the basis another implication $X \cup \{s\} \to b$, for some fixed attribute s. In this new basis, the attribute s may obtain an unnecessarily high measurement. As we pointed earlier in Sect. 2, both the D-basis and the canonical direct unit basis have the property of irreducibility for antecedents in their implications.

It would be interesting to compare the measurement of relevance parameters for individual attributes with respect to various bases and check whether the group of attributes with the high ranking will be independent of the choice of the basis.

4   Illustrating Example from Test Data on Ovarian Cancer

We will be illustrating our approach on a relatively small data set composed of genes found to be highly correlated with microRNA and DNA methylation in a common set of 291 serous ovarian tumor samples. Global gene expression, microRNA and DNA methylation data for each tumor sample were downloaded from The Cancer Genome Atlas (TCGA) along with meta-data that included censored time-to-death from all causes (survival) [18].

The resulting data matrices for each data type were jointly analyzed using matrix factorizations of rank-1 to identify a low-dimensional signature composed of genes, microRNA and DNA methylation loci that best represented the dominant source of variation in the data as a sparse linear model.

Hierarchical clustering and pathway analysis methods were then employed to identify an even smaller set of genes that continued to model the dominant signal as a sparse linear combination. We hypothesized that gene signatures obtained in this way would help to unravel the complex, inter-connected biology that drives the clinical trajectory of ovarian cancer. In particular, we focused on a gene expression signature composed of 21 genes (out of 16,000 interrogated) that were all direct down-stream targets of the OSM gene as determined by pathway analysis methods.

The gene expression profiles of the 21 genes were arranged in a binary table with 190 rows and 46 columns. The rows represent the ovarian cancer patients who participated in the study which observed their survival time for 2500 days after treatment with standard chemotherapy with cisplatin and paclitaxel.

The first 42 columns represent the indicator functions for the expression levels of the 21 genes (after quantile normalization). If patient y has relatively high expression of gene x, it will be marked by indicator 1 in column x, and when this patient shows relatively low levels of expression of gene x, the indicator value of 1 is put in column $21 + x$. Those patients whose gene expression is within some threshold around the average expression value in the group will have an indicator value of 0 in both columns x and $21 + x$.

The last four columns represent the survival groups within those 190 patients. Indicator is 1 in column 43 if a patient lived longer than 2000 days, and it is 1 in column 44 if she lived longer than 1300. Thus, the implication $43 \rightarrow 44$ holds in the binary table. The indicator is 1 in the 45th column if a patient lived less than 1300 days, and it is 1 in column 46 if she lived less than 850 days. Hence another implication $46 \rightarrow 45$ is also a part of the basis.

The cut-off thresholds for survival of 2000, 1300 and 850 days roughly correspond to quartiles for the entire group of 291 patients based on Kaplan-Meier analysis. Recall that the whole observation group was comprised of 291 patients, of which only 153 stayed in the study for the total period of 2500 days, while the remaining patients were observed for shorter periods. Nevertheless, 38 of them were observed long enough to include them into two upper quartiles, with some partial loss of information for those between 1300 and 2000 days of observation. A total of 101 patients were excluded from the testing related to survival for this test of D-basis extraction. These were patients who either left the study before 2000 days, or else had survived but for fewer than 2000 days at the end of the study, and hence could not be assigned to a survival cohort. Other ways of dealing with censoring that would include all the samples are discussed in the next section.

Let us illustrate the measurement of the implications in the D-basis of this matrix, when the target attribute b is 44, i.e., the indicator of longer surviving patients (at least 1300 days). In this particular study this is the group of 87 patients, which includes a subgroup of 31 patients who survived longer than 2000 days. Application of the algorithm from [1] with the request of $\beta(44)$ produces 1819 implications of the form $X \rightarrow 44$, including the expected implication $43 \rightarrow 44$.

It is possible to rank the implications in the retrieved basis by support. For example, among 1819 implications obtained, there is a single one with the highest cardinality of the support $= 9$: $\{16, 28\} \rightarrow 44$. (Here 16 represents high expression of the gene GBP2, while 28 is low expression of IL7.) There is also one with the support of 8: $\{14, 1, 3, 11\} \rightarrow 44$ where $14 = $ HLA-B, $1 = $ VDR, $3 = $ TRIM22, $11 = $ IL15. There are also 9 implications of support 7, and 17 implications of support 6. The great majority of implications have a support of 1 or 2, and it is hard to decide whether any implications from this large group could be of particular value.

With the new approach we were able to compute the relevance parameter for all the columns and choose the columns of the highest relevance. In our case, these were column 29 with relevance parameter 2.7894, column 9 with relevance value of 2.5137, and columns 1 and 4 with the relevance figures. 1.8702 and 1.8425, respectively. (Column 29 is low expression of IL4R, while 9 is high expression of IL1B, $1 = $ VDR, $4 = $ SELE.)

The value 2.7894 for column 29 can be interpreted as following: attribute 29 appeared in implications with the conclusion $b = 44$, i.e., indicator that the patient was in the longer surviving group (patients with the survival period longer than 1300 days), approximately 2.7 times more often than for the complement of this group (patients surviving less than 1300 days). According to the definition of tsup_{44} parameter, the contribution of each individual implication would be adjusted by the weight $\frac{\sup(X)}{|X|}$. In the $\beta(44)$ section of the D-basis the size of antecedent varied between 2 and 7, with most of implications having 3 or 4 attributes in their antecedent.

In any outcome of the testing, the follow-up validation of the discovery assumes the check on an additional data set of 99 ovarian cancer patients. The identified groups of parameters highly relevant for survival are validated, when they successfully separate the survival curves for both training and test data, based on Kaplan-Meier and Cox regression analysis.

The six genes (from this set of 21 targets of OSM) with the highest relevance to long survival (over 1300 days) turned out to be IL4R, IL1B, VDR, SELE, HLA-B, GBP2 and IL15RA. We did a Kaplan-Meier analysis of the signature on the 291 patient sample, and then tested it on the independent sample of 99 other ovarian cancer patients. The difference between the KM plots for the top and bottom quartiles of the 291 training samples ordered by the 6-gene D-basis signature are statistically significant in both the KM analysis (p = 0.00217) and Cox regression analysis (p = 0.0000269). The same analysis on the 99 validation samples gave p = 0.0176 for the KM analysis and p = 0.0419 for the Cox regression analysis. Thus the six-gene signature derived from the relevance parameter is associated with survival at a significant level.

5   The Larger Test Case of Ovarian Cancer

More comprehensive testing was done on a set of 40 genes that are downstream targets of IL4, identified by the combination of methods described in Sect. 4. This time all 291 patients were included into the testing, thus the size of the matrix was 291 × 84, where the first 80 columns represented the indicators for the high and low levels of expression for 40 genes, and the 4 columns represented 4 groups of patients based on the observed time to death due to all causes.

The 191 patients described in Sect. 4 had indicators placing them into longest, long, short and shortest surviving groups. The remaining 101 patients were coded by all zeros in 4 surviving groups columns. These patients were observed in the study for less than 1300 days, while their status after the last observation remained unknown: this is a group of so-called *censored* patients. Potentially, given longer term of observation, each patient from the censored group could appear in any of the surviving groups.

From the point of extraction of implications, inclusion of censored patients without marking them into survival groups results in the loss of a subset of implications that may belong to the D-basis. For example, if implication $X \to b$ holds in all the rows of 191 patients, with $b = 81$, which is a column of the longest surviving patients, and one of the patients from the group of 101, which was marked by all zeros, has all the parameters from X present, then the implication $X \to b$ fails in the row of matrix representing this patient, so that this implication is excluded from the output. Thus, the experiment on 291 × 84 matrix produces a subset of the basis for 190 × 84.

Another way to include the data of all 291 patients into the analysis is to assign 101 censored patients into 4 groups, based on the Kaplan-Meier analysis of 291 samples; see [13]. Potentially, marking censored patients into survival groups based on risk analysis may result in both excluding and adding some implications to the D-basis. This method will be used in some of our future tests.

We are planning to report on the full extent of this testing in a forthcoming publication. For the purposes of the current report, we outline the outcomes of the testing done on 291 × 84 matrix, when censored patients were marked by all zeros in the columns 81–84, representing the survival groups.

The D-basis was extracted with 4 requests, for $b = 81, 82, 83$ and 84. We used a feature of the algorithmic design, which may request only a portion of the basis, selecting the implication with minimum support parameter "minsup."

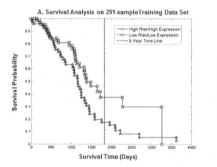
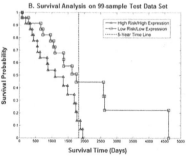

Fig. 1. Kaplan-Meier (KM) analysis of training and independent test data stratified by a 6-gene D-base expression signature (DBSig6). Panel A. KM plots of top and bottom quartiles of 291 ovarian tumors ordered by the 6-gene *DBSig6* expression signature. The KM plot marked by squares models the survival of the lowest quartile of patients ordered by *DBSig6* expression and KM plot of up-triangles models patients in highest quartile. The difference in KM plots is statistically significant (logrankP = 0.0143) and *DBSig6* expression is still associated with survival after adjustment for age and stage (CoxP = 0.0011). The intersection of the vertical dashed line with each KM curve gives the 5-year survival rate for each group of patients on the vertical axis. **Panel B.** KM plots on independent test data set composed of 99 ovarian tumors. The interpretation of the KM plots marked by squares and up-triangles and the vertical dashed line are the same as in Panel A. The difference in KM plots is statistically significant (logrankP = 0.0137) and *DBSig6* remains predictive of survival even after adjustment for age and stage (CoxP = 0.0086). Panels A and B demonstrate that the *DBSig6* expression signature is able to robustly identify "good" and "bad" reponders to standard chemotherapy for ovarian cancer.

When minsup $= k$, the program outputs only implications with the minimum support at least k. This considerably shortens the running time and the list of implications. For example, with $b = 83$ the algorithm produced 4325 implications of minimum support 3, in 91.94 s.

The six genes (from this set of 40 targets of IL4) with the highest relevance to long survival (over 1300 days) turned out to be FCGR2A, CD86, IFI30, CCL5, SELPLG and ICOS, all of which are associated with immune response and cancer. The results of the Kaplan-Meier and Cox regression analysis of this six-gene signature associated with IL4, on both the 291 patient training set and 99 sample validation data, are shown in Fig. 1. The difference between the KM plots for the top and bottom quartiles on the training data ordered by the 6-gene D-basis signature are statistically significant in both the KM analysis (p = 0.0143) and Cox regression analysis (p = 0.00112). The same analysis on the 99 validation samples gave p = 0.0137 for the KM analysis and p = 0.00858 for the Cox regression analysis. Again, this six-gene signature derived from the relevance parameter is associated with survival at a significant level.

On the other hand, the genes relevant to short survival did not separate the curves significantly, nor did the combined sets of genes for long and short survival. This may be an artifact of how we dealt with censoring. Other tests indicate that the 40-gene set associated with IL4 is much richer than the 21-gene set derived from OSM, and we anticipate that this will show up in further relevance experiments.

6   Analysis of Temporal Data for the Patients with Brain Tumor

The original set of data for 61 patients with brain tumors (astrocytomas, glioblastomas, and meningiomas) under new regimen of treatment was collected over the two years of observation in the hospital of Medical Holding in Astana, between 2012–2013. Patients were accepted into the experimental group after all options of standard chemotherapy were exhausted.

Three groups of parameters were regularly measured in patients. The first group was set of flow cytometry markers to identify major immune cell populations in peripheral blood: T helper cells, cytotoxic T cells, natural killers, B cells, and antigen-presenting cells. The second group was blood analysis for creatinine, bilirubin, calcium, protein, amylase, transferrin, C-reactive protein, immunoglobulins, lipase and iron. Finally, the third group of parameters was infectious markers including indicators for hepatitis A and C, cytomegalovirus, chlamydia, herpes simplex, Epstein-Barr virus, mycoplasma, ureaplasma, echinococcus, Helicobacter pylori, toxoplasma, and rubella.

The main read-out measurement of patient response to the new treatment was clinical assessment accompanied by the immune parameters, blood biochemistry parameters and the dynamic of infections.

All patients were divided into four groups depending on a clinical assessment. The first group included patients who did not survive cancer. The second group included patients succumbing to the illness. The third group included patients with stable health/tumor status. Finally, the fourth group incorporated patients with improving clinical assessments accompanied by reducing volume of tumor.

The goal of the study was to identify the elements of the treatment and measured parameters, which are associated to the positive response to the treatment.

The challenge of this data set was in the fact that each parameter was measured multiple times during the course of the treatment, while treatment and patient survival were of different time spans, resulting in significant variation in number of measurements. In order to take into consideration the dynamics of multiple parameters, and be able to analyze kinetics of very different length in similar consistent way, we transformed raw data into a set of increments, or differentials.

If the initial value of a parameter is V1, the value at the middle of the observation period is V2, and the final value during observation is V3, then the increment δ_{31} is defined as (V3-V1)/V1; increment δ_{21} is (V2-V1)/V1; increment δ_{32} is (V3-V2)/V2; Avg (average) is calculated average value for the parameter during the entire observation period. The analysis is performed in terms of (V1, Avg, $\delta_{21}, \delta_{32}, \delta_{31}$).

For conversion of real-valued data into binary form, the initial values V1 were taken into consideration only for part of the parameters. The full range of V1 parameter values was divided into quartiles and were coded into 4 columns representing the quartiles. The dynamic parameters were converted into two columns each, where the mark 1 in one of columns was an indicator of increasing of this parameter more than 10 % from average variation within the group, over the indicated period of treatment, while 1 in the other column was the indicator of decreasing by more than 10 %. Both columns would be marked by 0 when the parameter stayed within 10 % of average variation.

The 4 clinical assessment groups were combined into two: C1 group comprised the two groups of declining patients, and C2 group the stabilizing or improving patients.

The resulting binary table included 287 columns for 61 patients. An additional table was created for a subgroup of 33 patients with identical diagnosis of specific brain cancer, while the whole group incorporated patients with different sub-types of brain tumors.

The request for computation of the basis for a column that combined stabilizing and improved patients (C2) resulted in 1,138,518 implications computed in 39639 s, or just over 11 h. For the column indicating the group of declining patients (C1) the number of implications was 2,073,282, and it was computed in 170458 s, or 47.34 h.

The computation of the relevance parameter for all attributes, with respect to columns C1 and C2, provided the ranking of attributes in each case. For the computation of the most relevant implications, it was considered reasonable to make another run of the program to filter the implications first with respect to the minimum support parameter. It was established at the level minsup = 5, so that only those implications were produced whose antecedent held for at least 5 patients. The test for C1 now took only 1400 s and produced only 9,794 implications. The test for C2 took 345 s and produced 19,112 implications.

Similar tests were reproduced for the sub-group of 33 patients with the specific diagnosis of brain cancer. We observed much higher variation of the relevance parameter in the case of 61 patients. For example, several attributes showed relevance in the range of 1,000–10,000, mostly in the cases when $tsup_{\neg b}(a)$ for the attribute a had 0 value. Most of them got the relevance below 1 in the test of the sub-group of 33.

At the same time, most of the highly relevant attributes in the test on the subgroup of 33 patients showed their significance in the test for 61 as well.

On the set of 33 patients, the ranking of attributes by the relevance to column C1 revealed the attributes in ranking positions 1, 2 and 7, which correspond to dynamics of the same immune parameter: CD3+CD8+ cytotoxic T cells in the first, second halves of the treatment, and during the whole period, respectively. The dynamics was decreasing at the start, increasing in the second half, but still decreasing overall. In the C2 group, highly relevant were attributes in ranking positions 7 and 9 were decreasing dynamics of the presence of two specific viruses.

Table 2. Parameters associated with patients' clinical assessment.

Parameters *	Group 1	Group 2	Group 3	Group 4	P-value
INF_V1_HBsAg	0.540	0.573	0.572	0.427	0.015
INF_Avg_HBsAg	0.576	0.496	0.519	0.335	0.025
IMM_V1_CD3+CD8+	36.470	33.100	30.740	26.220	0.025
BLD_V1_IgG	9.910	7.100	12.660	9.070	0.007
BLD_V1_Fe_serum	8.290	16.570	13.205	24.030	0.030
BLD_δ31_triglycerids	0.175	-0.277	-0.222	-0.250	0.022
BLD_δ31_HDL	-0.330	0.052	0.030	0.011	0.002
BLD_δ31_LDL	-0.572	-0.230	-0.419	-0.116	0.011
BLD_δ31_Creatinin	-0.213	0.132	0.217	0.240	0.006
BLD_δ31_Total_protein	-0.109	0.012	0.016	-0.033	0.030
BLD_δ31_Albumin	-0.257	0.017	0.026	-0.130	0.035
BLD_δ31_CRP	4.399	0.337	-0.496	0.140	0.039
BLD_δ31_IgA	0.348	0.239	-0.092	-0.232	0.011
BLD_δ31_Lipase	-0.594	-0.294	0.127	nd	0.019
BLD_δ31_Fe_serum	-0.608	-0.437	0.251	-0.540	0.020
IMM_δ31_CD3-CD19+	-0.500	-0.300	0.000	-0.056	0.027
IMM_δ32_CD3+CD4+	-0.363	-0.106	0.076	0.176	0.013

* **BLD** – patients' blood parameters, **IMM** – immune parameters, **INF** – infection parameters. **HBsAg** - hepatitis B virus surface antigen, **CD3+CD8+** - cytotoxic T cells levels in peripheral blood, **IgG** – total immunoglobulins, **Fe_serum** - iron in blood, **HDL** – high density lipoproteins, **LDL** – low density lipoproteins, **CRP** - C-reactive protein, **IgA** – immunoglobulin of IgA isotype, **CD3-CD19+** - level of B cells in peripheral blood, **CD3+CD4+** - level of T helper cells. **Group** –median value of the parameter is presented for each group of patients. **P-value** – statistical significance of the Kruskal-Wallis non-parametric test.

Statistical analysis of the immune, biochemical and infection parameters: initial values, averages and increments (V1, Avg, δ_{21}, δ_{32}, δ_{31}) was also performed using non-parametric Spearman's correlation analysis [17] to find ties between parameters, see analogous use of increments in Adarichev et al. [3,4]. Global cross-correlation of all real-valued increments was performed for the data on 61 patients. Using analysis of correlation, the biases between only two parameters at a time could be studied, which is a major nuisance over the implication approach. For analysis of the parameters' difference between groups of patients, we used the non-parametric Kruskal-Wallis test in the R language for statistical computing [19]. This test is an equivalent of the one-way analysis of variance (ANOVA), but does not require assumption of the normal distribution of data.

Initial parameters (V1) most significantly associated with clinical assessment were HBsAg surface antigen of the hepatitis B virus ($p < 0.015$), amount of CD3+CD8+ cytotoxic T cells in peripheral blood ($p < 0.025$), total IgG immunoglobulins ($p < 0.007$), and higher concentration of blood iron ($p < 0.03$), see Table 2.

Change of blood parameters over the entire period of observation (δ_{31}) produced the longest list of significant associations: lipoproteins (triglycerides, HDL, LDL), creatinine, total protein, albumin, C-reactive protein, immunoglobulin IgA, lipase, and blood iron, refer to Table 2.

Importance of the total IgG immunoglobulins and specifically IgA isotype was in line with levels of B cells in peripheral blood (Table 3, $p < 0.02$), cells that produce immunoglobulins. Increase of T helper cells levels significantly correlated with good prognosis for survival ($p < 0.01$), see Table 2.

There were 151 attributes with the relevance above 1.5 associated with attribute C1, and 25 attributes associated with C2 with the same relevance threshold. Parallel analysis of same dataset using Kruskal-Wallis statistical test discovered 27 parameters at $p < 0.05$ significance level, they are presented in Table 3.

Table 3. Parameters discovered with both implications and statistical approaches.

Parameters *	relevance	group
BLD_δ31_Total_protein_inc10	4.12	C1
BLD_δ31_HDL_inc10	2.45	C1
BLD_δ31_CRP_inc10	2.33	C1
IMM_δ32_CD3+CD4+_inc10	2.12	C1
IMM_δ32_CD3-CD19+_inc10	1.87	C1
BLD_ δ31_Total_protein_dec10	1.85	C1
BLD_δ31_CRP_dec10	1.81	C1
BLD_δ31_LDL_dec10	1.62	C1
INF_Avg_HBsAg_dec10	1.59	C1
BLD_δ31_triglycerids_dec10	1.59	C1
BLD_δ31_Creatinin_inc10	1.53	C1
IMM_δ32_CD3-CD19+_dec10	1.52	C1
BLD_δ31_Albumin_dec10	6.87	C2
BLD_δ31_Fe_serum_inc10	4.43	C2
BLD_δ31_IgA_inc10	2.15	C2
BLD_δ31_Creatinin_dec10	1.91	C2
BLD_δ31_triglycerids_inc10	1.81	C2

Overlapping of these two analyses was compared using a Venn diagram. Twelve parameters were found both by implication and statistical approaches in C1, and five parameters were discovered in C2. Details of parameters are presented in Fig. 2.

The results presented in Tables 2 and 3 could be further corroborated by biological functionality of the discovered parameters. Parameter IMM_δ32_CD3-CD19+ reflects number of CD3-CD19+ B cells that produce antibodies. Correspondingly, blood parameter BLD_δ31_IgA is also in the list in the C2 group. Armed T cells, which are represented with IMM_δ32_CD3+CD4+ parameter could actually stimulate B cells to produce antibodies. Another group of parameters is related to blood lipoproteins of high and low density (HDL and LDL, respectively) and triglycerids. These parameters are known to be biased in the norm and pathology. Brain tumorigenesis and pharmaceutical intervention to fight cancer both lead to inflammatory reactions. Correspondingly, we found

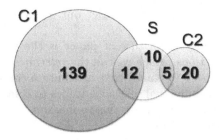

Fig. 2. Venn diagram for overlapping results of implication and statistical approaches. Circle **C1** - group of declining patients with total of $139 + 12 = 151$ implications. Circle **C2** - group of improving patients with total of $20 + 5 = 25$ implications. Threshold for relevance is set at 1.5. Circle **S** - results of statistical analysis with total of $12 + 10 + 5 = 27$ significant biases at $p < 0.05$ threshold. See Tables 2 and 3 for details.

inflammatory marker C-reactive protein in patients blood using both methods (parameter BLD_δ31_CRP). In brief, this set of parameter is functionally valid for the pathology under investigation. In the test on the binary conversion of this data, all blood parameters from the table showed high relevance of increasing trend for group C1, and four of them high relevance of decreasing trend in group C2, which confirms statistical observation.

7   Concluding Remarks and Future Research

This paper comes on the heels of a series of discoveries about computational complexity of extraction of the canonical basis of Guigues-Duquenne; see [6,10]. With the *D*-basis and algorithm for extracting implications derived in [1], the problem of handling relatively large data sets, which may include hundreds of attributes and objects, can be considered tamed. This brings researchers to a new challenge, already battled in data mining: the output of algorithms producing staggering amounts of association rules, for which further methods of analysis and filtering are needed.

Association rules were brought to consideration in analysis of transaction data. While the choice of purchases follow particular patterns, these are rather "soft" rules which do not need to hold even in a majority of all transactions. Whenever we come to analysis of biological data, the dependencies between attributes may reveal the laws of nature which are yet to be discovered. Thus, one may hope to find higher confidence levels of association rules discovered in this type of data.

The association rules of highest confidence ($=1$) are implications, and extraction of this special subset of association rules allows different approaches based on underlying structure of closure operators. On the other hand, the main metric for association rules, the confidence, is no longer a player for selection of important implications. This requires the development of new measurements for implications that would allow us to restrict our attention to those that may discover the hidden laws.

The main achievement of the current paper is the introduction of a new measurement for implications, which we call the *relevance*. It enables us to rank the attributes with respect to some fixed attribute b, in some given basis of implications, then apply individual relevance values to compute the relevance of implications. In particular, we can use this to measure the relevance of genetic or medical data to clinical outcomes.

The new approach was tested on two sources of medical data related to clinical assessment or survival of cancer patients. Our initial testing has already shown that the relevance parameter can be used to find genetic signatures associated with longer survival of ovarian cancer patients. The full analysis of these data sources will continue and we plan tuning the computation of relevance metric after validation of results of testing on additional data sets.

Acknowledgements. The C++ code for D-basis extraction on the binary table input used for testing in this project was created by undergraduate students of Yeshiva College in New York: Joshua Blumenkopf and Toviah Moldvin. We received the permission of Takeaki Uno, from the National Institute of Informatics in Tokyo, to implement the call to his subroutine performing the hypergraph dualization, within the structure of our programming code. We were assisted by colleagues Ulrich Norbisrath and Mark Sterling, from the Computer Science Department of School of Science and Technology of NU, when we needed tuning and debugging of the code, also to Rustam Bekishev and Anel Nurtay for assistance in the project. The first author is grateful to the bio-informatics group of the University of Hawaii Cancer Center, for the welcoming atmosphere and fruitful collaboration during her visit in June 2014, supported by Nazarbayev University grant N 13/42. The second author expresses his gratitude for support of his visit to Nazarbayev University in May–June 2013 and May 2014, which were partly funded by NU grant N 13/42 and grant N 0112PK02175 of Medical Holding of Astana. Tom Wenska, Ashkan Zeinalzadeh and Jenna Maligro contributed to the research and discussion in Honolulu.

References

1. Adaricheva, K., Nation, J.B.: Discovery of the D-basis in binary tables based on hypergraph dualization, Theoretical Computer Science (submitted to)
2. Adaricheva, K., Nation, J.B., Rand, R.: Ordered direct implicational basis of a finite closure system. Disc. Appl. Math. **161**, 707–723 (2013)
3. Adarichev, V.A., Vermes, C., Hanyecz, A., Mikecz, K., Bremer, E.G., Glant, T.T.: Gene expression profiling in murine autoimmune arthritis during the initiation and progression of joint inflammation. Arthritis Res. Ther. **7**, 196–207 (2005)
4. Adarichev, V.A., Vermes, C., Hanyecz, A., Ludanyi, K., Tunyogi-Csapó, M., Mikecz, K., Glant, T.T.: Antigen-induced differential gene expression in lymphocytes and gene expression profile in synovium prior to the onset of arthritis. Autoimmunity **39**, 663–673 (2006)

5. Agrawal, R., Mannila, H., Srikant, R., Toivonen, H., Verkamo, A.I.: Fast discovery of association rules. In: Fayyad, U.M., Piatetsky-Shapiro, G., Smyth, P., Uthurusamy, R. (eds.) Advances in Knowledge Discovery and Data Mining, pp. 307–328. AAAI Press, Menlo Park (1996)
6. Babin, M.A., Kuznetsov, S.O.: Computing premises of a minimal cover of functional dependencies is intractable. Disc. Appl. Math. **161**, 742–749 (2013)
7. Balcázar, J.L.: Redundancy, deduction schemes, and minimum-size bases for association rules. Log. Meth. Comput. Sci. **6**(2:3), 1–33 (2010)
8. Bertet, K., Monjardet, B.: The multiple facets of the canonical direct unit implicational basis. Theor. Comput. Sci. **411**, 2155–2166 (2010)
9. Boros, E., Elbassioni, K., Gurvich, V., Khachiyan, L.: Generating dual-bounded hypergraphs. Optim. Methods Softw. **17**, 749–781 (2002)
10. Distel, F., Sertkaya, B.: On the complexity of enumerating the pseudo-intents. Disc. Appl. Math. **159**, 450–466 (2011)
11. Fredman, M., Khachiyan, L.: On the complexity of dualization of monotone disjunctive normal forms. J. Algorithms **21**, 618–628 (1996)
12. Guigues, J.L., Duquenne, V.: Familles minimales d'implications informatives résultant d'une tables de données binares. Math. Sci. Hum. **95**, 5–18 (1986)
13. Kaplan, E.L., Meier, P.: Nonparametric estimation from incomplete observations. J. Amer. Statist. Assn. **53**(282), 457–481 (1958)
14. Kryszkiewicz, M.: Concise representations of association rules. In: Hand, D.J., Adams, N.M., Bolton, R.J. (eds.) Pattern Detection and Discovery. LNCS (LNAI), vol. 2447, p. 92. Springer, Heidelberg (2002)
15. Murakami, K., Uno, T.: Efficient algorithms for dualizing large scale hypergraphs. Disc. Appl. Math. **170**, 83–94 (2014)
16. Ryssel, U., Distel, F., Borchmann, D.: Fast algorithms for implication bases and attribute exploration using proper premises. Ann. Math. Art. Intell. **70**, 25–53 (2014)
17. Spearman, C.: The proof and measurement of association between two things. Amer. J. Psychol. **15**, 72–101 (1904)
18. Network, T.C.G.A.R.: The cancer genome atlas pan-cancer analysis project. Nat. Genet. **45**, 1113–1120 (2013)
19. R Core Team: R: a language and environment for statistical computing, R Foundation for Statistical Computing, Vienna, Austria (2013). URL http://www.R-project.org/

Survey

Formal Concept Analysis and Information Retrieval – A Survey

Victor Codocedo$^{(\boxtimes)}$ and Amedeo Napoli

LORIA - CNRS - INRIA - Université de Lorraine,
Nancy, France
{victor.codocedo,amedeo.napoli}@loria.fr

Abstract. One of the first models to be proposed as a document index for retrieval purposes was a lattice structure, decades before the introduction of Formal Concept Analysis. Nevertheless, the main notions that we consider so familiar within the community ("extension", "intension", "closure operators", "order") were already an important part of it. In the '90s, as FCA was starting to settle as an epistemic community, lattice-based Information Retrieval (IR) systems smoothly transitioned towards FCA-based IR systems. Currently, FCA theory supports dozens of different retrieval applications, ranging from traditional document indices to file systems, recommendation, multi-media and more recently, semantic linked data. In this paper we present a comprehensive study on how FCA has been used to support IR systems. We try to be as exhaustive as possible by reviewing the last 25 years of research as chronicles of the domain, yet we are also concise in relating works by its theoretical foundations. We think that this survey can help future endeavours of establishing FCA as a valuable alternative for modern IR systems.

1 Introduction

Surveying the intersection of Formal Concept Analysis (FCA) [33] and Information Retrieval [3] is not an easy task. The main complexity is that both domains have a an application range so wide that just getting a relevant set of articles to report about is a knowledge discovery process in itself. This is clearly exemplified by the survey presented by Poelmans et al. in 2012 [60] where FCA is used to report on 103 articles related to topics of FCA and IR in a period of only six years (2003–2009) crawled from the Web. In this paper we intend to approach the surveying in a more general and integral manner. We try to answer a very simple question. How have FCA and concept lattices been used in the context of IR applications? We answer this in a chronological narration, trying to cover the last 25 years of research since the first inception of the use of lattice structures to model the space of possible queries (or prescriptions, as they were called) to the last approaches, supporting file systems and semantic technologies.

As we can observe, most of the approaches presented here rest over a limited pool of *ideas and techniques* associated with FCA/IR but applied to a myriad of domains and applications. These ideas are:

© Springer International Publishing Switzerland 2015
J. Baixeries et al. (Eds.): ICFCA 2015, LNAI 9113, pp. 61–77, 2015.
DOI: 10.1007/978-3-319-19545-2_4

1. Using a concept lattice as a model of the description and document spaces
2. Enriching the description space through external knowledge sources
3. Enabling Relevance Feedback
   - Mixing querying and browsing
   - Query-by-navigation
   - Query-by-example
4. Using a concept lattice as a support for automatic retrieval

Our goal in this survey is to catalogue these ideas so future endeavours may have an easier way reaching further domains while developing new different and more interesting techniques. The remainder of this article is as follows: Sect. 2 introduces some context w.r.t. the use of lattice-based structures in the field of information retrieval. It also introduces the underlying model that generalizes the use of FCA for retrieval purposes. Section 3 describes the first approaches of FCA in the IR domain. Section 4 reviews works using background information to improve retrieval results. Section 5 reviews works based on the paradigm of relevance feedback and automatic document ranking. Section 6 lists the main applications and systems encompassing the ideas and notions described in the previous sections. Finally, Sect. 7 concludes the paper by introducing some concepts left out of the scope of this paper.

1.1 Related Work

Along with the work of Poelmans [60], there have been other important reviews of the literature regarding FCA and IR [13,64,68]. In 2005, Carpineto and Romano [13] described the main possible tasks that FCA could perform regarding querying and indexing by summarizing some of their work in the field. In 2007, Uta Priss [64] dedicated a full chapter to describe the state-of-the-art up to 2004 on FCA-based IR in her paper on *FCA and Information Sciences*. The last of these reviews was presented by Valverde and Peláez-Moreno in 2013 in the first (and sadly, the last) workshop on *Formal Concept Analysis meets Information Retrieval* in the context of the European Conference on Information Retrieval (ECIR 2013)[1]. This work differentiates between what is FCA *in* IR and what is FCA *for* IR, the latter of which refers to the possibility of *"augmenting IR with the methods and ideas of FCA"*. The authors describe these ideas in seven "affordances" of FCA for IR, classifying with them the body-of-work of FCA-based IR approaches.

1.2 Notation and Definitions

Formal Concept Analysis. For the sake of brevity, in this paper we assume a certain degree of familiarity with FCA. In what follows, we use the notation of [33]. A formal context is defined as $\mathcal{K} = (G, M, I)$ where G is a set of documents, M is a set of attributes or descriptors and I an incidence relation set indicating by gIm that document $g \in G$ *has* descriptor $m \in M$. Descriptors denote any

---

[1] http://fcair.hse.ru.

kind of metadata associated with documents, being *terms, phrases, symbols, authors, image features, etc.* For the sake of generality in this paper we will refer to M as *the set of descriptors*, unless indicated otherwise.

Boolean IR Model. The Boolean IR model is considered as the first and one of the simplest techniques to index and retrieve documents [3,47]. Given a collection of documents G, we can consider each document g as represented by a *conjunction* of Boolean descriptors $g' \subseteq M$, where M is the set of all descriptors (sometimes called "repertory" or "dictionary"). A query (or "request", or "prescription") is defined as a set of descriptors connected by a logical operator AND, OR, NOT. The simplest query is given by a set of descriptors connected by AND and is called a "conjunctive query". Given a conjunctive query Q_{and}, the set of relevant documents to be retrieved (Q'_{and}) are those that contain *at least* all the descriptors in the query. A disjunctive query (using OR) can always be split into its conjunctive parts and the set of relevant documents can be computed by the union of each separate set of relevant documents. A similar approach can be applied for NOT. In this work we will consider every query Q as being conjunctive, unless indicated otherwise.

A query $Q \subseteq M$ is a subset of descriptors usually provided by a given user. In this review we respect the original denominations given by different works to queries (requests, prescriptions, questions, etc.), however we indicate in parenthesis what denominations refer to. Finally, the *"space of documents"* is denoted as $(\wp(G), \subseteq)$ while *"the space of descriptors"* or *"the query space"* is indistinctly denoted as $(\wp(M), \subseteq)$.

2   Pre-FCA History - A Lattice to Model the Description and Document Spaces

Lattice structures were early adopted by information scientists as a model of document indexing [29,54]. As early as 1956, Robert A. Fairthorne [29] discussed how to model a library classification system by producing all possible requests (queries) as combinations of categories (descriptors) and logical connectors (AND, OR, NOT) and how this model could be compared to a "free distributive lattice". Some years later, Calvin Mooers [54] would consider two spaces for this model, namely the space of prescriptions P (descriptors) and the space of all possible documents subsets as $L = \wp(G)$.

He realised that L with the set inclusion operator $\subseteq$ was naturally a partially-ordered set (or poset) and that, under certain circumstances (actually when $P = \wp(M)$), P could also be modelled as such. With this, a retrieval system consists in a transformation $T : P \to L$ that is able to take a prescription (query) into the largest subset of documents that satisfies it (see Fig. 1).

It is important to note that Mooers did not describe an actual IR system, but a "model" for retrieval systems that would enable the comparison of different approaches. We can observe that FCA is an instance of this model, where the transformation T is naturally represented by a Galois connection defined between

Table 1. A document-term formal context.

	patient	laparoscopy	scan	user	medicine	response	time	MRI	practice	complication	arthroscopy	infection
d_1	×	×	×							×		×
d_2			×	×	×	×	×		×	×	×	
d_3	×				×			×				
d_4					×			×				×
d_5				×			×		×			
d_6										×		×
d_7										×	×	
d_8										×	×	×
d_9											×	×

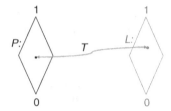

Fig. 1. Mooers' model: *"The space P of all possible retrieval prescriptions (queries), the space L of all possible document subsets, and the retrieval transformation T associating points of P with points of L."*

$\wp(G)$ and $\wp(M)$ and where the concept lattice is an elegant solution for the spaces P and L as it represents them in an integrated manner. Particularly, when this Galois connection is defined in terms of the derivation operator $((\cdot)')$, FCA becomes an implementation of the Boolean IR model.

2.1 The Underlying Model of FCA-IR

Let us introduce a general model of Boolean retrieval using the FCA framework with an example. In the following sections, we will re-use this model to explain how the tasks of browsing and querying can be performed using a concept lattice. Consider a formal context of documents and descriptors as the one shown in Table 1. Documents for a query $Q \subseteq M$ are retrieved through the derivation $Q' \subseteq G$ which works as the "transformation" T shown in Fig. 1. For example, the query $Q = \{arthroscopy, complication\}$ has as an answer documents in $Q' = \{d_2, d_7, d_8\}$.

Key Aspects: The query Q can be naturally *extended* to Q'', which of course, contains the same set of answers Q'. In the example, the query $Q = \{MRI\}$ extends to $Q'' = \{MRI, medicine\}$ and they both have the same answer $Q' = \{d_3, d_4\}$. This fact was already discussed by Mooers [54] and has been exhaustively exploited by FCA-IR approaches to provide context to user queries, in this case showing the user that his answer for MRI is within a *medical* context instead of several other possible interpretations[2]. The formal concept formed by (Q', Q'') has been called virtual node, virtual concept or *query concept*, and represents both, the extended query (intent), and the set of retrieved documents (extent). Notice that the latter can be an empty set if there are no documents satisfying the query (hence the name *virtual*). Finally, in this article we will make the distinction between "query extension" and "query expansion". The first of which refers to the *closure* of the query w.r.t. $(\cdot)'$. The second refers to an actual *modification* of the query by taking a set Q_1 where $Q_1'' \neq Q''$ and in general $Q_1 \cap Q \neq \emptyset$ (i.e. finding a query Q_1 related to Q which yields different results).

---

[2] http://en.wikipedia.org/wiki/MRI_(disambiguation).

Throughout all the approaches discussed in this survey, the underlying model described above has not varied much (notice that the book of Barbut and Monjardet which included what will be FCA later was published in 1970! [4,69]). This fact is in no way a negative point for FCA-based IR approaches, but actually a statement about the adequacy of the model to fit in different tasks and domains. On the other hand, this advantage of FCA is also one of its main drawbacks when dealing with modern IR systems.

The Boolean IR model was quickly regarded as too limited for the complex tasks involved in the retrieval of documents considering the size of modern document collections or the nature of their descriptions (e.g. numeric instead of Boolean). The IR community would shift to more complex models such as the vector space model (for ranking documents by "relevance" w.r.t. a query) or the probabilistic model (for predicting which are those more "relevant" for a user). Current introductory books on IR [3,47] do not mention lattice structures (not to say concept lattices) as valid IR models[3]. In [47], the chapter on the Boolean IR model finishes with the following quote attributed to Calvin Mooers in a book of Fairthorne (1961):

"It is a common fallacy (...) that the algebra of George Boole (1847) is the appropriate formalism for retrieval system design. This view is widely and uncritically accepted as it is wrong."

3 FCA Meets IR

The bad scenario for the Boolean retrieval model and its drawbacks did not stop many researchers from developing several different applications using this paradigm. In the '90s, the first FCA-based IR systems were developed, while several other systems based on the use of lattice structures became popular.

3.1 Non-FCA Lattice-Based IR Systems

Pedersen in [58] introduced BRAQUE (BRowse And QUery Environment) as a system that allowed the navigation of a document collection modelled as a *relationship lattice* [59], strongly resembling the features of a concept lattice. At the AT&T Bell labs, Ginsberg [34] introduced WorldViews, consisting *"of a system for automatic document indexing, an information retrieval system and a user interface"* using a taxonomy modelled as a lattice structure. In the work of Bosman et al. [5], a similar approach to Ginsberg's WorldLattice was presented for creating a "Hyperindex" of a *faceted hierarchical thesaurus* using a lattice structure. The lattice supported a "query-by-navigation" approach where the user could "refine" or "enlarge" a query. In the domain of software engineering, Mili et al. [53] proposed a lattice-based index of software descriptions for

---

[3] Actually, in [3] there is an entry of two paragraphs - in a 500 pages book - about lattices in Chap. 10 about user interfaces and visualization, referencing [9,58] as systems for query reformulation (expansion).

retrieval purposes based on software reuse needs. The authors describe two types of retrieval namely, *"exact"* which resembles the Boolean retrieval model and "approximate" measuring "proximity" w.r.t a given query.

3.2   FCA-Based IR Systems

The proposition of Godin et al. [36] revealed the capabilities of concept lattices for indexing and retrieval as an alternative to Boolean querying and hierarchical classifications. This work was built over the initial user interaction design proposed by the same authors years before [35,37]. A major highlight in this work is the efficient browsing capabilities generated from a document collection by the construction of a concept lattice which actually represents a query space. In this manner, the user can pose different queries without explicitly indicating a set of terms to be sought within documents. An important advantage of this model is that users do not have to be completely familiarised with the lexicon used for indexing.

In the same year, Carpineto and Romano presented their system GALOIS [7] for conceptual clustering[4] which would be later implemented for information retrieval purposes through a query browsing interface called ULYSSES [9,14]. ULYSSES develops further in the model for the unification of querying and browsing plus a third procedure called "bounding". The latter allows the user to restrict the search space within the concept lattice (deriving a sub-lattice) by including into the query sentences such as "all documents indexed by a given term m" (i.e. contained in formal concepts $(A, B) s.t. (A, B) \leq (m', m'')$) and "all documents not indexed by a given term m" (i.e. contained in formal concepts (A, B) such that $B \cap m'' = \emptyset$). Experimentation showed similar results to a plain Boolean retrieval system.

As Fairthorne proposed [29], in an ideal world we could take the descriptions of all the documents in a library and create a map of all the possible requests that could be made (this map would be the P space in Fig. 1). However, this is a rather an unlikely scenario as the size of such map grows *"faster than exponentially"* w.r.t. the number of categories [63]. Instead we would prefer to generate a smaller P space that represents the "most meaningful" queries[5]. For this reason, two main strategies were embraced. Firstly, the use of an authoritative source such as the thesaurus-based WordLattice in [34] which would model in a more concise manner the space P. Secondly, the elicitation of this space from document features (lexical properties [5], metadata [58] or terms [14,36]).

An Anecdote. Mooers described the size of the search space of a document collection (L in Fig. 1) of one million documents as the number of subsets we can construct from it, being the staggering figure of $10^{310,000}$ [54]. This remembered one of the authors the description of a googol (10^{100}), a number proposed in 1938 by mathematician Edward Kasner to exemplify the difference between "an

---

[4] Actually, GALOIS is an incremental algorithm for building a concept lattice.

[5] It is worth mentioning that the "meaningfulness" of a request is a matter of perspective. What is meaningful in a domain may not be in another. Meaning also changes with time.

unimaginably large number and infinity"[6]. While a googol is much larger than the number of particles in the observable universe[7], we can see that the L space is much larger than a googol. Apparently, we were not the first ones to step on this interesting fact. In 1997, a couple of entrepreneurs looking for a name for their search engine, in an attempt to represent the "indexing of an immense amount of data"[8], registered the misspelled version "Google".

4 Enriching the Description Space Through External Knowledge Sources

In the FCA-IR model explained in Sect. 2.1, attributes are descriptors obtained from the set of documents. As previously explained, this space (P) can be very large but other than that it can suffer from other problems. For example, it can be non-representative of the document set by different reasons (poor document description, poor vocabulary, incompleteness, etc.). Regarding these issues, it may be useful to use an external knowledge source to complement document descriptions. For example, if we are interested in considering synonymia for indexing (e.g. relating documents referring to *"concept lattices"* and *"Galois lattices"*) we may use a thesaurus. If we are interested in considering hierarchical relations (e.g. relating documents referring to *"monkeys"* with those referring to *"primates"*) we may use a taxonomy. If we are interested in considering logical implications (e.g. relating documents written by a French author to those written by a German author using the label *"European literature"*) we may use an ontology.

With these concerns, in 1996 Carpineto and Romano proposed a modified version of the GALOIS system to include "background information" in the form of a thesaurus for document indexing using FCA [8]. The modification was made in the order relation between formal concepts ($\leq_\mathcal{K}$) using the order between document descriptors ($\leq_\mathcal{T}$) induced by a thesaurus as follows:

$$(A_1, B_1) \leq_\mathcal{K} (A_2, B_2) \iff \forall m_2 \in B_2, \exists m_1 \in B_1 \; s.t. \; m_1 \leq_\mathcal{T} m_2$$

Furthermore, they redefined the intersection between two descriptor sets as:

$$B_1 \cap^* B_2 = \{m_i \mid m_i \geq_\mathcal{T} m_1, m_2, m_1 \in B_1, m_2 \in B_2, m_i \in \mathcal{T},$$
$$\nexists m_j \in \mathcal{T}, s.t. \; m_i \geq_\mathcal{T} m_j \geq_\mathcal{T} m_1, m_2\}$$

From the example in Table 1, consider a thesaurus $\mathcal{T}$ with the relations *arthroscopy, laparoscopy* $\leq_\mathcal{T}$ *endoscopy*[9]. Then, {*laparoscopy*} $\cap^*$ {*arthroscopy*} = {*endoscopy*} and we can build the formal concept $(\{d_1, d_2, d_3, d_6, d_7, d_8, d_9\}, \{endoscopy\})$.

---

[6] Wikipedia article - http://en.wikipedia.org/wiki/Googol.

[7] Video about googol from the University of Nottingham - https://www.youtube.com/watch?v=8GEebx72-qs.

[8] David Koller on the origin of the name "Google" http://graphics.stanford.edu/~dk/google_name_origin.html.

[9] Wikipedia categories http://en.wikipedia.org/wiki/Category:Endoscopy.

Consider this analogous to including in the formal context the attribute *endoscopy* and the relation where each document related either to *laparoscopy* or *arthroscopy* is also related to *endoscopy*.

The authors argue that this approach would lower the complexity associated to computing the concept lattice compared to the more simple approach of adding the thesaurus terms to the initial formal context. In 1997, Uta Priss presented several propositions for a FCA-based IR system in which three main components were discussed [62]. Firstly, a combined formal context comprising document descriptors and other metadata components (e.g. publisher, author, etc.). These kind of fields were coded by many-valued formal contexts which were later scaled (see attribute scaling in [33]). The second component described the inclusion of a thesaurus within the formal context by two approaches, namely by mapping document-descriptor pairs to thesaurus elements, and by constructing a combined formal context considering documents, descriptors and thesaurus elements in a relational concept analysis (RCA) manner (this RCA proposition is formally different from the one presented by Huchard et al. [40]). The third component referred to the use of "nested line diagrams" to represent in a better manner the combination of different concept lattices in an integrated view offering different description levels within a document collection.

Some of these ideas were later revisited by Cole and Eklund in 1999 [21] where the authors proposed an interactive e-mail retrieval system based on FCA. The formal context was built using "classifier outputs" as attributes which the user was asked to order in a hierarchy (G is a set of emails). *Conceptual scaling* was applied to many-valued attributes deriving views (sub-lattices) that were more manageable for the user to browse than the concept lattice of the entire email collection. In 2003, the authors (plus Gerd Stumme) would propose an extension of their work into a fully integrated system called "HIERMAIL" [22] in which nested-line-diagrams were used to represent conceptual scales (instead of sub-lattices) for knowledge discovery over an e-mail collection. Incidentally, Cole and Eklund had proposed a "folding" and "unfolding" mechanism (using the same notion of conceptual scales) for the concept lattice in a previous work oriented to model a document retrieval system in which documents were indexed by a medical thesaurus called SNOMED [20], although these procedures were not clearly defined.

A similar approach for domain-specific interactive FCA-based IR systems was presented by Mihye and Compton in 2001 [43] and later extended in [44]. An interesting point of this work is that it addresses the fact that taxonomies used to index documents are not static and should evolve through time. For this reason, the concept lattice is used not only to retrieve documents but also to aid users in the annotation of documents and in the evolution of the taxonomy.

5   Relevance Feedback and Automatic Retrieval

5.1   Relevance Feedback

Other than choosing and modelling the kind of data to be used as attributes in a formal context, an important factor in the efficiency of a retrieval system is to help the user closing what is usually called the knowledge or "the cognitive gap" [42]. The cognitive gap describes the distance between the space occupied by the

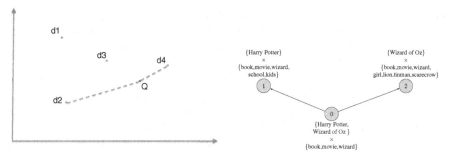

(a) Vector-space model example. Documents ($d1, d2, d3, d4$) and the query Q are represented as points. Axis are arbitrary description spaces. Dashed lines represent the distance between the query and documents $d2$ and $d4$. The latter is closer to the query than the former meaning that it is more "relevant" for retrieval purposes.

(b) Example of relevance feedback.

Fig. 2. Retrieval paradigm examples.

actual information needs of a user and the space occupied by its ability to describe its information needs. For example, consider a user searching for "the book which they made a film about and a wizard appears on it". Somebody could answer "Is it about a girl, a lion, a tin man and a scarecrow?" to which the user may answer "No, there are some kids in a school". Then, the answer could be narrowed down to the 7 books of the "Harry Potter" saga. Here we can see that the cognitive gap can be represented as the distance between the initial query, possibly with the keywords ``book film wizard´´, which the user is able to provide, and the query that he actually needs to provide which is ``Harry Potter book´´.

In 1971, Rocchio proposed his famous *relevance feedback* model to overcome this issue [65]. In a nutshell, we can see relevance feedback as a "query calibration" system using extra user inputs. In the previous example, the initial user query was very abstract. Somebody (possibly the librarian), with knowledge about fantasy books asked the user a question based on the assumption that the answer may be "The wizard of Oz". The negative answer provides a feedback of relevance (i.e. "The wizard of oz" is not relevant) which is used to generate the query: book film wizard school -``the wizard of oz´´[10].

In FCA terms, we can represent this scenario as the join of two object concepts as depicted in Fig. 2(b). The initial query yields concept 0 for which the system may propose concept 1 or concept 2.

This approach was proposed by Carpineto and Romano in 1998 through their system REFINER [10]. The user pose a query to which the system generates a "virtual concept in the lattice". By the use of the upper and the lower cover of the virtual concept, REFINER is able to propose minimal query refinements/enlargements (resp.) to the user. Experimental results showed significant

---

[10] In Google query syntax, '-' is used for excluding terms - http://goo.gl/7RZrQl.

better results in the search time employed by a user w.r.t. the Boolean IR model. In 2002, Grootjen et al. [38] proposed a similar rougher approach called "conceptual relevance feedback" further developed as a query expansion method [39] in the lines of pseudo-relevance feedback [47].

In 2007, Nauer and Toussaint [55] presented a model for "explicit relevance feedback"[11] over a standard Web search engine (such as Google) supported over a concept lattice. This model consisted of a constant iteration of the formal context by "extension" and "reduction" procedures. Extensions were made whenever the user submitted a new query or gave a positive assessment. Reductions were performed whenever the user gave a negative assessment. Explicit relevance feedback was also supported in a previous work by Martines and Loisant [49] for concept lattice-based image retrieval in a similar manner. Users were asked to evaluate images as "good" or "bad". An example of "implicit relevance feedback" can be found in the work of Ducrou et al. [26], supported by a procedure called "query-by-example". Instead of asking the user to give explicit relevance assessments, the query is modified by a sample set manually created by the user.

5.2   Ranking Documents

So far we have reviewed approaches that assist the user in navigating the query space and deciding what *is* or *is not* relevant. This is usually achieved by providing an interface that helps them retrieve parts of the concept lattice by the use of "query-by-navigation", "query-by-browsing", "relevance feedback" or "query-by-example". This however is not what we are used to when dealing with search engines. The "file search program" of any operating system, or the mechanics of traditional Web search engines follows a very simple scenario. The user inputs a query and the system outputs a list of documents already ordered by the "relevance" w.r.t. that query. Thus, the system is provided with the notion of *what is relevant* and *what is not.* For instance, in the vector-space model, documents and queries are represented by points in an arbitrary Euclidean-space. "Relevance" in this case may be represented by the distance between a query and a document (the closer the document, the more relevant it is w.r.t. the query) as shown in the example of Fig. 2(a) (explaining the meaning of the axis or why the documents and the query are located in the space as they are is out of the scope of this paper. For more information see [47]).

A similar notion was adopted by Carpineto and Romano in 2000 in what they called concept lattice-based ranking (CLR) [11] for a fully automated retrieval system. Using the REFINER model, the virtual concept representing the query is placed in the concept lattice and a series of "concentric rings" around the virtual concept yields a distance that allows to rank documents (e.g. in Fig. 2(b), documents in concepts 1 and 2 - and not in concept 0 - are at distance 1, while documents in their super-concepts would be at distance 2). Different measures are also introduced in the work of Ducrou et al. [26] where instead of using

---

[11] i.e. the user is explicitly asked to make relevance assessments in the system. Opposed to "implicit relevance feedback", where relevance assessments are "inferred" by the interaction of the user with the system.

the concept lattice structure, differences in extent and intent sets are taken into account. In 2014, Codocedo et al. [18] presented a system for lattice-based ranking using notions of case-based reasoning. This approach inspired in CLR uses the concept lattice to find suitable "query modifications" through pivotal elements called "cousin concepts". Query modifications are evaluated w.r.t. a semantic distance to the original query yielding automatic document ranking. Experimental evidence suggests that such an approach leads to better precision w.r.t. the Boolean querying model and CLR.

6 Applications and Systems

6.1 Applications

Semantic Retrieval. How to mix semantic technologies (what is known as semantic web) with IR techniques is still an open question. It is fair to say that modern IR systems are more focused on how to retrieve documents from very large collections than to provide reasoning or inferring capabilities to their engines. Nevertheless, this has not hindered the adoption of some of the semantic web notions such as the knowledge graph in the Google Web engine[12]. Regarding FCA-based IR approaches we can highlight the work of Messai et al. [50] presented in 2005 adapting the ideas of query refinement to support the use of ontologies for generalization purposes. In 2011, Codocedo et al. [17] presented an application of FCA to index songs using semantic similarity among keywords in a concept lattice. In 2012, Ferré et al. [30] introduced LISQL, a query language for logical information systems supporting complex relational properties among objects. These ideas were materialised in a geographical information system. Finally, in 2014 the work of Alam et al. [2] presents the concept lattice as a classification of SPARQL answers to provide views on linked open data retrieval system.

Recommender Systems (RSs). RSs have become increasingly popular at the point that currently, it is an independent research community. Nevertheless, RSs have their roots in IR sharing many notions such as indexing, retrieval and ranking. To phrase it in the terms of [68], an important *affordance* of FCA *for* RSs is the *characterization* it can provide to recommendations, i.e. it can explain why a certain item is being recommended, so the user can have a better experience with the system. This fact was addressed by [41] in 2008 which proposed a system for "well-interpretable recommendations based on FCA" for advertisement keywords using association rules. Previously, in 2006 [23] FCA was used as a method to pre-calculate groups of users that agree in certain groups of items. The notion of *query concept* is in here replaced by the "entry-level concept" of a user or an item. Experimental results suggest that FCA alleviates the otherwise hard task of finding the neighbourhood of a given item or user in

---

[12] http://www.google.com/insidesearch/features/search/knowledge.html.

the dataset. In 2013, Senatore et al. [66] proposed a recommender system based on an extension of FCA (namely, "Fuzzy FCA" or more precisely, FCA with fuzzy attributes) allowing to include *degrees of similarity* between users (i.e. not just Boolean relations for rating the same item) providing *ranked* recommended items. Finally, in 2014 Castellanos et al. [15] presented an approach based on [23] to extract preferences from a user activity log and derive semantically-enhanced item recommendations from them.

Others: For the sake of brevity, in here we give a summarised overview of some other applications of FCA-based IR systems. **File Systems (FS)** are an interesting application in low-level information retrieval (operating system level). FCA provides a more dynamic interaction with the file system structure where the FS can be represented as a lattice instead of a tree [31,48,67] **Source code location** is an important task in software engineering as it enables code refactoring, among other applications [1,53,61]. Other interesting applications are **mathematical expression search** [57] and **multimedia indexing** [26,49].

6.2   FCA-Based IR Systems

FaIR (2000) by Uta Priss [63]: A faceted IR system in which formal concepts of documents and descriptors are mapped to thesauri entries. It features a query language built on top of the set of formal concepts with the logical operators *AND, OR, NOT*.

CREDO (2004) by Carpineto and Romano [12]: CREDO works as the front-end of a Web search engine (such as Google or Yahoo). It implements some of their ideas in query expansion presented in REFINER providing context to an otherwise plain-list of ranked documents. Extensions of CREDO included its port to mobile devices (CREDINO and SmartCREDO [6]).

JBrainDead (2004) by Cigarrán et al. [16]: A FCA-based system that combines standard IR techniques such as term weighting and ranking for automated attribute selection. We highlight in this work the novel evaluation metrics considering the effort needed to find documents within a concept lattice derived from the number of concepts to visit and the percentage of those that represent relevant results.

Mail-Sleuth and the Sleuth Family (2004–2009), Ducrou, Eklund et al.: Building on previous work, the authors present a commercial tool called Mail-Sleuth [28], a system for searching and browsing personal email collections under the assumption that novice users are able to manage a line diagram of a lattice structure. The authors extended these ideas to different application domains: ImageSleuth [26] for image browsing and retrieval (discussed in the previous sections), DVDSleuth [24] for browsing Web catalogues, SearchSleuth [25] for browsing results from a standard Web search engine and AnnotationSleuth [27] a system designed for browsing a virtual-museum collection. In 2014, Wray and

Eklund presented the application *"A place for art"* [70] which followed in the steps of AnnotationSleuth with a much more elaborated user interface.

FooCA (2005) by Koester [45]: In the steps of CREDO, it also relied in the assumption that users can manage line diagrams of concept lattices, as well as interacting directly with the formal context.

BR-Explorer (2006) by Messai et al. [51]: An algorithm for document retrieval the notions of "query concept", "pivoting" and "ranking" for bioinformatic datasets.

Camelis (2007) by Sebastian Ferré: Based on "a generalization of FCA", named Logical Concept Analysis (LCA), where attributes are replaced by logical formulas. Designed to cover four main aspects: mixing query and navigation, expressive query language, genericity in data types, and efficiency for large collections. Camelis integrates several taxonomies different in nature, e.g. geographical ($Paris \sqsubseteq France$), numeric ($1999 \leq 2000$) and conceptual ($ICFCA \sqsubseteq Conference$), allowing complex querying and other tasks previously discussed, such as "query-by-navigation" and "query by example". An extension of Camelis called Sewelis (or Camelis 2) was introduced in [30] for "Query-based Faceted Search" on linked data, introducing an expressive query language called LISQL (Logical Information System Query Language).

CreChainDo (2007) by Nauer and Toussaint [56]: A FCA-based IR system supporting explicit relevance feedback (details in Sect. 5).

7   Conclusions

Two related topics have been left out of the scope of this review while they remain of extreme importance for FCA-based IR approaches. Firstly, the *use of complex data* for document indexing. Several approaches have proposed more sophisticated models than the standard Boolean retrieval model defined at the beginning of this article. Mainly, they rely in three FCA extensions for dealing with complex data, Logical Concept Analysis such as in [32], Fuzzy FCA such as the case of [66] and Pattern Structures, such as the case of [19] or [52] (the latter does not explicitly apply pattern structures, but the notions are very similar). Secondly, the application of FCA to large collections of documents or big data (an interesting discussion is provided in [46]). Both of these matters deserve a more extensive treatment than the one we could give them here.

Finally, this paper has presented an exhaustive review of FCA-based IR approaches focusing in the shared ideas and notions they share. We have shown how these ideas can be applied in a variety of domains and applications ranging from standard Boolean retrieval to semantic retrieval or file systems.

References

1. Al-Msie'Deen, R., Seriai, A., Huchard, M., Urtado, C., Vauttier, S., Salman, H.E.: Mining features from the object-oriented source code of a collection of software

variants using formal concept analysis and latent semantic indexing. In: The 25th International Conference on Software Engineering and Knowledge Engineering, Knowledge Systems Institute Graduate School, États-Unis, p. 8 (2013)

2. Alam, M., Napoli, A.: Defining views with formal concept analysis for understanding SPARQL query results. In: Proceedings of the 2014 International Conference on Concept Lattices and Their Applications (2014)

3. Baeza-Yates, R.A., Ribeiro-Neto, B.: Modern Information Retrieval. ACM Press, Boston (1999)

4. Barbut, M., Monjardet, B.: Ordre et Classification: Algebre et Combinatoire. Librairie Hachette, Paris (1970)

5. Bosman, F., Bouwman, R., Bruza, P.: The effectiveness of navigable information disclosure systems. In: Proceedings of the Informatiewetenschap (1991)

6. Carpineto, C., Mizzaro, S., Romano, G., Snidero, M.: Mobile information retrieval with search results clustering: prototypes and evaluations. J. Am. Soc. Inf. Sci. Technol. **60**(5), 877–895 (2009)

7. Carpineto, C., Romano, G.: Galois : an order-theoretic approach to conceptual clustering. In: Proceedings of the 10th International Conference on Machine Learning (ICML 1990) (1993)

8. Carpineto, C., Romano, G.: A lattice conceptual clustering system and its application to browsing retrieval. Mach. Learn. **24**(2), 95–122 (1996)

9. Carpineto, C., Romano, G.: Information retrieval through hybrid navigation of lattice representations. Int. J. Hum. Comput. Stud. **45**(5), 553–578 (1996)

10. Carpineto, C., Romano, G.: Effective reformulation of boolean queries with concept lattices. In: Andreasen, T., Christiansen, H., Larsen, H.L. (eds.) FQAS 1998. LNCS (LNAI), vol. 1495, pp. 83–94. Springer, Heidelberg (1998)

11. Carpineto, C., Romano, G.: Order theoretical ranking. J. Am. Soc. Inf. Sci. **51**(7), 587–601 (2000)

12. Carpineto, C., Romano, G.: Exploiting the potential of concept lattices for information retrieval with CREDO. J. Univ. Comput. Sci. **10**(8), 985–1013 (2004)

13. Carpineto, C., Romano, G.: Using concept lattices for text retrieval and mining. In: Ganter, B., Stumme, G., Wille, R. (eds.) Formal Concept Analysis. LNCS (LNAI), vol. 3626, pp. 161–179. Springer, Heidelberg (2005)

14. Carpineto, C., Romano, G.: ULYSSES: a lattice-based multiple interaction strategy retrieval interface. In: Blumenthal, B., Gornostaev, J., Unger, C. (eds.) EWHCI 1995. LNCS, vol. 1015, pp. 91–104. Springer, Heidelberg (1995)

15. Castellanos, A., García-Serrano, A., Cigarrán, J.: Linked data-based conceptual modelling for recommendation: a FCA-based approach. In: Hepp, M., Hoffner, Y. (eds.) EC-Web 2014. LNBIP, vol. 188, pp. 71–76. Springer, Heidelberg (2014)

16. Cigarrán, J.M., Gonzalo, J., Peñas, A., Verdejo, M.F.: Browsing search results via formal concept analysis: automatic selection of attributes. In: Eklund, P. (ed.) ICFCA 2004. LNCS (LNAI), vol. 2961, pp. 74–87. Springer, Heidelberg (2004)

17. Codocedo, V., Lykourentzou, I., Napoli, A.: A contribution to semantic indexing and retrieval based on FCA - an application to song datasets. In: Proceedings of the 2012 International Conference on Concept Lattices and their Applications (2012)

18. Codocedo, V., Lykourentzou, I., Napoli, A.: A semantic approach to concept lattice-based information retrieval. Ann. Math. Artif. Intell. **72**(1), 69–195 (2014)

19. Codocedo, V., Napoli, A.: A proposition for combining pattern structures and relational concept analysis. In: Glodeanu, C.V., Kaytoue, M., Sacarea, C. (eds.) ICFCA 2014. LNCS, vol. 8478, pp. 96–111. Springer, Heidelberg (2014)

20. Cole, R., Eklund, P.: Application of formal concept analysis to information retrieval using a hierarchically structured thesaurus. In: International Conference on Conceptual Graphs, ICCS 1996. University of New South (1996)

21. Cole, R., Eklund, P.: Analyzing an email collection using formal concept analysis. In: Żytkow, J.M., Rauch, J. (eds.) PKDD 1999. LNCS (LNAI), vol. 1704, pp. 309–315. Springer, Heidelberg (1999)

22. Cole, R.J., Eklund, P.W., Stumme, G.: Document retrieval for email search and discovery using formal concept analysis. J. Appl. Artif. Intell. (AAI) **17**(3), 257–280 (2003)

23. Duboucherryan, P., Bridge, D.: Collaborative recommending using formal concept analysis. Knowl. Based Syst. **19**(5), 309–315 (2006)

24. Ducrou, J.: DVDSleuth: a case study in applied formal concept analysis for navigating Web catalogs. In: Priss, U., Polovina, S., Hill, R. (eds.) ICCS 2007. LNCS (LNAI), vol. 4604, pp. 496–500. Springer, Heidelberg (2007)

25. Ducrou, J., Eklund, P.: SearchSleuth: the conceptual neighbourhood of a Web query. In: Proceedings of the 2007 International Conference on Concept Lattices and their Applications,CLA 2007 (2007)

26. Ducrou, J., Vormbrock, B., Eklund, P.: FCA-based browsing and searching of a collection of images. In: Schärfe, H., Hitzler, P., Øhrstrøm, P. (eds.) ICCS 2006. LNCS (LNAI), vol. 4068, pp. 203–214. Springer, Heidelberg (2006)

27. Eklund, P., Ducrou, J.: Navigation and annotation with formal concept analysis. In: Richards, D., Kang, B.-H. (eds.) PKAW 2008. LNCS, vol. 5465, pp. 118–121. Springer, Heidelberg (2009)

28. Eklund, P., Ducrou, J., Brawn, P.: Concept lattices for information visualization: can novices read line-diagrams? In: Eklund, P. (ed.) ICFCA 2004. LNCS (LNAI), vol. 2961, pp. 57–73. Springer, Heidelberg (2004)

29. Fairthorne, R.A.: The patterns of retrieval. Am. Doc. **7**(2), 65–70 (1956)

30. Ferré, S., Hermann, A.: Reconciling faceted search and query languages for the semantic web. IJMSO **7**(1), 37–54 (2012)

31. Ferré, S., Ridoux, O.: A file system based on concept analysis. In: Palamidessi, C., Moniz Pereira, L., Lloyd, J.W., Dahl, V., Furbach, U., Kerber, M., Lau, K.-K., Sagiv, Y., Stuckey, P.J. (eds.) CL 2000. LNCS (LNAI), vol. 1861, pp. 1033–1047. Springer, Heidelberg (2000)

32. Ferré, S., Ridoux, O.: A logical generalizationof formal concept analysis. In: Ganter, B., Mineau, G.W. (eds.) ICCS 2000. LNCS, vol. 1867, pp. 371–384. Springer, Heidelberg (2000)

33. Ganter, B., Wille, R.: Formal Concept Analysis: Mathematical Foundations. Springer, Heidelberg (1999)

34. Ginsberg, A.: A unified approach to automatic indexing and information retrieval. IEEE Expert **8**(5), 46–56 (1993)

35. Godin, R., Gecsei, J., Pichet, C.; Design of a browsing interface for information retrieval. In: Proceedings of the 12th Annual International ACM SIGIR Conference on Research and Development in Information Retrieval, SIGIR 1989, New York (1989)

36. Godin, R., Missaoui, R., April, A.: Experimental comparison of navigation in a Galois lattice with conventional information retrieval methods. Int. J. Man-Mach. Stud. **38**(5), 747–767 (1993)

37. Godin, R., Saunders, E., Gecsei, J.: Lattice model of browsable data spaces. Inf. Sci. Int. J. **40**(2), 89–116 (1986)

38. Grootjen, F.A., van der Weide, T.: Conceptual relevance feedback. In: 2002 IEEE International Conference on Systems, Man and Cybernetics, vol. 2, Oct 2002

39. Grootjen, F.A., van der Weide, T.P.: Conceptual query expansion. Data Knowl. Eng. **56**(2), 174–193 (2006)
40. Huchard, M., Hacene, M.R., Roume, C., Valtchev, P.: Relational concept discovery in structured datasets. Ann. Math. Artif. Intell. **49**(1–4), 39–76 (2007)
41. Ignatov D.I., Kuznetsov, S.O.: Concept-based recommendations for internet advertisement. In: Proceedings of the 2008 International Conference on Concept Lattices and Their Applications (2008)
42. Ingwersen, P.: Cognitive perspectives of information retrieval interaction: elements of a congnitive IR theory. J. Doc. **52**(1), 3–50 (1996)
43. Kim, M.H., Compton, P.: Formal concept analysis for domain-specific document retrieval systems. In: Stumptner, M., Corbett, D.R., Brooks, M. (eds.) Canadian AI 2001. LNCS (LNAI), vol. 2256, pp. 237–248. Springer, Heidelberg (2001)
44. Kim, M., Compton, P.: Evolutionary document management and retrieval for specialized domains on the web. Int. J. Hum. Comput. Stud. **60**(2), 201–241 (2004)
45. Koester, B.: Conceptual knowledge retrieval with FooCA: improving web search engine results with contexts and concept hierarchies. In: Perner, P. (ed.) ICDM 2006. LNCS (LNAI), vol. 4065, pp. 176–190. Springer, Heidelberg (2006)
46. Kuznetsov, S.O.: Fitting pattern structures to knowledge discovery in big data. In: Cellier, P., Distel, F., Ganter, B. (eds.) ICFCA 2013. LNCS, vol. 7880, pp. 254–266. Springer, Heidelberg (2013)
47. Manning, C.D., Raghavan, P., Schtze, H.: Introduction to Information Retrieval. Cambridge University Press, Cambridge (2008)
48. Martin, B.: Formal concept analysis and semantic file systems. In: Eklund, P. (ed.) ICFCA 2004. LNCS (LNAI), vol. 2961, pp. 88–95. Springer, Heidelberg (2004)
49. Martinez, J., Loisant, E.: Browsing image databases with galois' lattices. In: Proceedings of the 2002 ACM Symposium on Applied Computing, SAC 2002, New York (2002)
50. Messai, N., Devignes, M.-D., Napoli, A., Smaïl-Tabbone, M.: Querying a bioinformatic data sources registry with concept lattices. In: Dau, F., Mugnier, M.-L., Stumme, G. (eds.) ICCS 2005. LNCS (LNAI), vol. 3596, pp. 323–336. Springer, Heidelberg (2005)
51. Messai, N., Devignes, M.-D., Napoli, A., Smaïl-Tabbone, M.: BR-Explorer: an FCA-based algorithm for information retrieval. In: Fourth International Conference on Concept Lattices and Their Applications - CLA 2006, Hammamet/Tunisia (2006)
52. Messai, N., Devignes, M.-D., Napoli, A., Smaïl-Tabbone, M.: Many-valued concept lattices for conceptual clustering and information retrieval. In: Proceedings of the 2008 Conference on ECAI 2008: 18th European Conference on Artificial Intelligence, pp. 127–131, June 2008
53. Mili, A., Mili, R., Mittermeir, R.: Storing and retrieving software components: a refinement based system. In: Proceedings of the 16th International Conference on Software Engineering, ICSE 1994, Los Alamitos (1994)
54. Mooers, C.N.: A mathematical theory of language symbols in retrieval (1958)
55. Nauer, E., Toussaint, Y.: Dynamical modification of context fo an iterative and interactive information retrieval proess on the web. In: Proceedings of the 2007 International Conference on Concept Lattices and their Applications, CLA 2007 (2007)
56. Nauer, E., Toussaint, Y.: CreChainDo: an iterative and interactive web information retrieval system based on lattices. Int. J. Gen. Syst. **38**(4), 363–378 (2009)
57. Nguyen, T.T., Hui, S.C., Chang, K.: A lattice-based approach for mathematical search using formal concept analysis. Expert Syst. Appl. **39**(5), 5820–5828 (2012)

58. Pedersen, G.S.: A browser for bibliographic information retrieval, based on an application of lattice theory. In: Proceedings of the 16th Annual International ACM SIGIR Conference on Research and Development in Information Retrieval, SIGIR 1993, New York (1993)
59. Pedersen, G.S.: Relationship Lattices for Information Modelling. Information Modelling and Knowledge Bases. IOS Press, Amsterdam (1994)
60. Poelmans, J., Ignatov, D.I., Viaene, S., Dedene, G., Kuznetsov, S.O.: Text mining scientific papers: a survey on FCA-based information retrieval research. In: Perner, P. (ed.) ICDM 2012. LNCS, vol. 7377, pp. 273–287. Springer, Heidelberg (2012)
61. Poshyvanyk, D., Marcus, A.: Combining formal concept analysis with information retrieval for concept location in source code. In: 15th IEEE International Conference on Program Comprehension (ICPC 2007), June 2007
62. Priss, U.: A graphical interface for document retrieval based on formal concept analysis. In: Proceedings of the 8th Midwest Artificial Intelligence and Cognitive Science Conference (1997)
63. Priss, U.: Lattice-based information retrieval. Knowl. Organ. **27**, 132–142 (2000)
64. Priss, U.: Formal concept analysis in information science. Ann. Rev. Inf. Sci. Technol. **40**(1), 521–543 (2007)
65. Rocchio, J.J.: Relevance feedback in information retrieval. In: The Smart Retrieval System - Experiments in Automatic Document Processing (1971)
66. Senatore, S., Pasi, G.: Lattice navigation for collaborative filtering by means of (fuzzy) formal concept analysis. In: Proceedings of the 28th Annual ACM Symposium on Applied Computing, SAC 2013, New York (2013)
67. Shah, A., Caves, L.: ConceptOntoFs: a semantic file system for interno. In: First International Workshop on Plan 9 and Inferno (2006)
68. Valverde, F.J., Pelaez-Moreno, C.: System vs. methods: an analysis of the affordances of formal concept analysis for information retrieval. In: Proceedings of the Workshop Formal Concept Analysis Meets Information Retrieval (FCAIR 2013) (2013)
69. Wille, R.: Restructuring lattice theory: an approach based on hierarchies of concepts. In: Ferré, S., Rudolph, S. (eds.) ICFCA 2009. LNCS, vol. 5548, pp. 314–339. Springer, Heidelberg (2009)
70. Wray, T., Eklund, P.: Using formal concept analysis to create pathways through museum collections. In: Proceedings of the 3rd International Workshop "What can FCA do for Artificial Intelligence"? (2014)

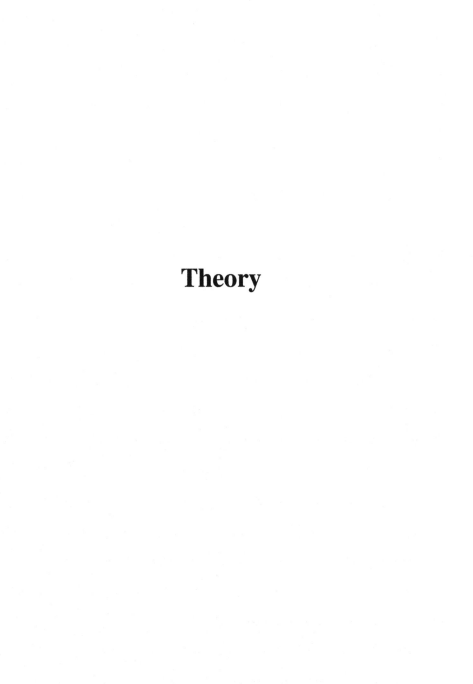

Theory

Bonds Between *L*-Fuzzy Contexts Over Different Structures of Truth-Degrees

Jan Konecny[(✉)]

Data Analysis and Modeling Lab,
Department of Computer Science, Palacky University,
Olomouc 17. Listopadu 12, 77146 Olomouc, Czech Republic
jan.konecny@upol.cz

Abstract. We consider the problem of bonds between *L*-fuzzy contexts over different complete residuated lattices. For this purpose we define (l, k)-connection and dual (l, k)-connection – pairs of mappings between the residuated lattices based on Krupka's results on factorizations of complete residuated lattices. We show that the bonds defined using the dual (l, k)-connection have very natural properties.

Keywords: Formal concept analysis · Galois connection · Bond · Factorization · Complete residuated lattice · Fuzzy set

1 Introduction

We study the problem of bonding formal fuzzy contexts over different structures of truth-degrees. This problem was addressed in [12][1] where the authors used residuation-preserving isotone Galois connections between complete residuated lattices to define bonds. We find the definition of residuation-preserving isotone Galois connection unnecessarily strict for its purpose and we take a new look at it.

Similarly as in [12] we look for an isotone Galois connection between two complete residuated lattices. We apply Krupka's results on factorization of residuated lattices [13] to find looser and more flexible requirements for the correspondence. As a result we obtain two interrelated correspondences between complete residuated concept lattices — (l, k)-connection and dual (l, k)-connection. Both of them can be considered to be a variant of the residuation-preserving isotone Galois connection from [12]. Using the dual (l, k)-connection we define bonds between formal fuzzy contexts over different complete residuated lattices.

The paper is organized as follows. In Sect. 2, we recall fundamental notions used in the paper. Sections 3 and 4 introduce the (l, k)-connection and dual the (l, k)-connection, respectively, and describe their properties. In Sect. 5 we utilize the new connections in formal concept analysis to define bonds between formal fuzzy contexts over different residuated lattices. Finally, Sect. 6 summarizes our conclusions and ideas for future research in this area.

---

[1] See [12] for motivations of the present research.

© Springer International Publishing Switzerland 2015
J. Baixeries et al. (Eds.): ICFCA 2015, LNAI 9113, pp. 81–96, 2015.
DOI: 10.1007/978-3-319-19545-2_5

2 Preliminaries

2.1 Residuated Lattices, Fuzzy Sets, and Fuzzy Relations

We use complete residuated lattices as basic structures of truth-degrees. A complete residuated lattice is a structure $\mathbf{L} = \langle L, \wedge, \vee, \otimes, \rightarrow, 0, 1 \rangle$ such that

(i) $\langle L, \wedge, \vee, 0, 1 \rangle$ is a complete lattice, i.e. a partially ordered set in which arbitrary infima and suprema exist;

(ii) $\langle L, \otimes, 1 \rangle$ is a commutative monoid, i.e. $\otimes$ is a binary operation which is commutative, associative, and $a \otimes 1 = a$ for each $a \in L$;

(iii) $\otimes$ and $\rightarrow$ satisfy adjointness, i.e. $a \otimes b \leqslant c$ iff $a \leqslant b \rightarrow c$.

0 and 1 denote the least and greatest elements. The partial order of $\mathbf{L}$ is denoted by $\leqslant$. Throughout this work, $\mathbf{L}$ denotes an arbitrary complete residuated lattice.

Elements a of L are called truth degrees. Operations $\otimes$ (multiplication) and $\rightarrow$ (residuum) play the role of (truth functions of) "fuzzy conjunction" and "fuzzy implication". Furthermore, we define the complement of $a \in L$ as

$$\neg a = a \rightarrow 0, \tag{1}$$

and binary operation of biresiduum $\leftrightarrow$ as

$$a \leftrightarrow b = (a \rightarrow b) \wedge (b \rightarrow a) \quad \text{for each } a, b \in L \tag{2}$$

An $\mathbf{L}$-set (or $\mathbf{L}$-fuzzy set) A in a universe set X is a mapping assigning to each $x \in X$ some truth degree $A(x) \in L$. The set of all $\mathbf{L}$-sets in a universe X is denoted L^X.

The operations with $\mathbf{L}$-sets are defined componentwise. For instance, the intersection of $\mathbf{L}$-sets $A, B \in L^X$ is an $\mathbf{L}$-set $A \cap B$ in X such that $(A \cap B)(x) = A(x) \wedge B(x)$ for each $x \in X$, etc. An $\mathbf{L}$-set $A \in L^X$ is also denoted $\{ ^{A(x)}\!/x \mid x \in X \}$. If for all $y \in X$ distinct from $x_1, x_2, \ldots, x_n$ we have $A(y) = 0$, we also write

$$\{ ^{A(x_1)}\!/x_1, {}^{A(x_2)}\!/x_2, \ldots, {}^{A(x_n)}\!/x_n \}.$$

An $\mathbf{L}$-set $A \in L^X$ is called crisp if $A(x) \in \{0, 1\}$ for each $x \in X$. Crisp $\mathbf{L}$-sets can be identified with ordinary sets. For a crisp A, we also write $x \in A$ for $A(x) = 1$ and $x \notin A$ for $A(x) = 0$. An $\mathbf{L}$-set $A \in L^X$ is called empty (denoted by $\emptyset$) if $A(x) = 0$ for each $x \in X$.

Binary $\mathbf{L}$-relations (binary $\mathbf{L}$-fuzzy relations) between X and Y can be thought of as $\mathbf{L}$-sets in the universe $X \times Y$. That is, a binary $\mathbf{L}$-relation $I \in L^{X \times Y}$ between a set X and a set Y is a mapping assigning to each $x \in X$ and each $y \in Y$ a truth degree $I(x, y) \in L$ (a degree to which x and y are related by I).

Various composition operators for binary $\mathbf{L}$-relations were extensively studied by [7]; we will use the following three composition operators, defined for relations $A \in L^{X \times F}$ and $B \in L^{F \times Y}$:

$$(A \circ B)(x, y) = \bigvee_{f \in F} A(x, f) \otimes B(f, y), \tag{3}$$

$$(A \triangleleft B)(x, y) = \bigwedge_{f \in F} A(x, f) \rightarrow B(f, y), \tag{4}$$

$$(A \triangleright B)(x, y) = \bigwedge_{f \in F} B(f, y) \rightarrow A(x, f). \tag{5}$$

All of them have natural verbal descriptions. For instance, $(A \circ B)(x, y)$ is the truth degree of the proposition "*there is a factor f such that f applies to object x and attribute y is a manifestation of f*"; $(A \triangleleft B)(x, y)$ is the truth degree of "*for every factor f, if f applies to object x then attribute y is a manifestation of f*". Note also that for $L = \{0, 1\}$, $A \circ B$ coincides with the well-known composition of binary relations.

2.2   Formal Fuzzy Concept Analysis

An **L-context** is a triplet $\langle X, Y, I \rangle$ where X and Y are (ordinary nonempty) sets and $I \in L^{X \times Y}$ is an **L**-relation between X and Y. Elements of X are called objects, elements of Y are called attributes, I is called an incidence relation. $I(x, y) = a$ is read: "The object x has the attribute y to degree a."

Consider the following pair $\langle \uparrow, \downarrow \rangle$ of operators $\uparrow : L^X \rightarrow L^Y$ and $\downarrow : L^Y \rightarrow L^X$ induced by an **L**-context $\langle X, Y, I \rangle$ as

$$A^{\uparrow}(y) = \bigwedge_{x \in X} A(x) \rightarrow I(x, y) \quad \text{and} \quad B^{\downarrow}(x) = \bigwedge_{y \in Y} B(y) \rightarrow I(x, y) \tag{6}$$

for all $A \in L^X$ and $B \in L^Y$.

Furthermore, denote the set of fixed points of $\langle \uparrow, \downarrow \rangle$ by $\mathcal{B}^{\uparrow\downarrow}(X, Y, I)$, i.e.

$$\mathcal{B}^{\uparrow\downarrow}(X, Y, I) = \{\langle A, B \rangle \in L^X \times L^Y \mid A^{\uparrow} = B, B^{\downarrow} = A\}. \tag{7}$$

The set of fixed points endowed with $\leqslant$, defined by

$$\langle A_1, B_1 \rangle \leqslant \langle A_2, B_2 \rangle \quad \text{if } A_1 \subseteq A_2 \text{ (equivalently } B_2 \subseteq B_1)$$

is a complete lattice [2,15], called a *standard* **L**-*concept lattice* associated with I, and its elements are called *formal concepts*. In a formal concept $\langle A, B \rangle$, the A is called an *extent*, and B is called an *intent*. The set of all extents and the set of all intents are denoted by $\text{Ext}^{\uparrow\downarrow}$ and $\text{Int}^{\uparrow\downarrow}$, respectively. That is,

$$\begin{aligned}
\text{Ext}^{\uparrow\downarrow}(X, Y, I) &= \{A \in L^X \mid \langle A, B \rangle \in \mathcal{B}^{\uparrow\downarrow}(X, Y, I) \text{ for some } B\}, \\
\text{Int}^{\uparrow\downarrow}(X, Y, I) &= \{B \in L^Y \mid \langle A, B \rangle \in \mathcal{B}^{\uparrow\downarrow}(X, Y, I) \text{ for some } A\}.
\end{aligned} \tag{8}$$

An **L**-relation $\beta \in L^{X_1 \times Y_2}$ is called an **L**-*bond*[2] from **L**-context $\langle X_1, Y_1, I_1 \rangle$ to **L**-context $\langle X_2, Y_2, I_2 \rangle$ if

$$\begin{aligned}
\text{Ext}^{\uparrow\downarrow}(X_1, Y_2, \beta) &\subseteq \text{Ext}^{\uparrow\downarrow}(X_1, Y_1, I_1), \\
\text{Int}^{\uparrow\downarrow}(X_1, Y_2, \beta) &\subseteq \text{Int}^{\uparrow\downarrow}(X_2, Y_2, I_2).
\end{aligned} \tag{9}$$

---

[2] The notion of **L**-bond was introduced in [11]; however we adapt its definition the same way as in [8,9].

3 (l, k)-Connections Between Complete Residuated Lattices

Similarly as in [12] we look for a pair of mappings $\lambda : \mathbf{L}_1 \to \mathbf{L}_2$ and $\kappa : \mathbf{L}_2 \to \mathbf{L}_1$ which form an isotone Galois connection. Set of its fixpoints with order defined as

$$\langle a_1, a_2 \rangle \leqslant \langle b_1, b_2 \rangle \quad \text{iff} \quad a_1 \leqslant b_1 \quad \text{(or equivalently } a_2 \leqslant b_2) \qquad (10)$$

is a complete lattice. We denote it as $\mathbf{L}_{\langle \lambda, \kappa \rangle}$. We need to assure that an adjoint pair exists in $\mathbf{L}_{\langle \lambda, \kappa \rangle}$ and this pair is related to adjoint pairs of both, $\mathbf{L}_1$ and $\mathbf{L}_2$. To this purpose we apply Krupka's results on factorization of residuated lattices [13]. In fact, the problem can be reformulated as finding an isomorphism between some factorizations of $\mathbf{L}_1$ and $\mathbf{L}_2$ as depicted in Fig. 1.

Let us recollect Krupka's approach to factorization of complete residuated lattices. Krupka defines the factorization by cuts of biresiduum as follows. Consider a complete residuated lattice $\mathbf{L}$, a truth degree $e \in L$, and mappings

$$a^e = \bigvee \{b \in L \mid a \leftrightarrow b \geqslant e\} = e \to a, \qquad (11)$$

$$a_e = \bigwedge \{b \in L \mid a \leftrightarrow b \geqslant e\} = e \otimes a. \qquad (12)$$

For each $a \in L$ define intervals

$$[a]_e = [a_e, (a_e)^e] = [e \otimes a, e \to (e \otimes a)],$$
$$[a]^e = [(a^e)_e, a^e] = [e \otimes (e \to a), e \to a].$$

Denote $L/e = \{[a]^e \mid a \in L\}(= \{[a]_e \mid a \in L\})$. Then we have the following result.

Theorem 1 ([13]). $L/e = \langle L/e, \wedge, \vee, \otimes, \to, 0, 1 \rangle$, where $\wedge$ and $\vee$ are given by the order

$$B_1 \leqslant B_2 \quad \text{iff} \quad \bigvee B_1 \leqslant \bigvee B_2$$

and

$$B_1 \otimes B_2 = [\bigvee B_1 \otimes \bigvee B_2]_e,$$
$$B_1 \to B_2 = [\bigvee B_1 \to \bigvee B_2]_e,$$
$$0 = [0, e \to 0],$$
$$1 = [e, 1]$$

for each $B_1, B_2 \in L/e$, is a complete residuated lattice.

Following lemma shows alternative ways to define $\otimes$ and $\to$ in $\mathbf{L}/e$

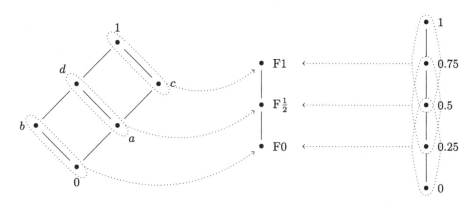

⊗	0	a	b	c	d	1
0	0	0	0	0	0	0
a	0	0	0	a	0	a
b	0	0	b	0	b	b
c	0	a	0	c	a	c
d	0	0	b	a	b	d
1	0	a	b	c	d	1

→	0	a	b	c	d	1
0	1	1	1	1	1	1
a	d	1	d	1	1	1
b	c	c	1	c	1	1
c	b	d	b	1	d	1
d	a	c	d	c	1	1
1	0	a	b	c	d	1

Fig. 1. Six-element residuated lattice, with ⊗ and → as showed in the bottom part (011010:00A0B0BCAB in [6]), factorized by c-cuts of biresiduum (left), five-element Łukasiewicz chain (111:000AB in [6]) factorized by 0.5-cuts of biresiduum (right), and their common lattice of factors (middle).

Lemma 1 ([13]). *For any $B_1, B_2 \in L/e$ we have*

$$\bigvee B_1 \otimes \bigwedge B_2 = \bigwedge B_1 \otimes B_2,$$
$$\bigvee B_1 \to \bigvee B_2 = \bigvee (B_1 \to B_2),$$
$$\bigwedge B_1 \to \bigwedge B_2 = \bigvee (B_1 \to B_2).$$

Note that the operators (11) and (12) form an isotone Galois connection on the complete residuated lattice **L**. We extend this approach to have an isotone Galois connection between two (different) complete residuated lattices.

Definition 1. *Let* $\mathbf{L}_1 = \langle L_1, \wedge_1, \vee_1, \otimes_1, \to_1, 0_1, 1_1 \rangle$, $\mathbf{L}_2 = \langle L_2, \wedge_2, \vee_2, \otimes_2, \to_2, 0_2, 1_2 \rangle$ *be complete residuated lattices, let* $l \in L_1, k \in L_2$ *and let* $\lambda : L_1 \to L_2, \kappa : L_2 \to L_1$ *be mappings, such that*

1. *$\langle \lambda, \kappa \rangle$ is an isotone Galois connection between $\mathbf{L}_1$ and $\mathbf{L}_2$,*
2. *$\kappa\lambda(a_1) = l \to_1 (l \otimes_1 a_1)$ for each $a_1 \in L_1$,*
3. *$\lambda\kappa(a_2) = k \otimes_2 (k \to_2 a_2)$ for each $a_2 \in L_2$.*

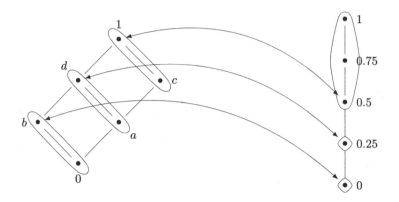

Fig. 2. $(c, 0.5)$-connection between the residuated lattices from Fig. 1.

We call $\langle \lambda, \kappa \rangle$ an (l, k)-connection from $\mathbf{L}_1$ to $\mathbf{L}_2$.

Figure 2 shows an example of (l, k)-connection corresponding to the factorizations in Fig. 1.

Remark 1.

(a) A pair of identities $\langle \mathrm{id}, \mathrm{id} \rangle$ on a complete residuated lattice $\mathbf{L}$ is $(1, 1)$-connection from $\mathbf{L}$ to $\mathbf{L}$.

(b) It is worth noting that an (l, k)-connection from $\mathbf{L}_1$ to $\mathbf{L}_2$ is not uniquely given by the pair of truth degrees $l \in L_1, k \in L_2$ as more than one isomorphism between $\mathbf{L}_1/l$ and $\mathbf{L}_2/k$ can exist. For example, consider four-element complete residuated lattice $\mathbf{L}$ in Fig. 3 (left) with $\otimes = \wedge$ and $\rightarrow$ as in Fig. 3 (right) and mapping $f : L \rightarrow L$ given by $f(0) = 0, f(a) = b, f(b) = a$, and $f(1) = 1$. Then $\langle \mathrm{id}_L, \mathrm{id}_L \rangle$ and $\langle f, f \rangle$ are both $(1, 1)$-connections from $\mathbf{L}$ to $\mathbf{L}$.

Utilizing Theorem 1 we can find particular adjoint pairs in the lattice of fixed points of $\langle \lambda, \kappa \rangle$.

Theorem 2. *Denote by* $L_{\langle \lambda, \kappa \rangle}$ *the set of all fixed points of* (l, k)-*connection* $\langle \lambda, \kappa \rangle$ *between* $\mathbf{L}_1$ *and* $\mathbf{L}_2$.

1. *The algebra* $\langle L_{\langle \lambda, \kappa \rangle}, \wedge, \vee, \otimes, \rightarrow, 0, 1 \rangle$ *where* $\wedge, \vee, 0,$ *and* 1 *are given by the order* (10) *and*

$$\langle a_1, a_2 \rangle \rightarrow \langle b_1, b_2 \rangle = \langle a_1 \rightarrow_1 b_1, \lambda(a_1 \rightarrow_1 b_1) \rangle,$$
$$\langle a_1, a_2 \rangle \otimes \langle b_1, b_2 \rangle = \langle l \rightarrow (l \otimes_1 a_1 \otimes_1 b_1), \lambda(a_1 \otimes_1 b_1) \rangle$$

is a complete residuated lattice.

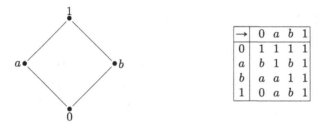

→	0	a	b	1
0	1	1	1	1
a	b	1	b	1
b	a	a	1	1
1	0	a	b	1

Fig. 3. Complete residuated lattice from Remark 1(b); (0:a0b in [6])

2. *The algebra* $\langle L_{\langle\lambda,\kappa\rangle}, \wedge, \vee, \boxtimes, \searrow, 0, 1\rangle$ *where* $\wedge$, $\vee$, 0, *and* 1 *are given by the order* (10) *and*

$$\langle a_1, a_2\rangle \searrow \langle b_1, b_2\rangle = \langle\kappa(k \otimes_2 (a_2 \to_2 b_2)), k \otimes_2 (a_2 \to_2 b_2)\rangle$$
$$= \langle\kappa((k \to_2 b_2) \to_2 (k \to_2 b_2)), (k \to_2 a_2) \to_2 (k \to_2 b_2)\rangle,$$
$$\langle a_1, a_2\rangle \boxtimes \langle b_1, b_2\rangle = \langle\kappa(a_2 \otimes_2 (k \to_2 b_2)), a_2 \otimes_2 (k \to_2 b_2)\rangle$$
$$= \langle\kappa((k \to_2 a_2) \otimes_2 b_2), (k \to_2 a_2) \otimes_2 b_2\rangle$$

is a complete residuated lattice.

Proof. Directly from Definition 1, Theorem 1 and Lemma 1.   □

Remark 2. For sake of completeness, we show how the meet, join, 0, and 1 given by the order (10) are defined in $L_{\langle\lambda,\kappa\rangle}$:

$$\langle a_1, a_2\rangle \wedge \langle b_1, b_2\rangle = \langle a_1 \wedge_1 b_1, k \otimes_2 ((k \to_2 a_2) \wedge_2 (k \to_2 b_2))\rangle, \tag{13}$$
$$\langle a_1, a_2\rangle \vee \langle b_1, b_2\rangle = \langle l \to_1 ((l \otimes_1 a_1) \vee_1 (l \otimes_1 b_1)), a_2 \vee_2 b_2\rangle, \tag{14}$$
$$0 = \langle l \to 0_1, 0_2\rangle, \tag{15}$$
$$1 = \langle 1_1, k\rangle. \tag{16}$$

It is easy to see, that the two ajdoint pairs, $\langle\otimes, \to\rangle$ and $\langle\boxtimes, \searrow\rangle$, from Theorem 2 can be different. As an example consider $\mathbf{L}_1$ being three-element Łukasiewicz chain, $\mathbf{L}_2$ being three-element Gödel chain and λ and κ being identities on $L_1 = L_2$. The related factorizations, $\mathbf{L}_1/1$ and $\mathbf{L}_2/1$ are the three-element Łukasiewicz chain and the three-element Gödel chain, respectively, again. Clearly, their adjoint pairs are different.

We call the (l, k)-connections whose factorizations produce the same adjoint pair *residuation-preserving*. The following corollary shows that for residuation-preserving (l, k)-connection $\langle\lambda, \kappa\rangle$ we can specify the adjoint pair on the lattice of its fixed points without the mappings $\langle\lambda, \kappa\rangle$.

Corollary 1. *Let* $\langle\lambda, \kappa\rangle$ *be a residuation-preserving* (l, k)-connection from $\mathbf{L}_1$ *to* $\mathbf{L}_2$. *The algebra* $\mathbf{L}_{\langle\lambda,\kappa\rangle} = \langle L_{\langle\lambda,\kappa\rangle}, \wedge, \vee, \otimes, \to, 0, 1\rangle$ *where* $\wedge, \vee, 0, 1$ *are given by* *the order* (10) *and*

$$\langle a_1, a_2 \rangle \to \langle b_1, b_2 \rangle = \langle a_1 \to_1 b_1, k \otimes_2 (a_2 \to_2 b_2) \rangle \tag{17}$$

$$= \langle a_1 \to_1 b_1, k \otimes_2 ((k \to_2 a_2) \to_2 (k \to_2 b_2))) \rangle, \tag{18}$$

$$\langle a_1, a_2 \rangle \otimes \langle b_1, b_2 \rangle = \langle l \to_1 (l \otimes_1 a_1 \otimes_1 b_1) a_2 \otimes_2 (k \to_2 b_2) \rangle \tag{19}$$

$$= \langle l \to_1 (l \otimes_1 a_1 \otimes_1 b_1), (k \to_2 a_2) \otimes_2 b_2 \rangle \tag{20}$$

is a complete residuated lattice.

Proof. Directly from Theorem 2 and the property of residuation-preservation, that is $\otimes = \boxtimes$ and $\to = \searrow$. $\qquad\square$

The following theorem provides more practical characterization of residuation-preserving (l, k)-connections.

Theorem 3. *Let $\langle \lambda, \kappa \rangle$ be an (l, k)-connection from $\mathbf{L}_1$ to $\mathbf{L}_2$. The following statements are equivalent*

(a) $\langle \lambda, \kappa \rangle$ is residuation-preserving.
(b) $\kappa(k \otimes_2 (\lambda(a) \to_2 \lambda(b))) = \kappa\lambda(a) \to_1 \kappa\lambda(b)$ holds true for any $a, b \in L_1$.
(c) $k \otimes_2 (\lambda\kappa(a) \to_2 \lambda\kappa(b)) = \lambda(\kappa(a) \to_1 \kappa(b))$ holds true for any $a, b \in L_2$.

Proof. (sketch) Follows from the fact, that pairs in $L_{\langle \lambda, \kappa \rangle}$ are exactly pairs $\langle \kappa\lambda(a_1), \lambda(a_1) \rangle$ for $a_1 \in L_1$ and exactly pairs $\langle \kappa(a_2), \lambda\kappa(a_2) \rangle$ for $a_2 \in L_2$. $\quad\square$

Note that left-hand sides of the equations in (b) and (c) of Theorem 3 contain an inconvenient multiplication by k. This leads to a quite cumbersome definition when we try to use them to define bonds between formal fuzzy context over different residuated lattices. In the next section we provide an alternative to (l, k)-connection which avoids this inconvenience.

4   Dual (l, k)-Connections Between Complete Residuated Lattices

We defined (l, k)-connections as an isotone Galois connection to assure that the set of its fixed points is a complete lattice and that it preserves order of both $\mathbf{L}_1$ and $\mathbf{L}_2$. But another property of isotone Galois connection, namely its non-duality, is undesired for our purpose, that is bonding fuzzy contexts over different residuated lattices. To fix this, we make a small trick with the (l, k)-connections. Instead of connecting upper bounds of intervals from $\mathbf{L}_1/l$ with lower bounds of intervals in $\mathbf{L}_2/k$, we simply connect upper bounds with upper bounds. To do that we need to drop the requirement of being an isotone Galois connection.

Definition 2. *Let $\mathbf{L}_1 = \langle L_1, \wedge_1, \vee_1, \otimes_1, \to_1, 0_1, 1_1 \rangle$, $\mathbf{L}_2 = \langle L_2, \wedge_2, \vee_2, \otimes_2, \to_2, 0_2, 1_2 \rangle$ be complete residuated lattices, let $l \in L_1, k \in L_2$ and let $\lambda' : L_1 \to L_2, \kappa' : L_2 \to L_1$ be mappings, such that*

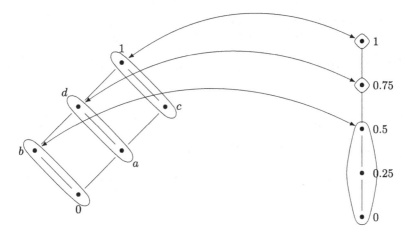

Fig. 4. Dual $(c, 0.5)$-connection between the residuated lattices from Fig. 1

- λ' and κ' are order-preserving,
- $\lambda'\kappa'\lambda'(a_1) = \lambda'(a_1)$ and $\kappa'\lambda'\kappa'(a_2) = \kappa'(a_2)$ for each $a_1 \in L_1$ and $a_2 \in L_2$,
- $\kappa'\lambda'(a_1) = l \to_1 (l \otimes_1 a_1)$ for each $a_1 \in L_1$,
- $\lambda'\kappa'(a_2) = k \to_2 (k \otimes_2 a_2)$ for each $a_2 \in L_2$.

We call the pair $\langle \lambda', \kappa' \rangle$[3] *a dual* (l, k)*-connection from* $\mathbf{L}_1$ *to* $\mathbf{L}_2$.

The notion of dual and non-dual (l, k)-connections are related in following way.

Theorem 4.

(a) For each (l, k)*-connection* $\langle \lambda, \kappa \rangle$ *from* $\mathbf{L}_1$ *to* $\mathbf{L}_2$ *there is a dual* (l, k)*-connection* $\langle \lambda', \kappa' \rangle$ *from* $\mathbf{L}_1$ *to* $\mathbf{L}_2$, *such that for each* $a_1 \in L_1, a_2 \in L_2$,

$$
\begin{aligned}
\langle a_1, a_2 \rangle \in L_{\langle \lambda, \kappa \rangle} \text{ implies } \langle a_1, k \to_2 a_2 \rangle \in L_{\langle \lambda', \kappa' \rangle}, \\
\langle a_1, a_2 \rangle \in L_{\langle \lambda', \kappa' \rangle} \text{ implies } \langle a_1, k \otimes_2 a_2 \rangle \in L_{\langle \lambda, \kappa \rangle}.
\end{aligned}
\tag{21}
$$

(b) For each dual (l, k)*-connection* $\langle \lambda', \kappa' \rangle$ *from* $\mathbf{L}_1$ *and* $\mathbf{L}_2$ *there is an* (l, k)*-connection* $\langle \lambda, \kappa \rangle$ *from* $\mathbf{L}_1$ *to* $\mathbf{L}_2$ *such that* (21) *is satisfied.*

Proof.

(a) Let $\langle \lambda, \kappa \rangle$ be an (l, k)-connection from $\mathbf{L}_1$ to $\mathbf{L}_2$. We show that $\langle \lambda', \kappa' \rangle$ defined as

$$
\lambda' = k \to_2 \lambda(a_1) \quad \text{and} \quad \kappa' = \kappa(k \otimes_2 a_2)
\tag{22}
$$

for each $a_1 \in L_1, a_2 \in L_2$ is a dual (l, k)-connection from $\mathbf{L}_1$ to $\mathbf{L}_2$ which satisfies (21). Since λ and κ are order-preserving and $\to$ and $\otimes$ are

---

[3] In this section, we consistently denote dual (l, k)-connections by prime, as $\langle \lambda', \kappa' \rangle$, to distinguish them from the non-dual (l, k)-connections introduced in the previous section.

both monotone in the second argument, the mapping λ' and κ' are order-preserving as well. We have for each $a_1 \in L_1, a_2 \in L_2$

$$\kappa'\lambda'(a_1) = \kappa(k \otimes_2 (k \rightarrow_2 (\lambda(a_1)))) = \kappa\lambda\kappa\lambda(a_1) = \kappa\lambda(a_1) = l \rightarrow_1 (l \otimes_1 a_1)$$

and

$$\lambda'\kappa'(a_2) = k \rightarrow_2 \lambda\kappa(k \otimes_2 a_2) = k \rightarrow_2 (k \otimes_2 (k \rightarrow_2 (k \otimes_2 a_2))) = k \rightarrow_2 (k \otimes_2 a_2).$$

Finally, we have for each $a_1 \in L_1, a_2 \in L_2$

$$\lambda'\kappa'\lambda'(a_1) = k \rightarrow_2 (k \otimes_2 (k \rightarrow_2 \lambda(a_1))) = k \rightarrow_2 \lambda(a_1) = \lambda'(a_1)$$

and

$$\kappa'\lambda'\kappa'(a_2) = \kappa'\lambda'(\kappa(k \otimes_2 a_2)) = \kappa\lambda\kappa(k \otimes_2 a_2) = \kappa(k \otimes_2 a_2) = \kappa'(a_2).$$

Thus, $\langle\lambda', \kappa'\rangle$ is a dual (l, k)-connection from $\mathbf{L}_1$ to $\mathbf{L}_2$. Now we show that $\langle\lambda', \kappa'\rangle$ satisfies (21). Let $\langle a_1, a_2\rangle \in L_{\langle\lambda,\kappa\rangle}$; from that we have

$$\begin{aligned}
\langle a_1, k \rightarrow a_2\rangle &= \langle\kappa(a_2), k \rightarrow_2 \lambda(a_1)\rangle \\
&= \langle\kappa\lambda(a_1), k \rightarrow_2 \lambda(a_1)\rangle \\
&= \langle\kappa'\lambda'(a_1), \lambda'(a_1)\rangle
\end{aligned}$$

showing $\langle a_1, k \rightarrow_2 a_2\rangle \in L_{\langle\lambda',\kappa'\rangle}$. The other part can be showed similarly.

(b) Similarly as in (a) we can show that $\langle\lambda, \kappa\rangle$ defined as

$$\lambda = k \otimes_2 \lambda'(a_1) \quad \text{and} \quad \kappa = \kappa'(k \rightarrow_2 a_2) \tag{23}$$

for each $a_1 \in L_1, a_2 \in L_2$ is a (l, k)-connection from $\mathbf{L}_1$ to $\mathbf{L}_2$ which satisfies (21).

$\square$

What we get from this trick are more convenient operations $\wedge$ and $\rightarrow$ in the complete residuated lattice $\mathbf{L}_{\langle\lambda,\kappa\rangle}$ of fixed points of $\langle\lambda, \kappa\rangle$. That is important for definition of bonds because concept-forming operators $\langle\uparrow, \downarrow\rangle$ are defined using the operations $\wedge$ and $\rightarrow$.

Theorem 5. *The (l, k)-connections from $\mathbf{L}_1$ to $\mathbf{L}_2$ are in one-to-one correspondence with dual (l, k)-connections from $\mathbf{L}_1$ to $\mathbf{L}_2$.*

Proof. From proof of Theorem 4 we have (22) and (23) providing ways to get a dual (l, k)-connection from an (l, k)-connection and *vice versa*. We only need to show, that they are mutually inverse. Let $\langle\lambda, \kappa\rangle$ be an (l, k)-connection from $\mathbf{L}_1$ and $\mathbf{L}_2$ and let $\langle\lambda', \kappa'\rangle$ be a dual (l, k)-connection from $\mathbf{L}_1$ to $\mathbf{L}_2$ defined by (22). Applying (23) we get

$$\lambda''(a_1) = k \otimes_2 \lambda'(a_1) = k \otimes_2 (k \rightarrow_2 \lambda(a_1)) = \lambda\kappa\lambda(a_1) = \lambda(a_1)$$

for each $a_1 \in L_1$ and

$$\kappa''(a_2) = \kappa'(k \rightarrow_2 a_2) = \kappa(k \otimes_2 (k \rightarrow_2 a_2)) = \kappa \lambda \kappa(a_2) = \kappa(a_2)$$

for each $a_2 \in L_2$. Similarly, the other composition can be showed to be an identity. □

From the above one-to-one correspondence we obtain the following theorem.

Theorem 6. *Let be $\langle \lambda', \kappa' \rangle$ dual (l, k)-connection from $\mathbf{L}_1$ to $\mathbf{L}_2$.*

1. *The algebra $\langle L_{\langle \lambda', \kappa' \rangle}, \wedge, \vee, \otimes_1, \rightarrow_1, 0, 1 \rangle$ where $\wedge$ and $\vee$ are given by the order (10) and*

$$\langle a_1, a_2 \rangle \rightarrow \langle b_1, b_2 \rangle = \langle a_1 \rightarrow_1 b_1, \lambda'(a_1 \rightarrow_1 b_1) \rangle$$
$$\langle a_1, a_2 \rangle \otimes \langle b_1, b_2 \rangle = \langle l \rightarrow_1 (l \otimes_1 a_1 \otimes_1 b_1), \lambda'(a_1 \otimes_1 b_1)))) \rangle$$

is a complete residuated lattice.
2. *The algebra $\langle L_{\langle \lambda', \kappa' \rangle}, \wedge, \vee, \boxtimes, \searrow, 0, 1 \rangle$ where $\wedge$ and $\vee$ are given by the order (10) and*

$$\langle a_1, a_2 \rangle \searrow \langle b_1, b_2 \rangle = \langle \kappa'(a_2 \rightarrow_2 b_2), a_2 \rightarrow_2 b_2 \rangle$$
$$\langle a_1, a_2 \rangle \boxtimes \langle b_1, b_2 \rangle = \langle \kappa'(a_2 \otimes_2 b_2), k \rightarrow_2 (k \otimes_2 a_2 \otimes_2 b_2) \rangle$$

is a complete residuated lattice.

Proof. Directly from Theorems 2 and 4 and its proof, and Theorem 5. □

For sake of completeness, we also show how $\wedge, \vee, 0$ and 1 are defined in the complete residuated lattice from the previous theorem:

$$\langle a_1, a_2 \rangle \wedge \langle b_1, b_2 \rangle = \langle a_1 \wedge_1 b_1, a_2 \wedge_2 b_2 \rangle, \tag{24}$$
$$\langle a_1, a_2 \rangle \vee \langle b_1, b_2 \rangle = \langle l \rightarrow_1 (l \otimes_1 (a_1 \vee_1 b_1)), k \rightarrow_2 (k \otimes_2 (a_2 \vee_2 b_2)) \rangle, \tag{25}$$

and $0 = \langle l \rightarrow 0_1, k \rightarrow 0_2 \rangle$, $1 = \langle 1_1, 1_2 \rangle$.

Again, we want the two adjoint pairs from Theorem 6 to be equal. We define the notion of residuation-preserving dual (l, k)-connection analogously, as in the non-dual case.

Theorem 7. *Let $\langle \lambda', \kappa' \rangle$ be a dual (l, k)-connection from $\mathbf{L}_1$ to $\mathbf{L}_2$. The following statements are equivalent*

(a) $\langle \lambda', \kappa' \rangle$ is residuation-preserving.
(b) $\kappa'(\lambda'(a_1) \rightarrow_2 \lambda'(b_1)) = \kappa' \lambda'(a_1) \rightarrow_1 \kappa' \lambda'(b_1)$ holds true for any $a_1, b_1 \in L_1$.
(c) $\lambda'(\kappa'(a_2) \rightarrow_1 \kappa'(b_2)) = \lambda' \kappa'(a_2) \rightarrow_2 \lambda' \kappa'(b_2)$ holds true for any for any $a_2, b_2 \in L_2$.

Proof. Similar as proof of Theorem 3. □

Theorem 8. *A dual (l, k)-connection $\langle \lambda', \kappa' \rangle$ from $\mathbf{L}_1$ to $\mathbf{L}_2$ is residuation-preserving if and only if its associated (l, k)-connection is residuation-preserving.*

Proof. We have

$$
\begin{aligned}
\kappa'(\lambda'(a_1) \rightarrow_2 \lambda'(b_1)) &= \kappa(k \otimes_2 ((k \rightarrow_2 \lambda(a_1)) \rightarrow_2 (k \rightarrow_2 \lambda(b_1)))) \\
&= \kappa(k \otimes_2 ((k \otimes_2 (k \rightarrow_2 \lambda(a_1))) \rightarrow_2 \lambda(b_1))) \\
&= \kappa(k \otimes_2 ((\lambda \kappa \lambda(a_1)) \rightarrow_2 \lambda(b_1))) \\
&= \kappa(k \otimes_2 (\lambda(a_1) \rightarrow_2 \lambda(b_1)))
\end{aligned}
$$

and

$$
\kappa' \lambda'(a_1) \rightarrow_2 \kappa' \lambda'(b_1) = \kappa \lambda(a_1) \rightarrow_2 \kappa \lambda(b_1)
$$

showing that the condition Theorem 7(b) is equivalent to Theorem 3(b). The statement of Theorem 8 then follows from Theorems 3 and 7. □

Remark 3. In the previous approach [12], the residuation-preserving Galois connections are defined as isotone Galois connections, as in the case of (l, k)-connections. In the same time, they have to satisfy conditions similar to Theorem 7(b) and (c), as in the case of dual (l, k)-connections. This is where we see the unnecessary strictness of the previous approach. Loosely speaking, the residuation-preserving isotone Galois connections were wanted to be both, (l, k)-connections and dual (l, k)-connections.

5 $\langle \lambda, \kappa \rangle$-Bonds

In this section, we define bond between formal fuzzy contexts over different complete residuated lattices $\mathbf{L}_1$ and $\mathbf{L}_2$ and describe their properties. More specifically, we propose new bonds, called $\langle \lambda, \kappa \rangle$-bonds, which are based directly on dual (l, k)-connections[4] from $\mathbf{L}_1$ to $\mathbf{L}_2$. In this section we omit proofs due to page limit.

Below, we define the $\langle \lambda, \kappa \rangle$-bonds as a special $\mathbf{L}_{\langle \lambda, \kappa \rangle}$-relation β between X_1 and Y_2 and we define concept-forming operators $\vartriangle \colon L_1^{X_1} \rightarrow L_2^{Y_2}$ and $\triangledown \colon L_2^{Y_1} \rightarrow L_2^{X_2}$ induced by $\langle \lambda, \kappa \rangle$-bond by[5]

$$
\begin{aligned}
A^{\vartriangle}(y_2) &= \bigwedge_2{}_{x_1 \in X_1} \lambda(A(x_1)) \rightarrow_2 \mathrm{proj}_2(\beta(x_1, y_2)), \\
B^{\triangledown}(x_1) &= \bigwedge_1{}_{y_2 \in Y_2} \kappa(B(y_2)) \rightarrow_1 \mathrm{proj}_1(\beta(x_1, y_2)).
\end{aligned}
\tag{26}
$$

Thus we can express the concept-forming operators $\langle \vartriangle, \triangledown \rangle$ using the classic ones, i.e. $\langle \uparrow, \downarrow \rangle$, as

$$
A^{\vartriangle} = (\lambda(A))^{\uparrow_{\mathrm{proj}_2(\beta)}} \quad \text{and} \quad B^{\triangledown} = (\kappa(B))^{\downarrow_{\mathrm{proj}_1(\beta)}}
$$

for each $A \in L_1^{X_1}$ and $B \in L_2^{Y_2}$.

---

[4] In this section $\langle \lambda, \kappa \rangle$ always denotes a dual (l, k)-connection.

[5] By proj_1 and proj_2 we denote projection of first and second entry of a pair, respectively; i.e. $\mathrm{proj}_1(\langle a_1, a_2 \rangle) \mapsto a_1, \mathrm{proj}_2(\langle a_1, a_2 \rangle) \mapsto a_2$.

Remark 4. The definition of concept-forming operators (26) actually follows as a corollary of particular setting in the framework of supremum-preserving aggregation structures. The framework was introduced in [3] and studied further in [4] (see also [1,5,10,14] for related works). We will bring detailed explanation in the full version of this paper.

Definition 3. *Let* $\mathbf{L}_1, \mathbf{L}_2$ *be complete residuated lattices,* $\langle \lambda, \kappa \rangle$ *be dual* (l, k)-*connection from* $\mathbf{L}_1$ *to* $\mathbf{L}_2$, *and let* $\langle X_1, Y_1, I_1 \rangle$ *and* $\langle X_2, Y_2, I_2 \rangle$ *be* $\mathbf{L}_1$-*context and* $\mathbf{L}_2$-*context, respectively. We call* $\beta \in L_{\langle \lambda, \kappa \rangle}^{X_1 \times Y_2}$ *a* $\langle \lambda, \kappa \rangle$-*bond from* $\langle X_1, Y_1, I_1 \rangle$ *to* $\langle X_2, Y_2, I_2 \rangle$ *if the following inclusions hold:*

$$Ext^{\triangle \triangledown}(X_1, Y_2, \beta) \subseteq Ext^{\uparrow \downarrow}(X_1, Y_1, \kappa \lambda(I_1)), \tag{27}$$

$$Int^{\triangle \triangledown}(X_1, Y_2, \beta) \subseteq Int^{\uparrow \downarrow}(X_2, Y_2, \lambda \kappa(I_2)). \tag{28}$$

Obviously, when $\mathbf{L}_1 = \mathbf{L}_2 = \mathbf{L}$ the pair of identities $\langle \mathrm{id}, \mathrm{id} \rangle$ on L is a $(1, 1)$-connection between them and the $\langle \mathrm{id}, \mathrm{id} \rangle$-bonds correspond with $\mathbf{L}$-bonds. The following theorem explains the relationship of $\langle \lambda, \kappa \rangle$-bonds with the $\mathbf{L}$-bonds more generally.

Theorem 9. *Let* $\beta \in L_{\langle \lambda, \kappa \rangle}^{X_1 \times Y_2}$. *The following statements are equivalent.*

(a) β *is a* $\langle \lambda, \kappa \rangle$-*bond from* $\langle X_1, Y_1, I_1 \rangle$ *to* $\langle X_2, Y_2, I_2 \rangle$;
(b) $\mathrm{proj}_1(\beta)$ *is a* $\mathbf{L}_1$-*bond from* $\langle X_1, Y_1, \kappa \lambda(I_1) \rangle$ *to* $\langle X_2, Y_2, \kappa(I_2) \rangle$;
(c) $\mathrm{proj}_2(\beta)$ *is a* $\mathbf{L}_2$-*bond from* $\langle X_1, Y_1, \lambda(I_1) \rangle$ *to* $\langle X_2, Y_2, \lambda \kappa(I_2) \rangle$;
(d) $\mathrm{proj}_1(\beta) = \lambda \kappa(I_1) \triangleright_1 S_i$ *and* $\mathrm{proj}_2(\beta) = S_e \triangleleft_2 \lambda \kappa(I_2)$ *for some* $S_e \in L_1^{X_1 \times X_2}$ *and* $S_i \in L_2^{Y_1 \times X_2}$.

From Theorem 9(a)$\Leftrightarrow$(d) we have the following corollary.

Corollary 2. *Set of all* $\langle \lambda, \kappa \rangle$-*bonds is an* $\mathbf{L}_{\langle \lambda, \kappa \rangle}$-*closure system.*

$\langle \lambda, \kappa \rangle$-direct products and regular $\langle \lambda, \kappa \rangle$-bonds

In this part, we assume that $\mathbf{L}_{\langle \lambda, \kappa \rangle}$ satisfies the *double negation law*, that is

$$(a \to 0) \to 0 = a \quad \text{for each } a \in L_{\langle \lambda, \kappa \rangle}.$$

Note that it means

$$\langle a_1, a_2 \rangle = (\langle a_1, a_2 \rangle \to \langle l \to_1 0_1, k \to_2 0_2 \rangle) \to \langle l \to_1 0_1, k \to_2 0_2 \rangle$$
$$= \langle (a_1 \to_1 (l \to_1 0)) \to_1 (l \to_1 0), (a_2 \to_2 (k \to_2 0_2)) \to_2 (k \to_2 0_2) \rangle$$

for each $\langle a_1, a_2 \rangle \in L_{\langle \lambda, \kappa \rangle}$.

Definition 4. *Let* $\mathbb{K}_1 = \langle X_1, Y_1, I_1 \rangle$ *be an* $\mathbf{L}_1$-*context,* $\mathbb{K}_2 = \langle X_2, Y_2, I_2 \rangle$ *be an* $\mathbf{L}_2$-*context, and* $\langle \lambda, \kappa \rangle$ *be a dual* (l, k)-*connection from* $\mathbf{L}_1$ *to* $\mathbf{L}_2$. *We define* $\langle \lambda, \kappa \rangle$-*direct product* $\mathbb{K}_1 \boxplus_{\langle \lambda, \kappa \rangle} \mathbb{K}_2$ *as* $\mathbf{L}_{\langle \lambda, \kappa \rangle}$-*context* $\langle X_2 \times Y_1, X_1 \times Y_2, \Delta \rangle$ *with*

$$\Delta(\langle x_2, y_1 \rangle, \langle x_1, y_2 \rangle) = \neg \langle \kappa \lambda I_1(x_1, y_1), \lambda I_1(x_1, y_1) \rangle \to \langle \kappa I_2(x_2, y_2), \lambda \kappa I_2(x_2, y_2) \rangle$$

for each $x_1 \in X_1, y_1 \in Y_1, x_2 \in X_2, y_2 \in Y_2$.

I_1	y_1	y_2	y_3
x_1	d	a	0
x_2	c	d	1

I_2	β_1	β_2	β_3
α_1	0.75	0.75	0.25
α_2	1	0.75	1

Δ	$\langle x_1,\beta_1\rangle$	$\langle x_1,\beta_2\rangle$	$\langle x_1,\beta_3\rangle$	$\langle x_2,\beta_1\rangle$	$\langle x_2,\beta_2\rangle$	$\langle x_2,\beta_3\rangle$
$\langle \alpha_1,y_1\rangle$	1	1	$\frac{1}{2}$	1	1	1
$\langle \alpha_1,y_2\rangle$	1	1	$\frac{1}{2}$	1	1	$\frac{1}{2}$
$\langle \alpha_1,y_3\rangle$	$\frac{1}{2}$	$\frac{1}{2}$	0	1	1	1
$\langle \alpha_2,y_1\rangle$	1	1	1	1	1	1
$\langle \alpha_2,y_2\rangle$	1	1	1	1	1	1
$\langle \alpha_2,y_3\rangle$	1	$\frac{1}{2}$	1	1	1	1

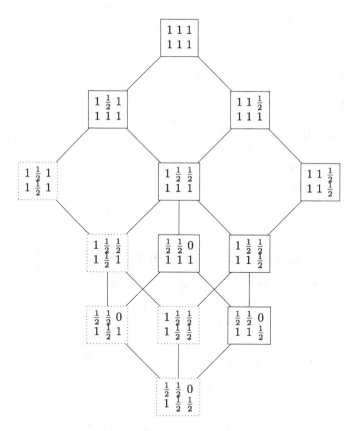

Fig. 5. A $\mathbf{L}_1$-context $\mathbb{K}_1$ and $\mathbf{L}_2$-context $\mathbb{K}_2$ (top left and top right) with $\mathbf{L}_1, \mathbf{L}_2$ as in Fig. 1; $\mathbb{K}_1 \boxplus_{\langle \lambda,\kappa\rangle} \mathbb{K}_2$ (middle) with $\langle \lambda,\kappa\rangle$ as in Fig. 4.; Lattice of all $\langle \lambda,\kappa\rangle$-bonds (bottom); the solid lined bonds are regular and the dotted lined bonds are irregular.

Extents of the $\langle \lambda, \kappa \rangle$-direct product are $\langle \lambda, \kappa \rangle$-bonds:

Theorem 10. *Let* $\mathbb{K}_1 \boxplus_{\langle \lambda, \kappa \rangle} \mathbb{K}_2 = \langle X_1 \times Y_2, X_2 \times Y_1, \Delta \rangle$ *be a* $\langle \lambda, \kappa \rangle$-*direct product. Extents in* $Ext^{\uparrow \downarrow}(X_1 \times Y_2, X_2 \times Y_1, \Delta)$ *are* $\langle \lambda, \kappa \rangle$-*bonds from* $\mathbb{K}_1$ *to* $\mathbb{K}_2$.

Analogously, with the **L**-bonds there exist $\langle \lambda, \kappa \rangle$-bonds which are not extents of the direct product $\mathbb{K}_1 \boxplus_{\langle \lambda, \kappa \rangle} \mathbb{K}_2$ (see Fig. 5). A $\langle \lambda, \kappa \rangle$-bond is called *regular* if it is extent of the direct product, otherwise it is called *irregular*.

6   Conclusions and Further Research

We revisited results on bonding formal fuzzy contexts in [12] and identified the main flaw: the residuation-preserving isotone Galois connections between complete residuated concept lattices had to fulfill two conflicting sets of requirements. In the present paper we studied two variants of residuation-preserving isotone Galois connections emerging by altering one of the two conflicting sets of requirements. One of the variants, namely dual (l, k)-connections, brought very convenient definition of bonds between formal fuzzy contexts with different structures of truth-degrees.

Our future research in this area includes:

- *Extension of the present results to homogeneous bonds wrt. isotone concept-forming operators and heterogeneous bonds studied in* [8,9]. We are going to generalize our previous results on bonds. Our preliminary observations show that (l, k)-connections will be useful for homogeneous bonds wrt. isotone concept-forming operators and for heterogeneous bonds.
- *Connections between complete residuated lattices based on antitone Galois connections.*

Acknowledgments. The author thanks to Ondrej Kridlo for valuable consultations. Supported by grant No. **15-17899S**, "Decompositions of Matrices with Boolean and Ordinal Data: Theory and Algorithms", of the Czech Science Foundation.

References

1. Bartl, E., Belohlavek, R.: Reducing sup-t-norm and inf-residuum to a single type of fuzzy relational equations. In: 2011 Annual Meeting of the North American Fuzzy Information Processing Society (NAFIPS), pp. 1–5 (2011)
2. Belohlavek, R.: Fuzzy Galois connections. Math. Log. Q. **45**(4), 497–504 (1999)
3. Belohlavek, R.: Optimal decompositions of matrices with entries from residuated lattices. J. Log. Comput. (2011)
4. Belohlavek, R.: Sup-t-norm and inf-residuum are one type of relational product: unifying framework and consequences. Fuzzy Sets Syst. **197**, 45–58 (2012)
5. Belohlavek, R., Vychodil, V.: What is a fuzzy concept lattice? In: Proceedings of CLA 2005, 3rd International Conference on Concept Lattices and Their Applications, pp. 34–45 (2005)

6. Belohlavek, R., Vychodil, V.: Residuated lattices of size ⩽ 12. Order **27**(2), 147–161 (2010)
7. Kohout, L.J., Bandler, W.: Relational-product architectures for information processing. Inf. Sci. **37**(1–3), 25–37 (1985)
8. Konecny, J.: Antitone L-bonds. In: IPMU, pp. 71–80 (2014)
9. Konecny, J., Ojeda-Aciego, M.: Isotone L-bonds. In: Ojeda-Aciego, M., Outrata, J., (eds.) CLA. CEUR Workshop Proceedings, vol. 1062, pp. 153–162 (2013). CEUR-WS.org
10. Krajči, S.: A generalized concept lattice. Log. J. IGPL **13**(5), 543–550 (2005)
11. Krídlo, O., Krajči, S., Ojeda-Aciego, M.: The category of L-chu correspondences and the structure of L-bonds. Fundam. Inform. **115**(4), 297–325 (2012)
12. Kridlo, O., Ojeda-Aciego, M.: CRL-chu correspondences. In: Ojeda-Aciego, M., Outrata, J., (eds.) CLA. CEUR Workshop Proceedings, vol. 1062, pp. 105–116 (2013). CEUR-WS.org
13. Krupka, M.: Factorization of residuated lattices. Log. J. IGPL **17**(2), 205–223 (2009)
14. Medina, J., Ojeda-Aciego, M., Ruiz-Calviño, J.: Formal concept analysis via multi-adjoint concept lattices. Fuzzy Sets Syst. **160**(2), 130–144 (2009)
15. Pollandt, S.: Fuzzy Begriffe: Formale Begriffsanalyse von unscharfen Daten. Springer, Heidelberg (1997)

The Linear Algebra in Formal Concept Analysis over Idempotent Semifields

Francisco J. Valverde-Albacete[1] and Carmen Peláez-Moreno[2]([⊠])

[1] Departamento de Lenguajes y Sistemas Informáticos, Universidad Nacional de Educación a Distancia, C/ Juan del Rosal 16., 28040 Madrid, Spain
fva@lsi.uned.es, franciscojvalverde@gmail.com
[2] Departamento de Teoría de la Señal y de las Comunicaciones, Universidad Carlos III de Madrid, 28911 Leganés, Spain
carmen@tsc.uc3m.es

Abstract. We report on progress in characterizing $\mathcal{K}$-valued FCA in algebraic terms, where $\mathcal{K}$ is an idempotent semifield. In this data mining-inspired approach, incidences are matrices and sets of objects and attributes are vectors. The algebraization allows us to write matrix-calculus formulae describing the polars and the fixpoint equations for extents and intents. Adopting also the point of view of the theory of linear operators between vector spaces we explore the similarities and differences of the idempotent semimodules of extents and intents with the subspaces related to a linear operator in standard algebra. This allows us to shed some new light into Formal Concept Analysis from the point of view of the theory of linear operators over idempotent semimodules.

1 Introduction

In [1] a generalization of Formal Concept Analysis was presented where incidences have values in a complete idempotent semifield $\mathcal{K}$. This is a complete idempotent semiring with a multiplicative semiring structure where the unit is distinct from top of the semiring, unlike in e.g. inclines. Logarithmic costs and amplifications are concrete instances of idempotent semifields, apart from their well attested use in morphological processing [2] and Markov chain decoding [3]. This setting was later generalized to the other four types of Galois connections or adjunctions arising from a single $\mathcal{K}$-valued incidence [4,5].

In [6] it was proven that, at least on a particular kind of dioids, the idempotent semifields, formal concepts are related to the eigenvectors of the unit in the semiring for certain matrices derived from the incidence, the *projector matrices*. The authors later went on to develop the spectral theory but never reviewed their original findings in two aspects as claimed in some of this work:

a. What is the actual relation between eigenvectors of the spectral projectors and extents or intents?

F.J. Valverde-Albacete—CPM has been supported by the Spanish Government-Comisión Interministerial de Ciencia y Tecnología project TEC2011-26807.

© Springer International Publishing Switzerland 2015
J. Baixeries et al. (Eds.): ICFCA 2015, LNAI 9113, pp. 97–113, 2015.
DOI: 10.1007/978-3-319-19545-2_6

b. What are the advantages of working in complete idempotent semifields?

In this paper we answer both of these questions, to wit:

a. the eigenvectors of the projectors generate the concepts by closure of their (extended) joins, and
b. all procedures related to FCA take the form of matrix-vector equations including, the polars, the closure operators and the generation.

This is possible due to the idempotent algebra analogues of vector spaces, the complete idempotent semimodules. Indeed, with the same tools we are able to extend the set of φ-formal concepts arising from a $\mathcal{K}$-valued formal context with a semimodule structure, including the multiplication of concepts by constants and an addition operation that takes into consideration the closure of the extents and intents obtained. We foresee that these will be useful tools for the development of data mining procedures.

2   Idempotent Semifields and Semimodules

A *semiring* is an algebra $\mathcal{S} = \langle S, \oplus, \otimes, \epsilon, e \rangle$ whose additive structure, $\langle S, \oplus, \epsilon \rangle$, is a commutative monoid and whose multiplicative structure, $\langle S \backslash \{\epsilon\}, \otimes, e \rangle$, is a monoid with multiplication distributing over addition from right and left and with additive neutral element absorbing for $\otimes$, i.e. $\forall a \in S$, $\epsilon \otimes a = \epsilon$.

Specifically, every commutative semiring accepts a canonical preorder, $a \leq b$ if and only if there exists $c \in D$ with $a \oplus c = b$. A *dioid* is a semiring $\mathcal{D}$ where this relation is actually an order. Dioids are zerosumfree and entire, that is they have no non-null additive or multiplicative factors of zero. Commutative complete dioids are already complete residuated lattices.

An *idempotent semiring* is a dioid whose addition is idempotent, and a *selective semiring* one where the arguments attaining the value of the additive operation can be identified.

Example 1. Examples of idempotent dioids are

1. The *Boolean lattice* $\mathbb{B} = \langle \{0, 1\}, \vee, \wedge, 0, 1 \rangle$
2. All fuzzy semirings, e.g. $\langle [0, 1], \max, \min, 0, 1 \rangle$
3. The *min-plus algebra* $\mathbb{R}_{\min,+} = \langle \mathbb{R} \cup \{\infty\}, \min, +, \infty, 0 \rangle$
4. The *max-plus algebra* $\mathbb{R}_{\max,+} = \langle \mathbb{R} \cup \{-\infty\}, \max, +, -\infty, 0 \rangle$ □

Of the semirings above, only the boolean lattice and the fuzzy semirings are complete dioids, since the rest lack the *top* element $\top$ as an adequate inverse for the bottom in the order.

2.1   Idempotent Semifields

A semiring is a *semifield* if there exists a multiplicative inverse for every element $a \in S$, except the null, notated as a^{-1}, and *radicable* if the equation $a^b = c$ can be solved for a. As exemplified above, idempotent semifields are incomplete in their natural order, but there are procedures for *completing* such structures [5] and we will not differentiate between *complete or completed* structures. Note, first, that in complete semifields $e \neq \top$ which distinguishes them from inclines, and also that the inverse for the null is prescribed as $\bot^{-1} = \top$.

Example 2. The max-plus $\mathbb{R}_{\max,+}$ and min-plus $\mathbb{R}_{\min,+}$ semifields can be completed as:

1. The *complete min-plus semifield* $\overline{\mathbb{R}}_{\min,+} = \langle \mathbb{R} \cup \{-\infty, \infty\}, \min, \dot{+}, -\cdot, \infty, 0 \rangle$.
2. The *complete max-plus semifield* $\overline{\mathbb{R}}_{\max,+} = \langle \mathbb{R} \cup \{-\infty, \infty\}, \max, +, -\cdot, -\infty, 0 \rangle$.

In this notation we have $\forall c, -\infty \dot{+} c = -\infty$ and $\infty \dot{+} c = \infty$, which solves several issues in dealing with the separately completed dioids. These two completions are inverses $\overline{\mathbb{R}}_{\min,+} = \overline{\mathbb{R}}_{\max,+}^{-1}$, hence order-dual lattices.   □

In fact, idempotent semifields $\mathcal{K} = \langle K, \oplus, \dot{\oplus}, \otimes, \dot{\otimes}, \cdot^{-1}, \bot, e, \top \rangle$, appear as enriched structures, the advantage of working with them being that meets can be expressed by means of joins and inversion as $a \dot{\oplus} b = (a^{-1} \oplus b^{-1})^{-1}$. On a practical note, residuation in complete commutative idempotent semifields can be expressed in terms of inverses, and this extends to eigenspaces.

2.2 Idempotent Semimodules

Let $\mathcal{D} = \langle D, +, \times, \epsilon_D, e_D \rangle$ be a commutative semiring. A $\mathcal{D}$-*semimodule* $\mathcal{X} = \langle X, \oplus, \odot, \epsilon_X \rangle$ is a commutative monoid $\langle X, \oplus, \epsilon_X \rangle$ endowed with a scalar action $(\lambda, x) \mapsto \lambda \odot x$ satisfying the following conditions for all $\lambda, \mu \in D$, $x, x' \in X$:

$$(\lambda \times \mu) \odot x = \lambda \odot (\mu \odot x) \qquad \lambda \odot (x \oplus x') = \lambda \odot x \oplus \lambda \odot x'$$
$$(\lambda + \mu) \odot x = \lambda \odot x \oplus \mu \odot x \qquad \lambda \odot \epsilon_X = \epsilon_X = \epsilon_D \otimes x$$
$$e_D \odot x = x \tag{1}$$

Matrices form a $\mathcal{D}$-semimodule $D^{g \times m}$ for given g, m. In this paper, we only use finite-dimensional semimodules where we can identify semimodules with column vectors, e.g. $\mathcal{X} \equiv \mathcal{D}^g$. If $\mathcal{D}$ is commutative, idempotent or complete, then $\mathcal{X}$ is also commutative, idempotent or complete. If $\mathcal{K}$ is a semifield, we may also define an inverse for the semimodule by the coordinate-wise inversion, $(x^{-1})_i = (x_i)^{-1}$.

For complete idempotent semifields, the following matrix algebra equations are proven in [7, Ch.8]:

Proposition 1. *Let $\mathcal{K}$ be an idempotent semifield, and $A \in \mathcal{K}^{m \times n}$. Then:*

1. $A \otimes (A^* \dot{\otimes} A) = A \dot{\otimes} (A^* \otimes A) = (A \dot{\otimes} A^*) \otimes A = (A \otimes A^*) \dot{\otimes} A = A$ and

$A^* \otimes (A \dot{\otimes} A^*) = A^* \dot{\otimes} (A \otimes A^*) = (A^* \dot{\otimes} A) \otimes A^* = (A^* \otimes A) \dot{\otimes} A^* = A^*.$

2. *Alternating $A - A^*$ products of 4 matrices can be shortened as in:*

$$A^* \overset{\cdot}{\otimes} (A \otimes (A^* \overset{\cdot}{\otimes} A)) = A^* \overset{\cdot}{\otimes} A = (A^* \overset{\cdot}{\otimes} A) \otimes (A^* \overset{\cdot}{\otimes} A)$$

3. *Alternating $A - A^*$ products of 3 matrices and another terminal, arbitrary matrix can be shortened as in:*

$$A^* \overset{\cdot}{\otimes} (A \otimes (A^* \overset{\cdot}{\otimes} M)) = A^* \overset{\cdot}{\otimes} M = (A^* \overset{\cdot}{\otimes} A) \otimes (A^* \overset{\cdot}{\otimes} M)$$

4. *The following inequalities apply:*

$$A^* \overset{\cdot}{\otimes} (A \otimes M) \geq M \qquad\qquad A^* \otimes (A \overset{\cdot}{\otimes} M) \leq M$$

2.3 Complete Congruences of Subsemimodules

The following approach is borrowed and adapted from [8]: Given a right $\mathcal{K}$-semimodule $\mathcal{X}$ a subset $\mathcal{W} \subset \mathcal{X}^2$ is called a *pre-congruence (of semimodules)*, if it is a subsemimodule $\langle W, \oplus, \odot \rangle$ such that $(x, x) \in W, \forall x \in X$, and if $(x_1, x_2) \in W(x_2, x_3) \in W$, then $(x_1, x_3) \in W$. Furthermore, it is a *congruence (of semimodules)* whenever $(x_1, x_2) \in W$ implies $(x_2, x_1) \in W$. So, these congruences are equivalences with a semimodule structure when thought of as a semimodule of $\mathcal{X}^2$.

On a complete (as a semimodule) pre-congruence $\mathcal{W} \subset \mathcal{X}^2$ for $x \in X$ set $\hat{x} = \vee\{x' \in X \mid (x', x) \in W\}$.

Lemma 1. *If $\mathcal{W}$ is a complete pre-congruence on $\mathcal{X}$, then $(x, \hat{x}) \in W$ and $x \leq \hat{x}$. Furthermore, if $\mathcal{W}$ is a complete congruence then: $\hat{x}$ is just the supremum in the equivalence class of $x \in X$ and it is a closure operator: $x \leq \hat{x} = \hat{\hat{x}}$ whence $x_1 \leq x_2$ implies $\hat{x}_1 \leq \hat{x}_2$, and in particular $\hat{x}_1 = \hat{x}_2$ if $(x_1, x_2) \in W$.*

Given a pre-dual pair $\mathcal{X}, \mathcal{Y}$ and a dot product $\langle X \mid Y \rangle$, we define the following correspondences between semimodules of $\mathcal{X}^2$ and $\mathcal{Y}$:

$$\begin{aligned} \mathcal{W} \subset \mathcal{X}^2 \mapsto \mathcal{W}^\top &= \{y \in Y \mid \langle x_1 \mid y \rangle = \langle x_2 \mid y \rangle, \forall (x_1, x_2) \in W\} \\ \mathcal{V} \subset \mathcal{Y} \mapsto \mathcal{V}^\perp &= \{(x_1, x_2) \in X^2 \mid \langle x_1 \mid y \rangle = \langle x_2 \mid y \rangle, \forall y \in V\} \end{aligned} \quad (2)$$

Note that $\mathcal{V}$ is a complete subsemimodule of $\mathcal{W}$ and $\mathcal{V} = (\mathcal{V}^\perp)^\top$ [8], and

Proposition 2. *Let $(\mathcal{X}, \mathcal{Y})$ be a pre-dual pair satisfying the property that if $\mathcal{W} \in \mathcal{X}^2$ is a complete congruence and $(s, t) \notin W$, then there exists a $y \in Y$ such that if $(x_1, x_2) \in W$ then $\langle x_1 \mid y \rangle = \langle x_2 \mid y \rangle$ and $\langle s \mid y \rangle \neq \langle t \mid y \rangle$. Then a subsemimodule $\mathcal{W} \subset \mathcal{X}^2$ is a complete congruence if and only if $\mathcal{W} = (\mathcal{W}^\top)^\perp$.*

2.4   Basic Spectral Theory Over Dioids

Let $\mathcal{M}_n(\mathcal{S})$ be the semiring of square matrices over a semiring $\mathcal{S}$ with the usual operations. Given $A \in \mathcal{M}_n(\mathcal{S})$ the *right (left) eigenproblem* is the task of finding the *right eigenvectors* $v \in S^{n \times 1}$ and *right eigenvalues* $\rho \in S$ (respectively *left eigenvectors* $u \in S^{1 \times n}$ and *left eigenvalues* $\lambda \in S$) satisfying:

$$u \otimes A = \lambda \otimes u \qquad\qquad A \otimes v = v \otimes \rho \qquad (3)$$

The left and right eigenspaces and spectra are the sets of these solutions:

$$\Lambda(A) = \{\lambda \in S \mid \mathcal{U}_\lambda(A) \neq \{\epsilon^n\}\} \qquad P(A) = \{\rho \in S \mid \mathcal{V}_\rho(A) \neq \{\epsilon^n\}\}$$
$$\mathcal{U}_\lambda(A) = \{u \in S^{1 \times n} \mid u \otimes A = \lambda \otimes u\} \quad \mathcal{V}_\rho(A) = \{v \in S^{n \times 1} \mid A \otimes v = v \otimes \rho\}$$
$$\mathcal{U}(A) = \bigcup_{\lambda \in \Lambda(A)} \mathcal{U}_\lambda(A) \qquad\qquad \mathcal{V}(A) = \bigcup_{\rho \in P(A)} \mathcal{V}_\rho(A) \qquad (4)$$

With so little structure it might seem hard to solve (3), but a very generic solution based in the concept of transitive closure of a matrix $A^+ = \sum_{i=1}^{\infty} A^i$ and transitive-reflexive closure $A^* = \sum_{i=0}^{\infty} A^i$ is given by the following theorem:

Theorem 1. *[9, Theorem 1] Let $A \in S^{n \times n}$. If A^* exists, the following two conditions are equivalent:*

1. *$A_{\cdot i}^+ \otimes \mu = A_{\cdot i}^* \otimes \mu$ for some $i \in \{1 \dots n\}$, and $\mu \in S$.*
2. *$A_{\cdot i}^+ \otimes \mu$ (and $A_{\cdot i}^* \otimes \mu$) is an eigenvector of A for e, $A_{\cdot i}^+ \otimes \mu \in \mathcal{V}_e(A)$.*

3   The Algebra in $\mathcal{K}$-Formal Concept Analysis

When $\overline{\mathcal{K}}$ is a completed idempotent semifield and $\mathcal{X} \equiv \overline{\mathcal{K}}^g$ and $\mathcal{Y} \equiv \overline{\mathcal{K}}^m$ are idempotent vectors spaces or semimodules, the definition of the Galois connection involves the use of a scalar product $\langle \cdot \mid R \mid \cdot \rangle : X \times Y \to K$ and a scalar $\varphi \in \overline{\mathcal{K}}$ [5]:

$$x_{R,\varphi}^\uparrow = \vee\{y \in Y \mid \langle x \mid R \mid y \rangle \leq \varphi\} \qquad y_{R,\varphi}^\downarrow = \vee\{x \in X \mid \langle x \mid R \mid y \rangle \leq \varphi\}$$

This definition is quite general and might even be valid for any dioid, but we now want to develop its *affordances* when the semiring has the richer algebraic structure of a complete idempotent semifield with its duality.

3.1   The Initial Scaling

In this case, we can reduce the studying of the connection with a generic φ to a much simpler setting: consider the dot product $\langle x \mid R \mid y \rangle = x^{\mathsf{T}} \otimes R \otimes y$, where $R \in \overline{\mathcal{K}}^{g \times m}$. Inspired by [10], we decompose in $\varphi = \gamma \overset{\cdot}{\otimes} \mu$ where $\{\gamma, \mu\} \subseteq \overline{\mathcal{K}}$.

When $\varphi \in (\bot, \top)$ so that $\{\gamma, \mu\} \subseteq (\bot, \top)$ then the normalizations $X \,\dot{/}\, \gamma = \widetilde{X}^\gamma$ and $Y \,\dot{/}\, \mu = \widetilde{Y}^\mu$ have the interpretation of (finite) scalings in the original spaces[1]:

$$x^\mathrm{T} \otimes R \otimes y \le \gamma \dot{\otimes} \mu \Leftrightarrow \gamma \dot{\backslash} (x^\mathrm{T} \otimes R \otimes y) \,\dot{/}\, \mu \le e$$

$$\Leftrightarrow (\gamma^{-1} \otimes x^\mathrm{T}) \otimes R \otimes (y \otimes \mu^{-1}) \le e \quad \Leftrightarrow (\widetilde{x}^\gamma)^\mathrm{T} \otimes R \otimes \widetilde{y}^\mu \le e,$$

whence we need only consider the case where $\varphi = e$. The following development presupposes this setting and we treat $x \in \widetilde{X}^\gamma$ and $y \in \widetilde{Y}^\mu$ simply as placeholders.

3.2   The Polars and the Galois Connection

Since $x^\mathrm{T} \otimes R \otimes y \le e \Leftrightarrow y^\mathrm{T} \otimes R^\mathrm{T} \otimes x \le e$, by using residuation we may write:

$$x_R^\uparrow = (x^\mathrm{T} \otimes R) \backslash e = R^* \dot{\otimes} x^{-1} \qquad y_R^\downarrow = (y^\mathrm{T} \otimes R^\mathrm{T}) \backslash e = R^{-1} \dot{\otimes} y^{-1} \qquad (5)$$

involving only transposition, inversion and operation in the dual semifield.

We recall the following proposition:

Proposition 3. $(\cdot_R^\uparrow, \cdot_R^\downarrow) : \widetilde{X}^\gamma \leftrightarroweq \widetilde{Y}^\mu$ *is a Galois connection between the semimodules* $\widetilde{X}^\gamma \cong \widetilde{(K^g)}^\gamma$ *and* $\widetilde{Y}^\mu \cong \widetilde{(K^m)}^\mu$: *for* $x \in X$, $y \in Y$, *we have* $y \le x_R^\uparrow \Leftrightarrow x \le y_R^\downarrow$.

Proof. We need only prove in one sense, since the other is similar. If $y \le x_R^\uparrow = R^* \dot{\otimes} x^{-1}$, then by inversion, $R^\mathrm{T} \otimes x \le y^{-1}$ whence, by residuation $x \le R^\mathrm{T} \backslash y^{-1} = R^{-1} \dot{\otimes} y^{-1} = y_R^\downarrow$. □

The diagram in Fig. 1 summarizes this Galois connection [5,11]. This immediately puts at our disposal a number of results which we collect in the following proposition:

Proposition 4. *Consider the Galois connection* $(\cdot_R^\uparrow, \cdot_R^\downarrow) : \mathcal{X} \leftrightarroweq \mathcal{Y}$. *Then:*

1. *The polars are antitone, join-inverting functions:*

$$(x_1 \oplus x_2)_R^\uparrow = x_{1R}^\uparrow \dot{\oplus} x_{2R}^\uparrow \qquad (y_1 \oplus y_2)_R^\downarrow = y_{1R}^\downarrow \dot{\oplus} y_{2R}^\downarrow. \qquad (6)$$

2. *The compositions of the polars:* $\pi_{R^\mathrm{T}} : X \to X, \pi_R : Y \to Y$

$$\pi_{R^\mathrm{T}}(x) = (x_R^\uparrow)_R^\downarrow = R^{-1} \dot{\otimes} (R^\mathrm{T} \otimes x) \qquad \pi_R(y) = (y_R^\downarrow)_R^\uparrow = R^* \dot{\otimes} (R \otimes y)$$

are closures, that is, extensive and idempotent operators.

$$\pi_{R^\mathrm{T}}(x) \ge x \qquad\qquad \pi_R(y) \ge y$$
$$\pi_{R^\mathrm{T}}(\pi_{R^\mathrm{T}}(x)) = \pi_{R^\mathrm{T}}(x) \qquad\qquad \pi_R(\pi_R(y)) = \pi_R(y)$$

---

[1] We leave the corner cases when $\varphi \in \{\bot, \top\}$ for later work.

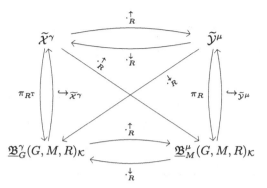

Fig. 1. $(\cdot{\overset{\uparrow}{_R}}, \cdot{\overset{\downarrow}{_R}}) : \tilde{\mathcal{X}}^{\gamma} \rightharpoonup \tilde{\mathcal{Y}}^{\mu}$, the Galois connection between **scaled** spaces. Refer to the text for the notation.

3. *The polars are mutual pseudo-inverses:*

$$(\cdot)_R^{\uparrow} \circ (\cdot)_R^{\downarrow} \circ (\cdot)_R^{\uparrow} = (\cdot)_R^{\uparrow} \qquad\qquad (\cdot)_R^{\downarrow} \circ (\cdot)_R^{\uparrow} \circ (\cdot)_R^{\downarrow} = (\cdot)_R^{\downarrow}$$

Proof. We prove for extents, since the proofs for intents are similar. Note that the techniques in this proof will be used many times subsequently. First:

$$(x_1 \oplus x_2)\overset{\uparrow}{_R} = R^* \mathbin{\dot\otimes} (x_1 \oplus x_2)^{-1} = R^* \mathbin{\dot\otimes} x_1^{-1} \dot\oplus R^* \mathbin{\dot\otimes} x_2^{-1} = x_1\overset{\uparrow}{_R} \dot\oplus x_2\overset{\uparrow}{_R}$$

From Proposition 1.4 we find that $\pi_R(y) = R^* \mathbin{\dot\otimes} (R \otimes y) \geq y$, that is, π_R is extensive. From Proposition 1.3 we know that $R^* \mathbin{\dot\otimes} (R \otimes (R^* \mathbin{\dot\otimes} M)) = R^* \mathbin{\dot\otimes} M$, whence

$$\pi_R(\pi_R(y)) = R^* \mathbin{\dot\otimes} (R \otimes (R^* \mathbin{\dot\otimes} (R \otimes y))) = R^* \mathbin{\dot\otimes} (R \otimes y) = \pi_R(y).$$

Finally,

$$((x_R^{\uparrow})_R^{\downarrow})_R^{\uparrow} = R^* \dot\otimes (R^{-1} \dot\otimes (R^* \dot\otimes x^{-1})^{-1})^{-1} = R^* \dot\otimes (R \otimes [R^* \dot\otimes x^{-1}]) = R^* \dot\otimes x^{-1} = x_R^{\uparrow}$$

where the reduction step also comes from Proposition 1.3.   □

One of the advantages of working in idempotent semimodules is that we can strengthen statement 1 in Proposition 4 to reveal that the polars are idempotent semimodule morphisms:

Proposition 5. *The polar of intents of the Galois connection transforms a $\overline{\mathcal{K}}$-semimodule of extents into a $\overline{\mathcal{K}}^{\mathrm{d}}$-semimodule of intents, and dually for the polar of the extents.*

Proof. For linearity, consider $x_1{\uparrow\atop R} = R^* \mathbin{\dot\otimes} x_1^{-1}$ and $x_2{\uparrow\atop R} = R^* \mathbin{\dot\otimes} x_2^{-1}$.

$$(\lambda_1 \otimes_{\cdot} x_1 \oplus_{\cdot} \lambda_1 \otimes_{\cdot} x_2)^{\uparrow}_R = R^* \mathbin{\dot\otimes} (\lambda_1 \otimes_{\cdot} x_1 \oplus_{\cdot} \lambda_1 \otimes_{\cdot} x_2)^{-1} =$$

$$= R^* \mathbin{\dot\otimes} (\lambda_1^{-1} \mathbin{\dot\otimes} x_1^{-1} \mathbin{\dot\oplus} \lambda_1^{-1} \mathbin{\dot\otimes} x_2^{-1}) =$$

$$= (\lambda_1^{-1} \mathbin{\dot\otimes} R^* \mathbin{\dot\otimes} x_1^{-1}) \mathbin{\dot\oplus} (\lambda_1^{-1} \mathbin{\dot\otimes} R^* \mathbin{\dot\otimes} x_1^{-1}) =$$

$$= (\lambda_1^{-1} \mathbin{\dot\otimes} x_1{\uparrow\atop R}) \mathbin{\dot\oplus} (\lambda_2^{-1} \mathbin{\dot\otimes} x_2{\uparrow\atop R}).$$

For the polar of extents the proof is similar.   □

Note that this is the $\mathcal{K}$-FCA analogue of the fact that the polars are join-inverting. But the novelty is that the scalings for one semimodule and the other are inverted. This theme will recur in our results: how to enrich the Galois connection in the setting of idempotent semimodules.

3.3   The Concept Equation and the Semimodules of Closures

Our next aim is to flesh out the dual isomorphism between the closed elements of the connection. Since we know that the Galois connection is concept-inducing, we consider (γ, μ)-formal concepts $(a, b) \in \underline{\underline{\mathfrak{B}}}^{\gamma \otimes \mu}(G, M, R)$ such that

$$a^{\uparrow}_R = R^* \mathbin{\dot\otimes} a^{-1} = b \qquad\qquad b^{\downarrow}_R = R^{-1} \mathbin{\dot\otimes} b^{-1} = a \qquad (7)$$

From these we can decouple extent and intent equations as in:

$$b^{-1} = R^{\mathrm{T}} \otimes_{\cdot} a \qquad\qquad a^{-1} = R \otimes_{\cdot} b$$

whence

$$\pi_{R^{\mathrm{T}}}(a) = R^{-1} \mathbin{\dot\otimes} (R^{\mathrm{T}} \otimes_{\cdot} a) = a \qquad\qquad \pi_R(b) = R^* \mathbin{\dot\otimes} (R \otimes_{\cdot} b) = b$$

These two, formally identical equations involve the closure operators making it explicit that the fixpoints are also closed elements, extents and intents, so call:

$$\mathrm{fix}(\pi_{R^{\mathrm{T}}}) = \{a \in \tilde{X}^{\gamma} \mid \pi_{R^{\mathrm{T}}}(a) = a\} \qquad \mathrm{fix}(\pi_R) = \{b \in \tilde{Y}^{\mu} \mid \pi_R(b) = b\},$$

the sets of fixpoints of each of these operators. Then:

Corollary 1. *The sets of fixpoints are the lattices of extents and intents:*

$$\mathrm{fix}(\pi_{R^{\mathrm{T}}}) = \underline{\underline{\mathfrak{B}}}_G^{\gamma}(G, M, R) \qquad\qquad \mathrm{fix}(\pi_R) = \underline{\underline{\mathfrak{B}}}_M^{\mu}(G, M, R).$$

Proof. The only thing left to prove is that the fixpoints are extents. Let $\pi_{R^{\mathrm{T}}}(a) = a$. Then:

$$b = a^{\uparrow}_R = R^* \mathbin{\dot\otimes} [R^{-1} \mathbin{\dot\otimes} (R^{\mathrm{T}} \otimes_{\cdot} a)]^{-1} = R^* \mathbin{\dot\otimes} (R \otimes_{\cdot} [R^* \mathbin{\dot\otimes} a]) = R^* \mathbin{\dot\otimes} a^{-1}$$

where the last step follows from the matrix equalities. So

$$b_R^\downarrow = R^{-1} \, \dot{\otimes} \, \left(R^* \, \dot{\otimes} \, a^{-1}\right)^{-1} = R^{-1} \, \dot{\otimes} \, (R^{\mathrm{T}} \otimes a) = a.$$

The proof for intents is dual.   □

Note that the Galois Connection theorem implies that the sets of extents and intents are dually isomorphic lattices through the polars. How is this expressed in our framework? Again, the advantages of working in complete idempotent semimodules make themselves evident in the following proposition:

Proposition 6. $\mathrm{fix}(\pi_{R^{\mathrm{T}}})$ and $\mathrm{fix}(\pi_R)$ are $\overline{\mathcal{K}}^{\mathrm{d}}$-subsemimodules of $\widetilde{X}^\gamma$ and $\widetilde{Y}^\mu$ respectively.

Proof. Consider a generic vector $z \in \widetilde{X}^\gamma$, then by Lemma 1.2 we have

$$\pi_R(R^* \, \dot{\otimes} \, z) = R^* \, \dot{\otimes} \, (R \otimes (R^* \, \dot{\otimes} \, z)) = R^* \, \dot{\otimes} \, z.$$

This means that any $\overline{\mathcal{K}}^{\mathrm{d}}$-combination of columns of R^* is a fixpoint of π_R, that is $\langle R^* \rangle_{\overline{\mathcal{K}}^{\mathrm{d}}} \subseteq \mathrm{fix}(\pi_R)$. Now, consider $b \in \mathrm{fix}(\pi_R)$. Then $R^* \, \dot{\otimes} \, (R \otimes b) = b$ whence $R^* \, \dot{\otimes} \, z = b$, so $\mathrm{fix}(\pi_R) \subseteq \langle R^* \rangle_{\overline{\mathcal{K}}^{\mathrm{d}}}$. Similarly $\mathrm{fix}(\pi_{R^{\mathrm{T}}}) = \langle R^{-1} \rangle_{\overline{\mathcal{K}}^{\mathrm{d}}}$.   □

This and the previous consideration yield the following Corollary:

Corollary 2. *The system of extents and intents of the Galois connection are* $\overline{\mathcal{K}}^{\mathrm{d}}$*-subsemimodules of* $\widetilde{X}^\gamma$ *and* $\widetilde{Y}^\mu$, *generated by the columns of* R^{-1} *and* R^*, *respectively.*

$$\underline{\mathfrak{B}}_G^\gamma(G, M, R) = \langle R^{-1} \rangle_{\overline{\mathcal{K}}^{\mathrm{d}}} \qquad \underline{\mathfrak{B}}_M^\mu(G, M, R) = \langle R^* \rangle_{\overline{\mathcal{K}}^{\mathrm{d}}}$$

Recall that $\mathcal{K}^{\mathrm{d}}$-semimodules are meet-semilattices for the order of $\mathcal{K}$, but also that they are complete with a top element, hence they are already complete lattices.

Although complete lattices, $\underline{\mathfrak{B}}_G^\gamma(G, M, R)$ and $\underline{\mathfrak{B}}_M^\mu(G, M, R)$ are only meet-*sub*semilattices of their ambient spaces $\widetilde{X}^\gamma$ and $\widetilde{Y}^\mu$ but not their join-*sub*semi-lattices. Luckily, the following Proposition allows us to characterize the lattices of extents and intents as $\mathcal{K}$-semimodules, that is, as complete join-semilattices, too. First consider the following structures on the set of extents and intents:

$$\mathcal{B}_G = \langle \underline{\mathfrak{B}}_G^\gamma(G, M, R), \tilde{\dot{\oplus}}, \tilde{\dot{\odot}}, \epsilon_G \rangle \qquad \mathcal{B}_M = \langle \underline{\mathfrak{B}}_M^\mu(G, M, R), \tilde{\dot{\oplus}}, \tilde{\dot{\odot}}, \epsilon_M \rangle \qquad (8)$$

with the two additions:

$$a_1 \tilde{\dot{\oplus}} a_2 = \pi_{R^{\mathrm{T}}}(a_1 \dot{\oplus} a_2) \qquad\qquad b_1 \tilde{\dot{\oplus}} b_2 = \pi_R(b_1 \dot{\oplus} b_2) \qquad (9)$$

two right traslations:

$$(a, \lambda) \mapsto \lambda \tilde{\otimes} a = \pi_{R^\mathsf{T}}(\lambda \otimes a) \qquad\qquad (b, \mu) \mapsto \mu \tilde{\otimes} b = \pi_R(\mu \otimes b) \qquad (10)$$

and neutral elements:

$$\epsilon_G = R^{-1} \dot{\otimes} \perp^M \qquad\qquad \epsilon_M = R^* \dot{\otimes} \perp^G \qquad (11)$$

We next prove that:

Proposition 7. *Let* $\lambda, \mu \in K$ *,* a_1 *and* a_2 *be extents of* $\underline{\mathfrak{B}}_G^\gamma(G, M, R)$ *and* b_1 *and* b_2 *be intents of* $\underline{\mathfrak{B}}_M^\mu(G, M, R)$ *. Then*

$$\lambda \tilde{\otimes} a_1 \tilde{\oplus} \mu \tilde{\otimes} a_2 = \pi_R(\lambda \otimes a_1 \oplus \mu \otimes a_2) \quad \lambda \tilde{\otimes} b_1 \tilde{\oplus} \mu \tilde{\otimes} b_2 = \pi_R(\lambda \otimes b_1 \oplus \mu \otimes b_2) \quad (12)$$

Proof. Call $a = \lambda \tilde{\otimes} a_1 \tilde{\oplus} \mu \tilde{\otimes} a_2$, then:

$$a = \left(R^{-1} \dot{\otimes} (R^\mathsf{T} \otimes (\lambda \otimes a_1)) \right) \tilde{\oplus} \left(R^{-1} \dot{\otimes} (R^\mathsf{T} \otimes (\mu \otimes a_2)) \right)$$

$$= R^{-1} \dot{\otimes} (R^\mathsf{T} \otimes [(R^{-1} \dot{\otimes} (R^\mathsf{T} \otimes (\lambda \otimes a_1))) \oplus (R^{-1} \dot{\otimes} (R^\mathsf{T} \otimes (\mu \otimes a_2)))])$$

Distributing R^T over $\oplus$ and applying the matrix equalities:

$$= R^{-1} \dot{\otimes} [R^\mathsf{T} \otimes (R^{-1} \dot{\otimes} [R^\mathsf{T} \otimes (\lambda \otimes a_1)]) \oplus R^\mathsf{T} \otimes (R^{-1} \dot{\otimes} [R^\mathsf{T} \otimes (\mu \otimes a_2)])]$$

$$= R^{-1} \dot{\otimes} [(R^\mathsf{T} \otimes (\lambda \otimes a_1)) \oplus (R^\mathsf{T} \otimes (\mu \otimes a_2))]$$

Redistributing and applying the definition of the closure we get:

$$= R^{-1} \dot{\otimes} (R^\mathsf{T} \otimes (\lambda \otimes a_1 \oplus \mu \otimes a_2)) = \pi_R(\lambda \otimes a_1 \oplus \mu \otimes a_2)$$

The proof for intents is similar.   □

We are now ready to prove the following proposition:

Proposition 8. $\mathcal{B}_G$ *and* $\mathcal{B}_M$ *are right complete idempotent* $\overline{\mathcal{K}}$*-semimodules.*

Proof. Consider first extents, and the neutral element as defined. Addition is clearly commutative and idempotent from the definition $a \tilde{\oplus} a = \pi_R(a \oplus a) = \pi_R(a) = a$. Associativity follows a pattern exploited in the rest of the proofs:

$$a_1 \tilde{\oplus} (a_2 \tilde{\oplus} a_3) = R^{-1} \dot{\otimes} (R^\mathsf{T} \otimes [a_1 \oplus (R^{-1} \dot{\otimes} [R^\mathsf{T} \otimes (a_2 \oplus a_3)])])$$

$$= R^{-1} \dot{\otimes} (R^\mathsf{T} \otimes a_1 \oplus R^\mathsf{T} \otimes [R^{-1} \dot{\otimes} (R^\mathsf{T} \otimes (a_2 \oplus a_3))])$$

$$= R^{-1} \dot{\otimes} (R^\mathsf{T} \otimes a_1 \oplus R^\mathsf{T} \otimes [a_2 \oplus a_3]) = R^{-1} \dot{\otimes} (R^\mathsf{T} \otimes [a_1 \oplus a_2 \oplus a_3])$$

$$= \pi_{R^\mathsf{T}}(a_1 \oplus a_2 \oplus a_3)$$

and the result follows by the associativity of $\oplus$, and the commutativity of $\tilde{\oplus}$ and $\dot{\oplus}$. The additive identity is correctly-defined:

$$
\begin{aligned}
\epsilon_G \tilde{\oplus} a &= R^{-1} \dot{\otimes} (R^{\mathrm{T}} \otimes [\epsilon_G \dot{\oplus} a]) = R^{-1} \dot{\otimes} (R^{\mathrm{T}} \otimes [R^{-1} \dot{\otimes} \perp^M \dot{\oplus} a]) \\
&= R^{-1} \dot{\otimes} ([R^{\mathrm{T}} \otimes (R^{-1} \dot{\otimes} \perp^M)] \dot{\oplus} [R^{\mathrm{T}} \otimes a]) = R^{-1} \dot{\otimes} (\perp^M \dot{\oplus} [R^{\mathrm{T}} \otimes a]) \\
&= R^{-1} \dot{\otimes} (R^{\mathrm{T}} \otimes a) = a
\end{aligned}
$$

where we have used that $\kappa_{R^{\mathrm{T}}}(x) = R^{\mathrm{T}} \otimes (R^{-1} \dot{\otimes} x) \le x$ is a kernel operator, that is a contractive, idempotent function, whence $R^{\mathrm{T}} \otimes (R^{-1} \dot{\otimes} \perp^M) = \perp^M$.

Only the external laws are left to be proven: first $\perp_{\mathcal{K}} \tilde{\odot} a = \pi_{R^{\mathrm{T}}}(\perp_{\mathcal{K}} \otimes a) = \pi_{R^{\mathrm{T}}}(\perp^G) = R^{-1} \dot{\otimes} (R^{\mathrm{T}} \otimes \perp^G) = R^{-1} \dot{\otimes} \perp^M = \epsilon_G$—indeed this might be taken for its definition. Next:

$$
\lambda \tilde{\odot} \epsilon_G = R^{-1} \dot{\otimes} (R^{\mathrm{T}} \otimes (\lambda \otimes R^{-1} \dot{\otimes} \perp^M)) = R^{-1} \dot{\otimes} (\lambda \otimes (R^{\mathrm{T}} \otimes (R^{-1} \dot{\otimes} \perp^M))).
$$

whence $\lambda \tilde{\odot} \epsilon_G = R^{-1} \dot{\otimes} (\lambda \otimes \perp^M) = R^{-1} \dot{\otimes} \perp^M = \epsilon_G$. From the definition of the scalar action $e_{\mathcal{K}} \tilde{\odot} a = \pi_{R^{\mathrm{T}}}(e_{\mathcal{K}} \odot a) = \pi_{R^{\mathrm{T}}}(a) = a$.

From (12) the rest of the laws follow by simple instantiation. The proof is identical for intents, *mutatis mutandis*.   □

We know consider the congruence on on $\mathcal{X}$ induced by the closure of extents and, dually, that of intents:

$$
\underline{\mathfrak{B}}_G^\gamma(G, M, R)^\perp = \{(x_1, x_2) \in X^2 \mid \pi_{R^{\mathrm{T}}}(x_1) = \pi_{R^{\mathrm{T}}}(x_2) = a, \forall a \in \underline{\mathfrak{B}}_G^\gamma(G, M, R)\}
$$
$$
\underline{\mathfrak{B}}_G^\gamma(G, M, R)^\perp = \{(y_1, y_2) \in Y^2 \mid \pi_R(y_1) = \pi_R(y_2) = b, \forall b \in \underline{\mathfrak{B}}_M^\mu(G, M, R)\}
$$

Clearly $(\underline{\mathfrak{B}}_G^\gamma(G, M, R)^\perp)^{\mathrm{T}} = \underline{\mathfrak{B}}_G^\gamma(G, M, R)$ and dually for intents.

When the connection between $\mathcal{X}$ and $\mathcal{Y}$ is an adjunction, Cuninghame-Green has proven that the Chebychev distance between x and its closure $\pi_{R^{\mathrm{T}}}(x)$ is minimal among all closures[2] [7]. For these reasons, Gaubert et al. have decided to call this the *orthogonal projection* with respect to the closure systems, and so the analogy goes on to define the *kernel* of π_{R_t} or π_R as:

$$
\begin{aligned}
\mathrm{KER}(\pi_{R^{\mathrm{T}}}) &= \underline{\mathfrak{B}}_G^\gamma(G, M, R)^\perp & \mathrm{KER}(\pi_R) &= \underline{\mathfrak{B}}_M^\mu(G, M, R)^\perp \\
\mathrm{KER}(\pi_{R^{\mathrm{T}}})^{\mathrm{T}} &= \underline{\mathfrak{B}}_G^\gamma(G, M, R) & \mathrm{KER}(\pi_R)^{\mathrm{T}} &= \underline{\mathfrak{B}}_M^\mu(G, M, R)
\end{aligned}
$$

---

[2] And this is also the case for the kernel operator in the adjunction.

3.4   The Semimodule of Formal Concepts

Thus each $\underline{\mathfrak{B}}_G^\gamma(G, M, R)$ and $\underline{\mathfrak{B}}_M^\mu(G, M, R)$ has a double semimodule structure:

1. A $\mathcal{K}^{\mathrm{d}}$-subsemimodule of their ambient spaces, as in Corollary 2.
2. A $\mathcal{K}$-semimodule as in Proposition 8. The semimodule structure is idiosyncratic in that it is defined with the help of the closure operators of each particular formal context.

Although the polars are bijective in the sets of extents and intents we do not have yet a full characterization in terms of semimodules. But:

Proposition 9. *The polars are isomorphisms from the $\mathcal{K}$- to $\mathcal{K}^{\mathrm{d}}$-semimodules.*

$$(\mathcal{B}_G)_R^\uparrow = \underline{\mathfrak{B}}_M^\mu(G, M, R) = \mathcal{B}_M \qquad (\mathcal{B}_M)_R^\downarrow = \underline{\mathfrak{B}}_G^\gamma(G, M, R) = \mathcal{B}_G$$

Proof. Let $\lambda, \mu \in K$, a_1 and a_2 be extents with b_1 and b_2 their intents. Then:

$$(\lambda \tilde{\otimes} a_1 \dot{\oplus} \mu \tilde{\otimes} a_2)_R^\uparrow = (\lambda \otimes a_1 \dot{\oplus} \mu \otimes a_2)^{\uparrow \downarrow \uparrow}_{RRR} = (\lambda \otimes a_1 \dot{\oplus} \mu \otimes a_2)_R^\uparrow = \lambda^{-1} \dot{\otimes} b_1 \dot{\oplus} \mu^{-1} \dot{\otimes} b_2$$

Since $\cdot^{-1}$ is a dual (auto)isomorphism of $\mathcal{K}$,

$$(\lambda \dot{\otimes} b_1 \dot{\oplus} \mu \dot{\otimes} b_2)_R^\downarrow = R^{-1} \dot{\otimes} (\lambda^{-1} \otimes b_1^{-1} \dot{\oplus} \mu^{-1} \otimes b_2^{-1})$$

$$= R^{-1} \dot{\otimes} (\lambda^{-1} \otimes R^{\mathrm{T}} \otimes a_1 \dot{\oplus} \mu^{-1} \otimes R^{\mathrm{T}} \otimes a_2)$$

$$= R^{-1} \dot{\otimes} (R^{\mathrm{T}} \otimes [\lambda^{-1} \otimes a_1 \dot{\oplus} \mu^{-1} \otimes a_2]) = \lambda^{-1} \tilde{\otimes} a_1 \dot{\oplus} \mu^{-1} \tilde{\otimes} a_2$$

On the other hand $(\mathcal{B}_G)_{RR}^{\uparrow \downarrow} = [\underline{\mathfrak{B}}_M^\mu(G, M, R)]_R^\downarrow = \mathcal{B}_G$. For the join semimodule of intents the proof is similar. $\square$

And it is not difficult to see how this translates into arbitrary joins in complete idempotent semifields—where instead of $\dot{\oplus}$ we write $\tilde{\sum}_{\bullet}$— to provide the basis for the following, restricted theorem of $\mathcal{K}$-FCA :

Theorem 2. *The (γ, μ)-concept lattice $\underline{\mathfrak{B}}^{\gamma \otimes \mu}(G, M, R)$ is a dually isomorphic pair of complete lattices in which infima and suprema are given by:*

$$\sum_{\bullet}{}_{i \in I} (a_i, b_i) = (\tilde{\sum}_{\bullet \, i \in I} a_i, \sum_{i \in I}^\bullet b_i) = \left[\left(\sum_{i \in I}^\bullet b_i\right)_R^\downarrow, \sum_{i \in I}^\bullet b_i\right]$$

$$\sum_{i \in I}^\bullet (a_i, b_i) = (\sum_{i \in I}^\bullet a_i, \tilde{\sum}_{\bullet \, i \in I} b_i) = \left[\sum_{i \in I}^\bullet a_i, \left(\sum_{i \in I}^\bullet a_i\right)_R^\uparrow\right]$$

Proof. A corollary of the Galois connection and the previous definitions. $\square$

Note that we do not use special notation for the meets and joins of concepts, as we did for the component lattices, and that the closure operations are hidden in the definitions of the new joins.

In fact, Proposition 6 says that the sets of extents and intents are idempotent semimodules, that is the idempotent analog of a vector space. Could we endow the set of concepts with a similar structure? Does this have practical consequences?

First, we next consider endowing the set of concepts with right upper and lower scalar actions:

$$\lambda \odot (a, b) = (\lambda \tilde{\odot} a, \lambda^{-1} \dot{\odot} b) \qquad \lambda \dot{\odot} (a, b) = (\lambda \dot{\odot} a, \lambda^{-1} \tilde{\odot} b) \qquad (13)$$

When $\lambda > e$ we call these operations *(concept) abstraction* and when $\lambda < e$ *(concept) concretion*.

The previous material leads to the following extended theorem of $\mathcal{K}$-FCA :

Theorem 3. *The (γ, μ)-concept double semimodule $\mathfrak{B}^{\gamma \tilde{\otimes} \mu}(G, M, R)$ is a dually isomorphic pair of complete idempotent semimodules in which infimum and supremum combinations are given by:*

$$\sum_{i \in I} \hspace{-0.5em}\raisebox{-0.3em}{$\cdot$}\; \lambda_i \otimes (a_i, b_i) = (\tilde{\sum}_{\raisebox{0.2em}{$\cdot$} i \in I} \lambda_i \tilde{\otimes} a_i, \sum_{i \in I}\hspace{-0.7em}\raisebox{0.6em}{$\cdot$}\; \lambda_i \dot{\otimes} b_i) = \left[\left(\sum_{i \in I}\hspace{-0.7em}\raisebox{0.6em}{$\cdot$}\; \lambda_i \dot{\otimes} b_i \right)^{\downarrow}_R, \sum_{i \in I}\hspace{-0.7em}\raisebox{0.6em}{$\cdot$}\; \lambda_i \dot{\otimes} b_i \right]$$

$$\sum_{i \in I}\hspace{-0.7em}\raisebox{0.6em}{$\cdot$}\; \lambda_i \dot{\otimes} (a_i, b_i) = (\sum_{i \in I}\hspace{-0.7em}\raisebox{0.6em}{$\cdot$}\; \lambda_i \dot{\otimes} a_i, \tilde{\sum}_{\raisebox{0.2em}{$\cdot$} i \in I} \lambda_i \tilde{\otimes} b_i) = \left[\sum_{i \in I}\hspace{-0.7em}\raisebox{0.6em}{$\cdot$}\; \lambda_i \dot{\otimes} a_i, \left(\sum_{i \in I}\hspace{-0.7em}\raisebox{0.6em}{$\cdot$}\; \lambda_i \dot{\otimes} a_i \right)^{\uparrow}_R \right]$$

Proof. This is a corollary of Proposition 9 and Theorem 2. □

3.5   Join-Dense and Meet-Dense Vectors

Standard concept lattices have "natural" building algorithms in terms of the object-intents and attribute-extents. We have just seen that $\mathcal{K}$-concept lattices are generated in terms of the dual semifield very straightforwardly, and in terms of the original semifield in a more convoluted way. Can we reconcile these two views? Certainly. First, we define concept-building operators from sets of objects and attributes respectively:

$$\gamma : \tilde{X}^{\gamma} \to \underline{\mathfrak{B}}^{\varphi}(G, M, R) \qquad\qquad \mu : \tilde{Y}^{\mu} \to \underline{\mathfrak{B}}^{\varphi}(G, M, R)$$

$$x \mapsto \gamma(x) = (\pi_{R^{\mathsf{T}}}(x), x_R^{\uparrow}) \qquad\qquad y \mapsto \mu(y) = (y_R^{\downarrow}, \pi_R(y))$$

Next, let I_G and I_M be the identity matrices of dimension $g \times g$ and $m \times m$ in $\overline{\mathcal{K}}$, whose columns are naturally conceived as the unitary vectors of objects and attributes, respectively.

Lemma 2. *Let I_G and I_M be the identity matrices of dimension $g \times g$ and $m \times m$ in $\overline{\mathcal{K}}$. Then the object- and attribute-concepts of the Galois connection are:*

$$\gamma_R(I_G) = (R^{-1} \dot{\otimes} R^{\mathsf{T}}, R^*) \qquad\qquad \mu_R(I_M) = (R^{-1}, R^* \dot{\otimes} R)$$

taken as pairs of co-indexed vectors.

Proof. Simple application of the polars to the identities.   □

Note how the matrix notation allows us to carry out multiple computations at the same time. We may now conclude the following:

Corollary 3. *Consider the Galois connection* $(\cdot^{\uparrow}_R, \cdot^{\downarrow}_R) : \tilde{X}^\gamma \leftthreetimes \tilde{Y}^\mu$. *Then, its system of extents is* $\overline{\mathcal{K}}^d$-*generated by the attribute-extents. Dually, its system of intents is* $\overline{\mathcal{K}}^d$-*generated by the object-intents.*

Proof. From Corollary 2 and Lemma 2.   □

This is a result that has a nice analogue with standard FCA where these sets are meet-dense, respectively. Furthermore, since these semimodules are complete and finitely-generated we can always find a subset of these $\dot{\oplus}$-dense sets. The schematic diagram of Fig. 2 makes these mechanisms evident.

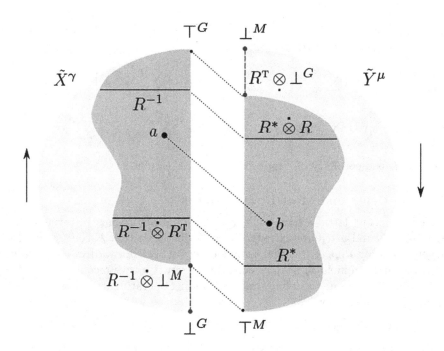

Fig. 2. Extended schematics of the Galois connection between $\tilde{X}^\gamma$ and $\tilde{Y}^\mu$ (outer clouds). Extents (left inner cloud) and intents (right inner cloud) are dually isomorphic $\mathcal{K}^d$-semimodules generated by R^{-1} and R^* respectively. Similarly, they are dually isomorphic $\mathcal{K}$-semimodules generated by $R^{-1} \dot{\otimes} R^{\mathrm{T}}$ and $R^* \dot{\otimes} R$ and the closure operators (see text).

Can we expect to find a similar mechanism for $\dot{\oplus}$-dense sets, that is object-extents and attribute intents? The answer suggestively blends the spectral theory of matrices and $\mathcal{K}$-FCA. First, consider a property of the object-extents $P_G = R^{-1} \dot{\otimes} R^{\mathrm{T}}$ and attribute intents $P_M = R^* \dot{\otimes} R$.

Proposition 10. *The object extents (respectively, attribute extents) are fundamental eigenvectors of P_G (respectively, P_M) for the eigenvalue e.*

Proof. Consider the power $(R^* \dot{\otimes} R) \otimes (R^* \dot{\otimes} R) = R^* \dot{\otimes} R$, where the equality comes from the matrix product laws. It is easy to see by induction that $(R^* \dot{\otimes} R)^n = R^* \dot{\otimes} R$. Furthermore, its diagonal only has the elements $\{e, \top\}$ wherefore:

$$P_M^{\dot{*}} = \sum_{n=0}^{\infty} {}_{\bullet} P_M^n = I_M \oplus \sum_{n=1}^{\infty} {}_{\bullet} P_M^n = P_M^{\dot{+}}$$

hence by Theorem 1 we know that the columns of P_M are all eigenvectors of P_M for e.   □

From this we obtain that:

Proposition 11. *The eigenspace of P_M (resp. P_G) generates the set of extents (resp. intents.)*

$$\underline{\mathfrak{B}}_G^{\gamma}(G, M, R) = (\mathcal{V}_e(R^* \dot{\otimes} R))_R^{\downarrow} \qquad \underline{\mathfrak{B}}_M^{\mu}(G, M, R) = (\mathcal{V}_e(R^{-1} \dot{\otimes} R^{\mathrm{T}}))_R^{\uparrow}$$

Proof. We prove it for intents: recall that the eigenspace of e generated by the columns of $R^* \dot{\otimes} R$, $\mathcal{V}_e(R^* \dot{\otimes} R) = \langle R^* \dot{\otimes} R \rangle_{\mathcal{K}}$. Furthermore, we know that the eigenspaces are $\mathcal{K}$-semimodules, that the polars transform $\mathcal{K}$-semimodules into $\mathcal{K}^{\mathrm{d}}$-semimodules, and $(R^* \dot{\otimes} R)_R^{\downarrow} = R^{-1}$, hence

$$(\mathcal{V}_e(R^* \dot{\otimes} R))_R^{\downarrow} = (\langle R^* \dot{\otimes} R \rangle_{\mathcal{K}})_R^{\downarrow} = \langle (R^* \dot{\otimes} R)_R^{\downarrow} \rangle_{\mathcal{K}^{\mathrm{d}}} = \langle R^{-1} \rangle_{\mathcal{K}^{\mathrm{d}}} = \underline{\mathfrak{B}}_G^{\gamma}(G, M, R).$$

For extents the procedure is similar.   □

Corollary 4. *Both the set of extents and intents are generated from the object extents, and dually from the attribute intents.*

$$\underline{\mathfrak{B}}_G^{\gamma}(G, M, R) = \langle R^{-1} \rangle_{\mathcal{K}^{\mathrm{d}}} = \pi_{R^{\mathrm{T}}}(\langle R^{-1} \dot{\otimes} R^{\mathrm{T}} \rangle_{\mathcal{K}})$$

$$\underline{\mathfrak{B}}_M^{\mu}(G, M, R) = \langle R^* \rangle_{\mathcal{K}^{\mathrm{d}}} = \pi_R(\langle R^* \dot{\otimes} R \rangle_{\mathcal{K}})$$

Proof. We directly use the intent polar (and dually for extents):

$$\underline{\mathfrak{B}}_M^{\mu}(G, M, R) = (\underline{\mathfrak{B}}_G^{\gamma}(G, M, R))_R^{\uparrow} = ((\mathcal{V}_e(R^* \dot{\otimes} R))_R^{\downarrow})_R^{\uparrow} = \pi_R(\langle R^* \dot{\otimes} R \rangle_{\mathcal{K}}).$$

□

4   Discussion

Our main result is that $\mathcal{K}$-concept lattices are both $\mathcal{K}^{\mathrm{d}}$-subsemimodules and $\mathcal{K}$-semimodules, as well as lattice-ordered, and that the polars are also dual isomorphisms of semimodules, on top of dual lattice isomorphisms.

Cuninghame-Green [7, Ch. 22] already developed a construction similar to that of Sect. 3 but describing an adjunction (a neighbourhood lattice) hence his failure to give primary status to the pairs of closed elements that constitute concepts: this is the merit of FCA. On the other hand, our scaling is reminiscent of that of his (although inspired by [10]), apart from the technicalities, although the interpretation as abstraction and concretion only can be contextualized in terms of similar concepts of FCA, like the joins representing a generalization, and the meet a specialization. Of course, the data analysis in the presence of infinite elements is all new.

Since $\mathcal{K}$ is an idempotent *semifield* where an inversion is available, as it is in the ambient spaces, we might wonder whether the semimodules of extents and intents had a similar inversion available. In general this is not the case, since $a^{-1} = (a^*)^{\mathrm{T}}$ we have for $a \in \mathfrak{B}_G^{\gamma}(G, M, R)$ that $a^{-1} = R \otimes (R^* \overset{\cdot}{\otimes} a^{-1})$. That is, the inverses of extents are actually fixpoints of a kernel operator, and likewise for intents. Dually, if the inverse a^{-1} were an extent, then a would also have to be a fixpoint of that kernel. This would only happen on very particular extents.

References

1. Valverde-Albacete, F.J., Peláez-Moreno, C.: Towards a generalisation of formal concept analysis for data mining purposes. In: Missaoui, R., Schmidt, J. (eds.) ICFCA 2006. LNCS (LNAI), vol. 3874, pp. 161–176. Springer, Heidelberg (2006)
2. Serra, J., Soille, P. (eds.): Mathematical Morphology and its Application to Image Processing. Computational Imaging and Vision. Kluwer Academic, Dordrecht (1994)
3. Deller, J.R., Proakis, J.G., Hansen, J.H.: Discrete-Time Processing of Speech Signals. Prentice-Hall, Englewood Cliffs (1987)
4. Valverde-Albacete, F.J., Peláez-Moreno, C.: Further Galois connections between semimodules over idempotent semirings. In: Diatta, J., Eklund, P., (eds.): In: Proceedings of the 4th Conference on Concept Lattices and Applications (CLA 07), Montpellier, France, 199–212 (2007)
5. Valverde-Albacete, F.J., Peláez-Moreno, C.: Extending conceptualisation modes for generalised formal concept analysis. Inf. Sci. **181**, 1888–1909 (2011)
6. Valverde-Albacete, F.J., Peláez-Moreno, C.: Spectral lattices of $(R_{max,+})$-formal contexts. In: Medina, R., Obiedkov, S. (eds.) ICFCA 2008. LNCS (LNAI), vol. 4933, pp. 124–139. Springer, Heidelberg (2008)
7. Cuninghame-Green, R.: Minimax Algebra. Number 166 in Lecture notes in Economics and Mathematical Systems. Springer, Heidelberg (1979)
8. Gaubert, S., Katz, R.D.: The tropical analogue of polar cones. Linear Algebra Appl. **431**, 608–625 (2009)

9. Gondran, M., Minoux, M.: Valeurs propres et vecteurs propres dans les dioïdes
   et leur interprétation en théorie des graphes. EDF, Bulletin de la Direction des
   Etudes et Recherches, Serie C, Mathématiques Informatique **2**, 25–41 (1977)
10. Bělohlávek, R.: Fuzzy Relational Systems. Foundations and Principles. IFSR Inter-
    national Series on Systems Science and Engineering, vol. 20. Kluwer Academic,
    New York (2002)
11. Erné, M.: Adjunctions and Galois connections: origins, history and development.
    In: [12], pp. 1–138
12. Denecke, K., Erné, M., Wismath, S. (eds.): Galois Connections and Applications.
    Mathematics and Its Applications, vol. 565. Springer, The Netherlands (2004)

On Closure Systems and Adjunctions Between Fuzzy Preordered Sets

F. García-Pardo, I.P. Cabrera$^{(\boxtimes)}$, P. Cordero, and M. Ojeda-Aciego

Andalucía Tech, Universidad de Málaga, Málaga, Spain
{fgarciap,ipcabrera,pcordero,aciego}@uma.es

Abstract. The aim of this work is providing a characterization in terms of closure systems, for the construction, given a mapping $f \colon \mathbb{A} \to B$ from a fuzzy preordered set $\mathbb{A}$ into an unstructured set B, of a suitable fuzzy preordering on B for which there exists a mapping $g \colon B \to A$ such that the pair (f, g) constitutes an adjunction (isotone Galois connection). This contribution continues our research line on the construction of adjunctions in which the theory of fuzzy closure systems is used in order to provide a more meaningful framework for the extension to the fuzzy case of previous results.

Keywords: Galois connection · Adjunction · Preorder · Fuzzy sets

1 Introduction

Adjunctions (also called isotone Galois connection) between two mathematical structures provide a means of linking both theories allowing for mutual cooperative advantages.

A number of results can be found in the literature concerning sufficient or necessary conditions for adjunctions between ordered structures to exist. Likewise, this paper is related to the existence and construction of the right adjoint to a given mapping f, but *in a more general framework*. Our abstract setting is to consider a mapping $f \colon \mathbb{A} \to B$ from an "ordered" set $\mathbb{A}$ into an unstructured set B, and then characterize those situations in which B can be "adequately ordered" and a mapping $g \colon B \to A$ can be built so that the pair (f, g) is an adjunction [12].

Our specific framework focuses on a fuzzy setting, so both the ordering and the notion of adjunction need to be adequately generalized. Concerning the ordering, the natural choice is that of fuzzy ordering leading to the so-called fuzzy posets; however, taking into account several approaches which suggest to consider even more general structures (some authors [8] suggest dropping reflexivity, whereas others argue that reflexivity and antisymmetry are conflicting properties [4]), our choice in this case has been to ignore antisymmetry and, therefore, consider fuzzy preorders.

Partially supported by the Spanish Science Ministry projects TIN12-39353-C04-01 and TIN11-28084.

J. Baixeries et al. (Eds.): ICFCA 2015, LNAI 9113, pp. 114–127, 2015.
DOI: 10.1007/978-3-319-19545-2_7

Several papers on fuzzy Galois connections or fuzzy adjunctions have been written since its introduction by Bělohlávek in [1]; consider for instance [2,9,15,17] for some recent extensions. Some authors have introduced alternative approaches guided by the intended applications: for instance, Shi et al. [16] introduced a definition of fuzzy adjunction for its use in fuzzy mathematical morphology. Our approach in this paper is more in consonance with Bělohlávek's logic approach, but in terms of the generalization provided by Yao and Li [17] within the framework of fuzzy posets and fuzzy closure operators.

In order to study the existence of adjoint, we will rely on the well-known relationship between closure operators (and closure systems) and antitone Galois connections; this relationship can be defined in categorical terms [7], but is much better known in the framework of domain theory and denotational semantics [13]. Essentially, the link can be stated as follows: if (f, g) is an antitone Galois connection, then the compositions fg and gf satisfy the conditions of a closure operator. In the case of an adjunction between two posets, just the mapping gf is a closure operator whereas fg is a kernel operator.

Specifically, in this paper we characterize the situations in which given a mapping $f \colon A \to B$ from a fuzzy preordered set A into an unstructured set B, there exists a fuzzy preorder on B and a mapping $g \colon B \to A$ such that (f, g) is an adjunction in terms of the existence of a closure system in A with suitable properties.

The structure of this work is the following: in the next section, we introduce the preliminary definitions and results, essentially notions related to fuzzy preorderings and to Galois connections, and some results which will be later needed. Section 4 introduces the notion of closure system and closure operator which better serves our purpose, together with several lemmas which allow to simplify the presentation of the proof of the main result in Sect. 4, where the actual construction of the right adjoint is given under adequate conditions. The final section includes some final comments and prospects for future work.

2   Preliminaries

Although the present work is developed in a fuzzy setting, the basic definitions of Galois connections and adjunctions on the crisp case as well as the different characterizations and equivalencies from them will be recalled for the paper to be more self-contained. For this we formulate the results in the general setting of preordered sets that are sets endowed with a reflexive and transitive binary relation. For any preordered set $\mathbb{A} = (A, \leq)$, we consider its dual preordered set $\mathbb{A}^{op} = (A, \geq)$. Also we call, $a^{\downarrow} = \{x \in A : x \leq a\}$ and $a^{\uparrow} = \{x \in A : x \geq a\}$. Let $f \colon (A, \leq) \to (B, \leq)$ be a Galois connection.

- f is *isotone* if, for all $a, b \in A$, $a \leq b$ implies $f(a) \leq f(b)$.
- f is *antitone* if, for all $a, b \in A$, $a \leq b$ implies $f(b) \leq f(a)$.

In the particular case in which $A = B$,

- f is *inflationary* (also called extensive) if $a \leq f(a)$ for all $a \in A$.

- f is *deflationary* if $f(a) \leq a$ for all $a \in A$.
- f is *idempotent* if $f \circ f = f$.
- f a *closure operator* if it is inflationary, isotone and idempotent.
- f a *kernel operator* if it is deflationary, isotone and idempotent.

For a more detailed study of closure and kernel operators we refer to [6].

Definition 1 (Galois Connections/Adjunctions). *Let* $\mathbb{A} = (A, \leq)$ *and* $\mathbb{B} = (B, \leq)$ *be preordered sets, and* $f : A \to B$ *and* $g : B \to A$ *be two mappings. The pair (f,g) is called a*

- **Right Galois Connection between** $\mathbb{A}$ **and** $\mathbb{B}$, *denoted by* $(f, g) : \mathbb{A} \leftharpoondown\!\!\!\rightarrow \mathbb{B}$, *if*

$$\text{for all } a \in A \text{ and } b \in B, \quad a \leq g(b) \text{ if only if } b \leq f(a).$$

- **Left Galois Connection between** $\mathbb{A}$ **and** $\mathbb{B}$, *we write* $(f, g) : \mathbb{A} \leftarrow\!\!\!\rightharpoonup \mathbb{B}$, *if*

$$\text{for all } a \in A \text{ and } b \in B, \quad g(b) \leq a \text{ if only if } f(a) \leq b.$$

- **Adjunction between** $\mathbb{A}$ **and** $\mathbb{B}$, *denoted by* $(f, g) : \mathbb{A} \leftrightharpoons \mathbb{B}$, *if*

$$\text{for all } a \in A \text{ and } b \in B, \quad a \leq g(b) \text{ if only if } f(a) \leq b.$$

- **Co-Adjunction between** $\mathbb{A}$ **and** $\mathbb{B}$, *denoted by* $(f, g) : \mathbb{A} \rightleftharpoons \mathbb{B}$, *if*

$$\text{for all } a \in A \text{ and } b \in B, \quad g(b) \leq a \text{ if only if } b \leq f(a).$$

Observe that the difference among the four definitions above is not significant because one can change from one version to another by swapping between A and A^{op} and/or B and B^{op}.

Theorem 1. *Let* $\mathbb{A} = (A, \leq)$ *and* $\mathbb{B} = (B, \leq)$ *be preordered sets, and* $f : A \to B$ *and* $g : B \to A$ *be two mappings. Then, the following conditions are equivalent*

1. $(f, g) : \mathbb{A} \leftharpoondown\!\!\!\rightarrow \mathbb{B}$.
2. $(f, g) : \mathbb{A}^{op} \leftarrow\!\!\!\rightharpoonup \mathbb{B}^{op}$.
3. $(f, g) : \mathbb{A} \leftrightharpoons \mathbb{B}^{op}$.
4. $(f, g) : \mathbb{A}^{op} \rightleftharpoons \mathbb{B}$.

Hereafter, we will work with adjunctions, but as a direct consequence of the previous theorem, any property about adjunctions can be extended by duality to the other kind of connections or to co-adjunctions.

Proposition 1. *Given two preordered sets* $\mathbb{A} = (A, \leq_A)$ *and* $\mathbb{B} = (B, \leq_B)$, *and two mappings* $f : A \to B$ *and* $g : B \to A$. *The following conditions are equivalent:*

1. $(f, g) : \mathbb{A} \leftrightharpoons \mathbb{B}$.
2. f *and* g *are isotone,* $g \circ f$ *is inflationary, and* $f \circ g$ *is deflationary.*

As a consequence, given $(f, g) : \mathbb{A} \leftrightarrows \mathbb{B}$, then the mapping gf is a closure operator and fg is a kernel operator.

In order to introduce fuzziness in this setting, the most usual underlying structure for considering fuzzy extensions of Galois connections/adjunctions is that of residuated lattice, $\mathbb{L} = (L, \vee, \wedge, \top, \bot, \otimes, \rightarrow)$. An $\mathbb{L}$-fuzzy set is a mapping from the universe set to the membership values structure $X : U \rightarrow L$ where $X(u)$ means the degree in which u belongs to X. Given X and Y two $\mathbb{L}$-fuzzy sets, X is said to be *included in* Y, denoted as $X \subseteq Y$, if $X(u) \leq Y(u)$ for all $u \in U$.

An $\mathbb{L}$-*fuzzy binary relation* on U is an $\mathbb{L}$-fuzzy subset of $U \times U$, that is $\rho_U : U \times U \rightarrow L$, and it is said to be:

- *Reflexive* if $\rho_U(a, a) = \top$ for all $a \in U$.
- *Transitive* if $\rho_U(a, b) \otimes \rho_U(b, c) \leq \rho_U(a, c)$ for all $a, b, c \in U$.
- *Symmetric* if $\rho_U(a, b) = \rho_U(b, a)$ for all $a, b \in U$.

Definition 2 (Fuzzy Preordered Set). *An $\mathbb{L}$-fuzzy preordered set is a pair* $\mathbb{U} = (U, \rho_U)$ *in which ρ_U is a reflexive and transitive $\mathbb{L}$-fuzzy relation on U.*

From now on, when no confusion arises, we will omit the prefix "$\mathbb{L}$-".

The extensions to the fuzzy setting of the notions of upset and downset of an element $a \in A$ are defined by $a^\uparrow, a^\downarrow : A \rightarrow L$ where $a^\downarrow(x) = \rho_A(x, a)$ and $a^\uparrow(x) = \rho_A(a, x)$ for all $x \in A$.

Definition 3. *Let $\mathbb{A} = (A, \rho_A)$ be a fuzzy preordered set.*
An element $m \in A$ is a p-minimum for a fuzzy set X if

1. $X(m) = \top$, *and*
2. $X \subseteq m^\uparrow$, *i.e.,* $X(x) \leq \rho_A(m, x)$, *for all $x \in A$.*

An element $M \in A$ is a p-maximum for a fuzzy set X if

1. $X(M) = \top$, *and*
2. $X \subseteq M^\downarrow$, *i.e.,* $X(x) \leq \rho_A(x, M)$, *for all $x \in A$.*

Notice that, due to the absence of any kind of antisymmetry, there exists a crisp set of p-minima (resp. maxima) for X, which is not necessarily a singleton, which we will denote p-min(X) (resp., p-max(X)).

3   Closure Systems and Closure Operators

The notion of closure system on a fuzzy preordered set that we will use is a natural extension of the classical closure system on a crisp partial ordered set. In fact, the definition is formulated in the same terms, though we will use an alternative characterization that is easier to handle with.

Definition 4. *Let $\mathbb{A} = (A, \rho_A)$ be a fuzzy preordered set and let $S \subseteq A$ be a crisp subset of A. Then S is said to be a closure system if the set p-min$(a^\uparrow \cap S)$ is non-empty, for all $a \in A$.*

Other definitions of closure system in a fuzzy setting can be found in the literature. It is remarkable the one given by Bělohlávek in [3], where the notions of L_K-closure operator and L_K-closure system on L-ordered sets were introduced, where K means a filter of the residuated lattice L. In Bělohlávek's definition, fuzzy closure systems are fuzzy sets, whereas in our definition above, fuzzy closure systems are crisp subsets.

There exists another definition similar in spirit to the one we propose, which was introduced in the framework of the so-called L-ordered sets. In the following result we state an alternative characterization of the notion of closure system based on ideas from [14].

Proposition 2. *Let $\mathbb{A} = (A, \rho_A)$ be a fuzzy preordered set. A non-empty subset $S \subseteq A$ is a closure system on A if and only if for any $a \in A$, there exists $m_a \in S$ such that*

1. $\rho_A(a, m_a) = \top$ and
2. $\rho_A(s_1, m_a) \otimes \rho_A(a, s_2) \leq \rho_A(s_1, s_2)$ for any $s_1, s_2 \in S$.

Proof. Let S be a subset of A and suppose that for an element $a \in A$, the set p-min$(a^\uparrow \cap S)$ is non-empty. Let us see that any element $m_a \in$ p-min$(a^\uparrow \cap S)$ satisfies Conditions 1 and 2 above. By definition of the set p-min$(a^\uparrow \cap S)$, since

$$(a^\uparrow \cap S)(x) = (a^\uparrow)(x) \wedge S(x) = \begin{cases} \bot & \text{if } x \notin S \\ (a^\uparrow)(x) = \rho_A(a, x) & \text{if } x \in S \end{cases}$$

it is clear that $m_a \in S$, $\rho_A(a, m_a) = \top$ and, furthermore, $\rho_A(a, s) \leq \rho_A(m_a, s)$ for all $s \in S$. Then, for all $s_1, s_2 \in S$,

$$\rho_A(s_1, m_a) \otimes \rho_A(a, s_2) \leq \rho_A(s_1, m_a) \otimes \rho_A(m_a, s_2) \leq \rho_A(s_1, s_2)$$

Conversely, given $a \in A$, let $m_a \in S$ be an element satisfying Conditions 1 and 2 and let us prove that $m_a \in$ p-min$(a^\uparrow \cap S)$. Condition 1 is common to the notion of p-minimum, on the other hand, by reflexivity and Condition 2, for all $s \in S$ we have that $\rho_A(a, s) = \rho_A(m_a, m_a) \otimes \rho_A(a, s) \leq \rho_A(m_a, s)$.   □

The previous result can be further improved by providing a new characterization which involves just one condition.

Theorem 2. *Let $\mathbb{A} = (A, \rho_A)$ be a fuzzy preordered set. $S \subseteq A$ is a closure system if and only if for all $a \in A$ there exists $m_a \in S$ satisfying $\rho_A(a, u) = \rho_A(m_a, u)$ for all $u \in S$.*

Proof. Assume that S is a closure system, by Proposition 2, for all $a \in A$ there exists $m_a \in S$ such that

1. $\rho_A(a, m_a) = \top$ and
2. $\rho_A(s_1, m_a) \otimes \rho_A(a, s_2) \leq \rho_A(s_1, s_2)$ for any $s_1, s_2 \in S$.

Using Condition 1, we can write $\rho_A(m_a, u) = \rho_A(a, m_a) \otimes \rho_A(m_a, u) \leq \rho_A(a, u)$. Using Condition 2 above, using m_a for s_1 and u for s_2, by reflexivity and transitivity we obtain $\rho_A(a, u) \leq \rho_A(m_a, u)$. Hence, we have $\rho_A(a, u) = \rho_A(m_a, u)$.

Conversely, it is enough to prove that $m_a \in \text{p-min}(a^\uparrow \cap S)$. As $\rho_A(a, u) = \rho_A(m_a, u)$ for all $u \in S$, in particular $\rho_A(a, m_a) = \rho_A(m_a, m_a) = \top$. For $x \notin S$, we have $(a^\uparrow \cap S)(x) = \bot \leq m_a^\uparrow(x)$; for $x \in S$, we have $(a^\uparrow \cap S)(x) = \rho_A(a, x) = \rho_A(m_a, x) = m_a^\uparrow(x)$. $\qquad\square$

As an easy consequence of this theorem we obtain a constructive version of the sets p-min$(a^\uparrow \cap S)$ when S is a closure system.

Corollary 1. *Let $\mathbb{A} = (A, \rho_A)$ be a fuzzy preordered set. If $S \subseteq A$ is a closure system then* $\text{p-min}(a^\uparrow \cap S) = \{s \in S \mid \rho_A(a, u) = \rho_A(s, u) \text{ for all } u \in S\}$, *for $a \in A$.*

It is well-known that closure systems and closure operators in the classical setting are different approaches to the same phenomenon. We will focus now on the development of the link between these two notions on fuzzy preordered sets. We first recall some basic definitions.

Definition 5. *Let $\mathbb{A} = (A, \rho_A)$ and $\mathbb{B} = (B, \rho_B)$ be fuzzy preordered sets. A mapping $f: A \to B$ is said to be* isotone *if $\rho_A(a_1, a_2) \leq \rho_B(f(a_1), f(a_2))$ for each $a_1, a_2 \in A$.*
A mapping $f: A \to A$ is said to be

- inflationary *if $\rho_A(a, f(a)) = \top$ for all $a \in A$.*
- deflationary *if $\rho_A(f(a), a) = \top$ for all $a \in A$.*

Definition 6. *Let $\mathbb{A} = (A, \rho_A)$ be a fuzzy preordered set. A mapping $c: A \to A$ is said to be a* closure operator *if it is isotone, inflationary and satisfies $\rho_A(c(c(a)), c(a)) = \top$ for all $a \in A$.*

The following lemma states that the notions of closure system and closure operator keep being interdefinible in the framework of fuzzy preordered sets.

Lemma 1. *Let $\mathbb{A} = (A, \rho_A)$ be a fuzzy preordered set.*

(i) *If $S \subseteq A$ is a closure system, then any mapping $c: A \to A$ such that $c(a) \in$ p-min$(a^\uparrow \cap S)$ is a closure operator.*
(ii) *If $c: A \to A$ is a closure operator, then $S = \{a \in A : \rho_A(c(a), a) = \top\}$ is a closure system.*

Proof. (i) Assume that $S \subseteq A$ is a closure system and let $c: A \to A$ be a map such that $c(a) \in \text{p-min}(a^\uparrow \cap S)$ for all $a \in A$. In particular, $c(a) \in S$, thus, by Corollary 1, we have that $\rho_A(a, c(a)) = \rho_A(c(a), c(a)) = \top$, which implies that c is inflationary. Similarly, $\rho_A(c(a), c(a)) = \rho_A(c(c(a)), c(a)) = \top$. Finally, c is isotone as

$$\rho_A(a_1, a_2) = \rho_A(a_1, a_2) \otimes \rho_A(a_2, c(a_2)) \leq \rho_A(a_1, c(a_2)) = \rho_A(c(a_1), c(a_2))$$

(ii) The subset $S = \{a \in A : \rho_A(c(a), a) = \top\}$ is a closure system because $c(a) \in$ p-min$(a^{\uparrow} \cap S)$, for all $a \in A$. Firstly notice that, as $\rho_A(c(c(a)), c(a)) = \top$, then $c(a) \in S$. Let us show that $\rho_A(a, u) = \rho_A(c(a), u)$ for all $u \in S$. Since c is isotone and $u \in S$, we deduce that

$$\rho_A(a, u) \leq \rho_A(c(a), c(u)) = \rho_A(c(a), c(u)) \otimes \rho_A(c(u), u) \leq \rho_A(c(a), u)$$

Now, as c is inflationary, $\rho_A(c(a), u) = \rho_A(a, c(a)) \otimes \rho_A(c(a), u) \leq \rho_A(a, u)$. □

In the following, the constructions given in the different items of the previous lemma will be called, respectively, *the closure operator associated to S* (denoted c_S) and *a closure system associated to c* (denoted S_c).

It is well-known that, in (crisp) posets, there exists a one-to-one correspondence between closure operator and closure systems (for every closure operator $c = c_{S_c}$ and for any closure system $S = S_{c_S}$). The relationship between both notions is weaker when the underlying structure is a fuzzy preordered set.

Proposition 3. *Let* $\mathbb{A} = (A, \rho_A)$ *be a fuzzy preordered set.*

1. *If* $c: A \to A$ *is a closure operator, then* $\rho_A(c(a), c_{S_c}(a)) = \rho_A(c_{S_c}(a), c(a)) = \top$ *for all* $a \in A$.
2. *If* S *is a closure system then* $S \subseteq S_{c_S}$ *and for all* $s_1 \in S_{c_S}$ *there exists* $s_2 \in S$ *such that* $\rho_A(s_1, s_2) = \rho_A(s_2, s_1) = \top$.

Proof. Firstly, for any closure operator $c: A \to A$, the closure system S_c is defined as $\{a \in A: \rho_A(c(a), a) = \top\}$. Consider a closure operator $c_{S_c}: A \to A$ such that $c_{S_c}(a) \in$ p-min$(a^{\uparrow} \cap S_c)$. By the definition of closure operator, $\rho_A(c(c(a)), c(a)) = \top$ and therefore $c(a) \in S_c$. Moreover, since $\rho_A(a, c(a)) = \top$, we also have $c(a) \in$ p-min$(a^{\uparrow} \cap S_c)$. Finally, by definition of p-minimum, $\rho_A(c(a), c_{S_c}(a)) = \rho_A(c_{S_c}(a), c(a)) = \top$.

To prove the second item, consider a closure system S, a closure operator $c_S: A \to A$ associated to S (i.e., satisfying $c_S(a) \in$ p-min$(a^{\uparrow} \cap S)$ for all $a \in A$) and the closure system $S_{c_S} = \{a \in A : \rho_A(c_S(a), a) = \top\}$. For any $a \in S$, it is straightforward that $a \in$ p-min$(a^{\uparrow} \cap S)$ and since $c_S(a) \in$ p-min$(a^{\uparrow} \cap S)$ as well, then $\rho_A(c_S(a), a) = \top$. Therefore, $a \in S_{c_S}$. Thus, $S \subseteq S_{c_S}$ is proved. Furthermore, if $s_1 \in S_{c_S}$ then $\rho_A(c_S(s_1), s_1) = \top$. On the other hand, since $c_S(s_1) \in$ p-min$(s_1^{\uparrow} \cap S)$, then $c_S(s_1) \in S$ and $\rho_A(s_1, c_S(s_1)) = \top$. □

The rest of the section is devoted to define the notion of a closure system compatible with an arbitrary fuzzy equivalence relation (a reflexive, symmetric and transitive fuzzy relation) and the particular case of the so-called *kernel relation*.

Definition 7. *Let* $\mathbb{A} = (A, \rho_A)$ *be a fuzzy preordered set and let* $\sim$ *be a fuzzy equivalence relation on* A.

(i) *A closure operator* $c: A \to A$ *is said to be* compatible with the relation $\sim$ *if* $(a_1 \sim a_2) \leq \rho_A(c(a_1), c(a_2))$, *for all* $a_1, a_2 \in A$.

(ii) A closure system $S \subseteq A$ is said to be compatible *with $\sim$ if any closure operator associated to S is compatible with $\sim$.*

Lemma 2. *Let $\mathbb{A} = (A, \rho_A)$ be a fuzzy preordered set and a fuzzy equivalence relation $\sim$ on A. Then, a closure system S is compatible with $\sim$ if and only if*

$$\rho_A(a, s) \leq \bigwedge_{u \in A} ((a \sim u) \to \rho_A(u, s)) \tag{1}$$

for all $s \in S$ and $a \in A$.

Proof. Firstly, note that $\rho_A(a, s) \leq \bigwedge_{u \in A} ((a \sim u) \to \rho_A(u, s))$ for all $s \in S$ and $a \in A$ is equivalent to $\rho_A(a, s) \otimes (a \sim u) \leq \rho_A(u, s)$ for all $s \in S$ and $a, u \in A$, according to adjointness property of residuated lattices.

Assume that S is a closure system compatible with $\sim$ and consider $c \colon A \to A$ a closure operator associated to S, that is, $c(a) \in$ p-min$(a^\uparrow \cap S)$ for all $a \in A$ and $(a_1 \sim a_2) \leq \rho_A(c(a_1), c(a_2))$, for all $a_1, a_2 \in A$. Let $s \in S$ and $a, u \in A$, then, by Corollary 1, we have $\rho_A(a, s) \otimes (a \sim u) = \rho_A(c(a), s) \otimes (a \sim u)$. By compatibility and symmetry, $(a \sim u) = (u \sim a) \leq \rho_A(c(u), c(a))$ and by monotonicity of $\otimes$, we obtain $\rho_A(c(a), s) \otimes (a \sim u) \leq \rho_A(c(a), s) \otimes \rho_A(c(u), c(a)) \leq \rho_A(c(u), s)$. Applying that c is inflationary, $\rho_A(c(u), s) = \rho_A(u, c(u)) \otimes \rho_A(c(u), s) \leq \rho_A(u, s)$. Therefore, we obtain $\rho_A(a, s) \otimes (a \sim u) \leq \rho_A(u, s)$ for all $s \in S$ and $a, u \in A$.

Conversely, let S be a closure system such that $\rho_A(a, s) \otimes (a \sim u) \leq \rho_A(u, s)$ for all $s \in S$ and $a, u \in A$. Let us prove that any closure operator associated to S is compatible with $\sim$, that is, for $c \colon A \to A$ such that $c(a) \in$ p-min$(a^\uparrow \cap S)$ for all $a \in A$, one can obtain that $(a_1 \sim a_2) \leq \rho_A(c(a_1), c(a_2))$, for all $a_1, a_2 \in A$. In effect, $(a_1 \sim a_2) = (a_2 \sim a_1) = \rho_A(a_2, c(a_2)) \otimes (a_2 \sim a_1)$. By the hypothesis, and Corollary 1 (since $c(a_1) \in$ p-min$(a_1^\uparrow \cap S)$ and $c(a_2) \in S$) one obtains, $\rho_A(a_2, c(a_2)) \otimes (a_2 \sim a_1) \leq \rho_A(a_1, c(a_2)) = \rho_A(c(a_1), c(a_2))$. $\square$

Later, we will use the previous lemma on the particular case of the fuzzy equivalence relation $\sim$ being the kernel relation $\equiv_f$ associated to a crisp mapping f defined for $a_1, a_2 \in A$ as follows:

$$(a_1 \equiv_f a_2) = \begin{cases} \bot & \text{if } f(a_1) \neq f(a_2) \\ \top & \text{if } f(a_1) = f(a_2) \end{cases}$$

Corollary 2. *Let $\mathbb{A} = (A, \rho_A)$ be a fuzzy preordered set, consider a crisp mapping $f \colon A \to B$, and let $\equiv_f$ be the kernel relation associated to f. A closure system $S \subseteq A$ is compatible with the kernel relation if and only if $\rho_A(a, s) = \rho_A(u, s)$ for all $s \in S$ and $a, u \in A$ such that $f(a) = f(u)$.*

Proof. It is straightforward, due to the fact that if $f(a) = f(u)$ then $\top \to \rho_A(u, s) = \rho_A(u, s)$ and if $f(a) \neq f(u)$ then $\bot \to \rho_A(u, s) = \top$ for all $u \in A$. $\square$

Adjunctions Between Fuzzy Preordered Sets: Previous Results. It is well-known that there exists a close relationship between closure operators and adjunctions. In this section, we focus on this relation in the more general setting of fuzzy preordered sets.

To begin with, the definition of adjunction that we will use in this context is the natural generalization of the classical one.

Definition 8 (Fuzzy Adjunction Between $\mathbb{A}$ and $\mathbb{B}$.). *Consider two fuzzy preordered sets $\mathbb{A} = (A, \rho_A)$ and $\mathbb{B} = (B, \rho_B)$ and two mappings $f \colon A \to B$ and $g \colon B \to A$. The pair (f, g) forms an adjunction between A and B, denoted $(f, g) \colon \mathbb{A} \leftrightharpoons \mathbb{B}$ if $\rho_A(a, g(b)) = \rho_B(f(a), b)$ for all $a \in A$ and $b \in B$.*

As in the crisp case, it is possible to characterize a fuzzy adjunction in many different ways. Below we recall just one alternative option which will be used in the next section.

Theorem 3 [10]. *Consider fuzzy preordered sets $\mathbb{A} = (A, \rho_A)$ and $\mathbb{B} = (B, \rho_B)$, and two mappings $f \colon A \to B$ and $g \colon B \to A$. The following conditions are equivalent:*

1. $(f, g) \colon \mathbb{A} \leftrightharpoons \mathbb{B}$.
2. f and g are isotone, $g \circ f$ is inflationary, and $f \circ g$ is deflationary.

Given a mapping f from a fuzzy preordered set (A, ρ_A) to any set B, we aim to introduce conditions which allow for defining a fuzzy preordering on B and a mapping from B to A such that the pair (f, g) constitutes a fuzzy adjunction. In a previous work, [11], we solved this problem in the case of (A, ρ_A) also satisfies a certain kind of antisymmetry, that is $\rho_A(a_1, a_2) = \rho_A(a_2, a_1) = \top$ implies $a_1 = a_2$, for all $a_1, a_2 \in A$. The obtained characterization for the case of fuzzy posets is the following:

Theorem 4 [11]. *Given a fuzzy poset (A, ρ_A) and a mapping $f \colon A \longrightarrow B$, let $[a]_f$ be the equivalence class of an element $a \in A$ with respect to the kernel relation $a \equiv_f b \iff f(a) = f(b)$. Then, there exists a fuzzy order ρ_B in B and a map $g \colon B \longrightarrow A$ such that $A \leftrightharpoons B$ if and only if the following conditions hold:*

1. There exists $\max[a]_f$ for all $a \in A$.
2. $\rho_A(a_1, a_2) \leq \rho_A(\max[a_1]_f, \max[a_2]_f)$, for all $a_1, a_2 \in A$.

Next section provides a characterization of the existence of (fuzzy ordered structure in B and) adjunction associated to a mapping $f \colon (A, \rho_A) \to B$ which, as we will see, considerably differs from that obtained in the framework of fuzzy posets.

4   The Characterization Result

In order to address this problem, we will proceed by proving a number of preliminary results which will pave the way for the characterization.

As one would expect, the mere existence of the adjunction induces a closure system in A which, moreover, is compatible with the kernel relation associated to the mapping $\equiv_f$. Therefore, the existence of such a closure systems turns out to be a necessary condition for the problem under study.

Proposition 4. *Consider two fuzzy preordered sets* $\mathbb{A} = (A, \rho_A)$ *and* $\mathbb{B} = (B, \rho_B)$, *and two mappings* $f\colon A \to B$ *and* $g\colon B \to A$ *such that* $(f, g)\colon \mathbb{A} \leftrightarrows \mathbb{B}$ *forms a fuzzy adjunction. Then the set* $g(f(A))$ *is a closure system in* $\mathbb{A}$ *which is compatible with the kernel relation* $\equiv_f$.

Proof. Firstly, let us prove that $S = g(f(A))$ is a closure system, that is $g(f(a)) \in$ p-min$(a^\uparrow \cap S)$ for all $a \in A$.

By Theorem 2, it suffices to prove that $\rho_A(a, u) = \rho_A(g(f(a)), u)$ for all $u \in S$. On the one hand, since $g \circ f$ is inflationary, we have that $\rho_A(a, g(f(a))) = \top$, therefore

$$\rho_A(g(f(a)), u) = \rho_A(a, g(f(a))) \otimes \rho_A(g(f(a)), u) \leq \rho_A(a, u)$$

for all $u \in A$. On the other hand, for $u = g(f(x)) \in S$, we have also

$$\rho_A(a, u) = \rho_A(a, g(f(x))) = \rho_B(f(a), f(x))$$
$$\leq \rho_A(g(f(a)), g(f(x))) = \rho_A(g(f(a)), u).$$

Now, notice that for $x, a, u \in A$ such that $f(a) = f(u)$ and $s = g(f(x)) \in S$, we have the following chain of equalities

$$\rho_A(a, s) = \rho_A(a, g(f(x))) = \rho_B(f(a), f(x))$$
$$= \rho_B(f(u), f(x)) = \rho_A(u, g(f(x))) = \rho_A(u, s)$$

According to Corollary 2, the closure system S is compatible with the kernel relation. □

In order to actually build the fuzzy preorder on the codomain B, we will make use of a suitable preorder between subsets. The idea is to consider the Hoare preorder, which was introduced in the study of the semantics of non-determinism and its definition is the following:

Given two subsets X, Y of a crisp poset $(A, \leq)$, the subset X is said to be Hoare-smaller than Y, denoted $X \sqsubseteq_H Y$, if for every $x \in X$ there exists $y \in Y$ such that $x \leq y$.

If we consider a fuzzy preorder ρ_A instead of the crisp ordering $\leq$, it is possible to extend the notion of Hoare ordering to a fuzzy preorder between crisp subsets. Formally, the extension is given below:

Definition 9. *Let* (A, ρ_A) *be a fuzzy preordered set, and consider* C, D *crisp subsets of* A, *the fuzzy relation* $\sqsubseteq_H$ *is defined as*

$$(C \sqsubseteq_H D) = \bigwedge_{c \in C} \bigvee_{d \in D} \rho_A(c, d)$$

Proposition 5. *The relation* $\sqsubseteq_H$ *is a fuzzy preorder in the powerset of* A.

Proof. Reflexivity is obvious.

For transitivity, given $C, D, E \subseteq A$, we have to prove that

$$(C \sqsubseteq_H D) \otimes (D \sqsubseteq_H E) \leq (C \sqsubseteq_H E)$$

We will use the distributive properties of t-norms with respect to suprema and infima, together with the transitivity of ρ_A, in the following form

$$\bigvee_{d \in D} (\rho_A(c,d) \otimes \rho_A(d,e)) \leq \rho_A(c,e)$$

for all $c \in C$, $d \in D$, $e \in E$:

$$
\begin{aligned}
(C \sqsubseteq_H D) \otimes (D \sqsubseteq_H E) &= \bigwedge_{c \in C} \bigvee_{d \in D} \rho_A(c,d) \otimes \bigwedge_{d \in D} \bigvee_{e \in E} \rho_A(d,e) \\
&\leq \bigwedge_{c \in C} \left(\bigvee_{d \in D} \rho_A(c,d) \otimes \bigwedge_{d \in D} \bigvee_{e \in E} \rho_A(d,e) \right) \\
&= \bigwedge_{c \in C} \bigvee_{d \in D} \left(\rho_A(c,d) \otimes \bigwedge_{d \in D} \bigvee_{e \in E} \rho_A(d,e) \right) \\
&\leq \bigwedge_{c \in C} \bigvee_{d \in D} \left(\rho_A(c,d) \otimes \bigvee_{e \in E} \rho_A(d,e) \right) \\
&= \bigwedge_{c \in C} \bigvee_{d \in D} \bigvee_{e \in E} (\rho_A(c,d) \otimes \rho_A(d,e)) \\
&= \bigwedge_{c \in C} \bigvee_{e \in E} \bigvee_{d \in D} (\rho_A(c,d) \otimes \rho_A(d,e)) \\
&\leq \bigwedge_{c \in C} \bigvee_{e \in E} \rho_A(c,e) = (C \sqsubseteq_H E)
\end{aligned}
$$

$\square$

The fuzzy preorder defined above will be used in the construction of the fuzzy preorder on B needed in order to define the right adjoint g. It is remarkable that $\sqsubseteq_H$ will be used just on (crisp) subsets $X \subseteq A$ with a particular property; namely, for all $x_1, x_2 \in X$ we have $\rho_A(x_1, x_2) = \top$. We will say a subset to be *cyclic* if it satisfies the previous property.

The following lemma states that, for the specific case of this kind of sets, the fuzzy relation $\sqsubseteq_H$ can be very easily computed.

Lemma 3. *Consider a fuzzy preordered set (A, ρ_A), and X, Y two crisp cyclic subsets of A, then $X \sqsubseteq_H Y = \rho_A(x, y)$ for any $x \in X$ and $y \in Y$.*

Proof. It is enough to show that $\rho_A(x_1, y_1) = \rho_A(x_2, y_2)$ for all $x_1, x_2 \in X$, $y_1, y_2 \in Y$. Indeed,

$$\rho_A(x_1, y_1) \geq \rho_A(x_1, x_2) \otimes \rho_A(x_2, y_1)$$

$$= \top \otimes \rho_A(x_2, y_1) \geq \rho_A(x_2, y_2) \otimes \rho_A(y_2, y_1) = \rho_A(x_2, y_2)$$

Analogously, $\rho_A(x_2, y_2) \geq \rho_A(x_1, y_1)$. $\square$

The result below actually builds a fuzzy preorder relation on B; in the definition the previous lemma is used for the particular case of the sets of p-minima of a fuzzy subset, which turn out to be cyclic (this is just a straightforward consequence of the definition).

Lemma 4. *Let $\mathbb{A} = (A, \rho_A)$ be a fuzzy preordered set together with a mapping $f \colon A \to B$ and $S \subseteq A$ a closure system compatible with the kernel relation $\equiv_f$. For every $a_0 \in A$, the fuzzy relation $\rho_B^{a_0} \colon B \times B \to L$ defined as*

$$\rho_B^{a_0}(b_1, b_2) = \left((\text{p-min}(a_1^\uparrow \cap S)) \sqsubseteq_H (\text{p-min}(a_2^\uparrow \cap S)) \right)$$

where $a_i \in f^{-1}(b_i)$ if $f^{-1}(b_i) \neq \varnothing$ and $a_i = a_0$ otherwise, for each $i \in \{1, 2\}$, is a fuzzy preordering on B.

Proof. The definition does not depend on the choice of preimages, because given $x, y \in f^{-1}(b)$ we have $\text{p-min}(x^\uparrow \cap S) = \text{p-min}(y^\uparrow \cap S)$. In effect, consider $s \in \text{p-min}(x^\uparrow \cap S)$ then $\rho_A(x, u) = \rho_A(s, u)$, for all $u \in S$. Furthermore, as $f(x) = f(y)$ and S is compatible with the kernel relation then, by Corollary 2, $\rho_A(x, u) = \rho_A(y, u)$ for all $u \in S$. As a consequence, we have that $\rho_A(y, u) = \rho_A(s, u)$ and, hence, $s \in \text{p-min}(y^\uparrow \cap S)$. As the roles of x and y in the previous argument are interchangeable this proves the equality of both sets.

By Lemma 3, we have that

$$\rho_B^{a_0}(b_1, b_2) = \rho_A(x, y)$$

for any $x \in \text{p-min}(a_1^\uparrow \cap S)$ and $y \in \text{p-min}(a_2^\uparrow \cap S)$ and it has been proved that the value is independent from the choice of x and y.

From the reflexivity of ρ_A, it is straightforward that $\rho_B^{a_0}$ is reflexive as $\rho_B^{a_0}(b, b) = \rho_A(x, x) = \top$ for $x \in \text{p-min}(a^\uparrow \cap S))$ where $a \in f^{-1}(b)$ if $f^{-1}(b) \neq \varnothing$ and $a = a_0$ otherwise.

Similarly, notice that

$$\rho_B^{a_0}(b_1, b_2) \otimes \rho_B^{a_0}(b_2, b_3) = \rho_A(x, y) \otimes \rho_A(y, z) \leq \rho_A(x, z) = \rho_B^{a_0}(b_1, b_3)$$

for $x \in \text{p-min}(a_1^\uparrow \cap S)), y \in \text{p-min}(a_2^\uparrow \cap S))$ and $z \in \text{p-min}(a_3^\uparrow \cap S))$ where $a_i \in f^{-1}(b_i)$ if $f^{-1}(b_i) \neq \varnothing$ and $a_i = a_0$ otherwise, for each $i \in \{1, 2, 3\}$. This implies that $\rho_B^{a_0}$ is transitive. $\square$

We can now focus on the definition of suitable mappings $g \colon B \to A$ such that (f, g) forms an adjoint pair.

Proposition 6. *Let $\mathbb{A} = (A, \rho_A)$ be a fuzzy preordered set, $f \colon A \to B$ be a mapping and $S \subseteq A$ be a closure system compatible with the kernel relation $\equiv_f$. Then, there exists a fuzzy preordering ρ_B on B and a mapping $g \colon B \to A$ such that (f, g) constitutes a fuzzy adjunction.*

Proof. Firstly, the existence of a fuzzy preordering on B is given by Lemma 4 and observe that this fuzzy relation is not unique, as it depends on an element $a_0 \in A$. Assume that the element $a_0 \in A$ has been fixed and consider the fuzzy preordering $\rho_B^{a_0}$ defined on B. There is a number of suitable definitions of $g \colon B \to A$, and all of them can be specified as follows:

(C1) If $b \in f(A)$, then $g(b) \in \text{p-min}(x_b^\uparrow \cap S)$ for some $x_b \in f^{-1}(b)$.
(C2) If $b \notin f(A)$, then $g(b) \in \text{p-min}(a_0^\uparrow \cap S)$.

The existence of g is clear by the axiom of choice, since for all $b \in f(A)$, the sets $f^{-1}(b)$ are nonempty (so x_b can be chosen for all $b \in f(A)$) and, moreover, $\text{p-min}(x_b^\uparrow \cap S)$ and $\text{p-min}(a_0^\uparrow \cap S)$ are nonempty as well because S is a closure system.

Now, consider any g satisfying the conditions $(C1)$ and $(C2)$ above. Let us prove that g is a right adjoint to f, that is, $\rho_B^{a_0}\big(f(a), b\big) = \rho_A\big(a, g(b)\big)$ for all $a \in A$ and $b \in B$.

By definition of $\rho_B^{a_0}$ (see Lemma 4), we have that

$$\rho_B^{a_0}(f(a), b) = \Big((\text{p-min}(a^\uparrow \cap S)) \sqsubseteq_H (\text{p-min}(w^\uparrow \cap S))\Big)$$

where w satisfies either $w \in f^{-1}(b)$ if $b \in f(A)$ (therefore, we can choose w to be x_b, see (C1) above) or, otherwise, $w = a_0$. According to Lemma 3 and the definition of g, we have

$$\rho_B^{a_0}(f(a), b) = \rho_A(x, g(b)) \quad \text{for any } x \in \text{p-min}(a^\uparrow \cap S)$$

The proof will be finished if we show that $\rho_A(x, g(b)) = \rho_A(a, g(b))$ for any $x \in \text{p-min}(a^\uparrow \cap S)$. By Corollary 1 for $x \in \text{p-min}(a^\uparrow \cap S)$ we have $\rho_A(a, u) = \rho_A(x, u)$ for all $u \in S$. Both, in case (C1) and in case (C2), we have $g(b) \in S$ hence, $\rho_A(x, g(b)) = \rho_A(a, g(b))$. $\qquad\qquad\square$

The previous results can be now summarized in the theorem below:

Theorem 5. *Consider a fuzzy p reordered set $\mathbb{A} = (A, \rho_A)$ and a map $f \colon A \to B$. There exists a fuzzy preordering ρ_B on B and a map $g \colon B \to A$ such that (f, g) constitutes a fuzzy adjunction if and only if there exists $S \subseteq A$ a closure system compatible with the kernel relation $\equiv_f$.*

5   Conclusions and Future Work

We have characterized the situations for the existence of right adjunction for a mapping $f \colon (A, \rho_A) \to B$ where (A, ρ_A) is a preordered set and B is an unstructured codomain, in terms of closure systems.

When focusing on fuzzy extensions of order relations one can find some interesting developments on the study of both fuzzy partial orders and fuzzy preorders, see [4,5] for instance. In these works, it is noticed that the versions of antisymmetry and reflexivity commonly used are too strong and, as a consequence, the resulting fuzzy partial orders are very close to the classical case. Accordingly, one interesting line of future work will be the adaptation of the current results to these alternative weaker definitions.

Another source of future work could be the definition of alternative interpretations of the notion of adjunction between multivalued functions (i.e., relations) both in crisp and fuzzy frameworks, with the aim of building a right adjoint for a given multivalued function.

References

1. Bělohlávek, R.: Fuzzy Galois connections. Math. Logic Q. **45**(4), 497–504 (1999)
2. Bělohlávek, R., Osička, P.: Triadic fuzzy Galois connections as ordinary connections. Fuzzy Sets Syst. **249**, 83–99 (2014)
3. Bělohlávek, R.: Fuzzy closure operators. J. Math. Anal. Appl. **262**, 473–489 (2001)
4. Bodenhofer, U.: A similarity-based generalization of fuzzy orderings preserving the classical axioms. Int. J. Uncertainty Fuzziness Knowl. Based Syst. **8**(5), 593–610 (2000)
5. Bodenhofer, U., De Baets, B., Fodor, J.: A compendium of fuzzy weak orders: representations and constructions. Fuzzy Sets Syst. **158**(8), 811–829 (2007)
6. Davey, B., Priestley, H.: Introduction to Lattices and Order, 2nd edn. Cambridge University Press, Cambridge (2002)
7. Dikranjan, D., Tholen, W.: Categorical Structure of Closure Operators. Kluwer Academic Publishers, Dordrecht (1995)
8. Fodor, J., Roubens, M.: Fuzzy Preference Modelling and Multicriteria Decision Support. Kluwer Academic Publishers, Dordrecht (1994)
9. Frascella, A.: Fuzzy Galois connections under weak conditions. Fuzzy Sets Syst. **172**(1), 33–50 (2011)
10. García-Pardo, F., Cabrera, I.P., Cordero, P., Ojeda-Aciego, M.: On Galois connections and soft computing. In: Rojas, I., Joya, G., Cabestany, J. (eds.) IWANN 2013, Part II. LNCS, vol. 7903, pp. 224–235. Springer, Heidelberg (2013)
11. García-Pardo, F., Cabrera, I.P., Cordero, P., Ojeda-Aciego, M.: On the construction of fuzzy Galois connections. In: Proceedings of XVII Spanish Conference on Fuzzy Logic and Technology, pp. 99–102 (2014)
12. García-Pardo, F., Cabrera, I.P., Cordero, P., Ojeda-Aciego, M., Rodríguez, F.J.: On the definition of suitable orderings to generate adjunctions over an unstructured codomain. Inf. Sci. **286**, 173–187 (2014)
13. Gierz, G., et al.: Continuous Lattices and Domains. Cambridge University Press, Cambridge (2003)
14. Guo, L., Zhang, G.-Q., Li, Q.: Fuzzy closure systems on L-ordered sets. Math. Logic Q. **57**(3), 281–291 (2011)
15. Konecny, J.: Isotone fuzzy Galois connections with hedges. Inf. Sci. **181**(10), 1804–1817 (2011)
16. Shi, Y., Nachtegael, M., Ruan, D., Kerre, E.: Fuzzy adjunctions and fuzzy morphological operations based on implications. Int. J. Intell. Syst. **24**(12), 1280–1296 (2009)
17. Yao, W., Lu, L.-X.: Fuzzy Galois connections on fuzzy posets. Math. Logic Q. **55**(1), 105–112 (2009)

Extensional Confluences and Local Closure Operators

Henry Soldano[1,2]([⊠])

[1] L.I.P.N UMR-CNRS 7030, Université Paris 13,
Sorbonne Paris Cité, 93430 Villetaneuse, France
henry.soldano@lipn.univ-paris13.fr
[2] Atelier de BioInformatique, Université Pierre et Marie Curie, 75005 Paris, France

Abstract. This work is motivated by knowledge discovery in attributed graphs. Our approach consists in extending the methodology of frequent closed pattern mining, as developed in Formal Concept Analysis (FCA), to the case where the objects in which attribute patterns may occur are the vertices of a graph, typically representing a social network. For that purpose we extend the framework of abstract concept lattices, in which the extensional space is a pointed join-subsemilattice of the powerset X of the object set, by considering as the extensional space a weaker structure called a *confluence* of X. Confluences were recently investigated as intensional spaces in FCA. In this article we show that when the intensional space is a lattice L and the extensional space is a confluence F of X, that leads to a set of closure operators, called local closure operators, whose union form the set of intensions of F. We investigate the structure of the set of (extension,intension) pairs, i.e. the set of local concepts built on (L, F) and related local implications. As an example, we consider the detection of all frequent k-communities in an attributed network.

1 Introduction

A way recently proposed to search for frequent closed patterns in attributed graphs is to define a restricted extensional space, i.e. a lattice obtained by applying a graph abstraction operator to vertex subsets. The idea of such an operator is to minimally reduce a vertex subset until the reduced vertex subset satisfies some connectivity property within the corresponding induced subgraph [1]. This approach, based on a previous work on abstraction in Formal Concept Analysis [2] extracts attribute patterns whose support vertex sets induce subgraphs made of dense parts[1]. However there are also recent works in data mining that are interested in local patterns made of a constraint on a subset of attributes together with a density constraint on a vertex subset, and this using various notions of maximality [3,4]. We are interested here in defining *local closed patterns* corresponding to maximal attribute patterns each associated to one dense

---

[1] In data mining the *support set* of a pattern is the extension of this pattern in a set of objects

© Springer International Publishing Switzerland 2015
J. Baixeries et al. (Eds.): ICFCA 2015, LNAI 9113, pp. 128–144, 2015.
DOI: 10.1007/978-3-319-19545-2_8

subgraph, allowing to extract knowledge, and implication rules, particular to specific dense groups of objects. For that purpose we propose to extend Formal Concept Analysis (FCA) to take into account this notion of locality.

In the framework we propose, several closure operators may be applied to the same pattern: a closed pattern will then be local as the closure will depend on which region of the extensional space is concerned. The simplest example is an extension made of various connected components, each leading to a local closed pattern.

Formally, the dense vertex subsets we consider form a partial order, included in the powerset of the vertex set. Such a partial order, called a confluence in a recent investigation in Formal Concept Analysis [5][2] is close to the notion of a *confluent familiy* recently investigated in [6] and that have been shown, with a mild restriction, as the structure of a family of itemsets in which a closure operator can be defined with respect to any object set. The structure of the set of closed patterns, when the language is a confluence, has been shown in [5] to be a more general structure called a pre-confluence, thus leading to define intensional Galois pre-confluences.

We show here that whenever we consider the symmetric case where the extensional space, i.e. the set of dense vertex subsets allowed, is a confluence of the powerset of the object set O, and the pattern language a lattice L, there exists a set of local closure operators, one for each minimal element of F such that the set of intensions int$[F]$ is the union of the corresponding closed patterns. We also show that there is a unique closure operator h on the extensional confluence. As a whole, we obtain a set of pairs (e, l) where e is a closed element of the extensional space F and l is closed with respect to a local closure operator. This set of *local* concepts, ordered following the partial order on F, is a pre-confluence, which is called an extensional Galois pre-confluence.

This leads to define also a set of local implications, written $\Box_e c \to \Box_e l$ or equivalently $\Box_m c \to \Box_m l$ where m is any minimal element of F such that $m \le e$. Such a local implication means that the local extension e of the closed pattern c is the same as the local extension of l, the corresponding local closed pattern.

As a direct application of these results we can compute extensional Galois pre-confluences where F is the set of vertex subsets inducing connected subgraphs of an attributed graph $G = (O, E)$. However we can enlarge our purpose as follows: we can use this methodology by considering a derived graph $G_T = (T, E_T)$ where T is a family of subsets of the object set O and where each new vertex $t \in T$ is labelled by the most specific pattern common to the vertices of t in 2^O. When considering T as the set of k-cliques of an attributed graph $G = (O, E)$, this allows to enumerate the k-communities [7], of size at least s, in all subgraphs induced by extensions of patterns in $G = (O, E)$, as we will see in Sect. 4. We then also obtain local implications, each indexed by some k-clique, such as $\Box_{t_1} a_1 \to \Box_{t_1} a_3 a_7$ stating that the members of any k-clique belonging, in the extension of attribute a_1, to the same k-community as t_1, also

---

[2] In that article, these structures were called confluence's. We use here a more standard terminology.

share attributes a_3 and a_7. This way, we gather some knowledge which is *local* as it is only valid in a vicinity of particular vertices. In such a local rule, the left part is a closed pattern c whose extension, in the k-clique graph, is split into several connected components, each corresponding to a k-community in the original graph, and the right part is the local closed pattern representing the most specific pattern shared by all the members of this k-community.

In Sect. 2 we recall the basic definition and properties related to closure operators and confluences. In Sect. 3 we define and investigate the set of closure operators $\{f_m\}$ relating a pattern lattice to a confluence $F \subseteq 2^O$, where m belongs to the set $\min[F]$ of minimal elements of F. In Sect. 4 we recall that given a graph $G_T = (T, E_T)$, the family of vertex subsets inducing connected subgraphs of G_T is a confluence of 2^T and consider the case in which T is a set of subsets of the vertices of some attributed graph and investigate the k-communities of subgraphs induced by extensions of patterns. Section 5 briefly suggests an efficient algorithm to compute frequent local closed patterns and related local implication rules.

2   Closure Operators in Pre-confluences and Confluences

We first recall standard definitions and results from FCA about closure operators in lattice, using the formulation of the T.S. Blyth's book [8] and then definitions and results concerning a structure weaker than a lattice and called a pre-confluence, from a recent work in Formal Concept Analysis [5]. Our purpose is here to extend the standard result about the range of closure operators in lattices to pre-confluences, in order to further present, in Sect. 3, the main contribution of this article, i.e. the definition and investigation of local concepts pre-confluences as a generalization of extensional abstract concept lattices [2].

All ordered sets considered here are finite. All lattices we consider are then *bounded lattices*, the top element $\top$ is the meet of the empty subset and the join of the element set, while the bottom element $\bot$ is the meet of the element set and the join of the empty subset. We will also further need *topped* $\wedge$-*subsemilattices of a lattice* X, i.e. bounded lattices with same meet and same greatest element as X. Dually, we consider *pointed* $\vee$-*subsemilattices* of a lattice X, i.e. bounded lattices with same join and same bottom element as X. A *pointed* $\vee$-*subsemilattices* is also further called, following [2], an *abstraction*. All along the article the set of upper bounds of some element x in an ordered set E is denoted by the up set $E^x = \{y \mid y \geq x\}$. In the same way, the set of lower bounds of x is denoted by the down set $E_x = \{y \mid y \leq x\}$. Closure operators and dual closure operators are defined as follows:

Definition 1. *Let E be an ordered set and $f : E \to E$ a self map such that for any $x, y \in E$, f is monotone, i.e. $x \leq y \implies f(x) \leq f(y)$ and idempotent, i.e. $f(f(x)) = f(x)$, then:*

- *if f is extensive, i.e. $f(x) \geq x$, f is called a closure operator*
- *if f is intensive, i.e. $f(x) \leq x$, f is called a dual closure or an interior operator, or also a projection.*

In the first case, an element such that $x = f(x)$ is called a closed element.

A *closure subset of an ordered set* E is defined as the range $f[E]$ of a closure operator on E while a *dual closure subset* is the range $p[E]$ of an interior operator. We then obtain the following well known result when the ordered set is a lattice [8]:

Proposition 1. *Let X be a lattice. A subset C of X is a closure subset if and only if C is closed under meet. The closure $f : X \to X$ is then unique and defined as $f(x) = \bigwedge_{\{c \in C \cap X^x\}} c$.*

By "C is closed under meet" we intend here that the meet of any subset c, including the empty subset $\emptyset$, belongs to C. Therefore $\top = \bigwedge_{\emptyset} c$ belongs to C, and C is a topped $\wedge$-subsemilattice of X. We will also further need the dual proposition which states that a subset A of X is a *dual closure subset* whenever A is closed under joins i.e. A is a pointed $\vee$-subsemilattices of X, i.e. an abstraction of X. The associated interior operator $p : X \to X$ is then defined as $p(x) = \bigvee_{\{a \in A \cap X_x\}} a$. In particular when X is a powerset 2^K, $p(x) = \bigcup_{\{a \in A | a \subseteq x\}} a$.

We are interested now in pre-confluences which are structures weaker than lattices introduced in [5], and recall a theorem extending Proposition 1.

Definition 2. *Let F be a finite ordered set such that for any $t \in F$, F^t is a lattice. F is called a pre-confluence, $x \wedge_t y$ is a local infimum or local meet, and $\top_t$ a local top.*

The two following Lemmas allows a better understanding of what is a preconfluence, emphasizing, first, that there exists a partial join operator $x \vee_F y$ on F, and then, that we only need minimal elements of F to characterize a pre-confluence.

Lemma 1. *Let F be a pre-confluence, then for any t in F and $x, y \in F^t$*

1. F^t has as join, denoted by $x \vee_F y$, the least element of $F^x \cap F^y$
2. Let $t' \geq t$ then $F^{t'}$ is a sublattice of F^t.

Lemma 2. *F is a pre-confluence if and only for any $m \in \min(F)$, F^m is a lattice.*

Clearly, according to the latter Lemma, a pre-confluence F with a minimum element $\perp$ is a lattice as $F^\perp = F$. We define hereunder what means for a subset of a pre-confluence to be "closed under local meet", and the resulting fundamental theorem:

Definition 3. *A subset C of a pre-confluence F is said closed under local meet whenever for any element t and any $C' \subseteq C \cap F^t$ we have*

$$\bigwedge_t {}_{\{c \in C'\}} c \quad belongs \ to \ C.$$

Theorem 1. *Let F be a pre-confluence. A subset C of F is a closure subset if and only if C is closed under local meet. The closure $f : F \to F$ is then defined as $f(x) = \wedge_{t\{c \in C \cap F^x\}} c$ and $C = f[F]$ is a pre-confluence.*

Whenever F is a lattice, "closed under local meet", as defined in Definition 3, simply means "closed under meet" and Theorem 1 comes down to Proposition 1.

It was previously shown that when restricting the standard extensional space 2^O, where O is the set of objects, to a subset which is an abstraction, we generalize concept lattices and obtain *abstract concept lattices* [2]. In order to generalize abstractions of a lattice, and later on generalize concept lattices and abstract concept lattices to local concept pre-confluences, we need to consider particular pre-confluences F, part of some host lattice X and sharing the same join operator.

Definition 4. *Let X be a lattice and $F \subseteq X$ a pre-confluence with as join $\vee_F = \vee$, F is called a confluence of X.*

Again it should be clear that confluences of a lattice X generalize abstractions of X: an abstraction of X is a confluence of X containing the minimum of X. We have the following characterization of confluences which is close to, but differs from, the definition of a confluent family given by M. Boley and co-authors [6]:

Proposition 2. *Let X be a lattice and $F \subseteq X$, F is a confluence of X if and only if for any x, y, t in F with $x \geq t$ and $y \geq t$, we have that $x \vee y$ belongs to F.*

The following Lemmas show that a confluence is associated to a set of interior operators on its host lattice and that we only need the interior operators associated to the minimal elements of F to represent all interior operators.

Lemma 3. *Let F be a confluence of X, the mapping $p_t : X^t \to X^t$ defined by $p_t(x) = \vee_{q \in F^t \cap X_x} q$ is an interior operator on X^t and $p_t[X^t] = F^t$.*

Lemma 4. *Let F is a confluence of X, then if $q \leq t$, and $x \in X^t$, then $p_t(x) = p_q(x)$.*

Again, by considering abstractions of X as confluences containing the minimum of X, we find a previous result, here the dual of Proposition 1 that characterizes abstractions as ranges of dual closure operators on lattices: whenever the confluence F is an abstraction, there is only one interior operator to consider, namely $p_\perp$ which is such that $p_\perp[X] = F$.

In what follows, we consider confluences of some powerset of objects 2^O, so generalizing extensional abstractions into extensional confluences. More precisely, we are interested in reducing the extension $ext(t)$ of some pattern t in such a way that the reduced extension belong to a confluence F. In the example that follows, and that will be the core of Sect. 4, we will define a confluence by considering a non-directed graph $G = (O, E)$.

Example 1. Let $O = \{1, 2, 3, 4\}$, $G = (O, E)$ be a graph whose vertex set is O and edge set is E. Let $F \subseteq 2^O$ be the set of vertex subsets inducing connected subgraphs of G. F is a confluence whose set of minimal elements is $M = \{\{1\}, \{2\}, \{3\}, \{4\}\}$, i.e. the set of singletons of 2^O. The union of two vertex subsets each inducing a connected subgraph of G that contains a given vertex s is a vertex subset obviously inducing a connected subgraph of G: s connects the two subgraphs, and therefore F is a confluence of 2^O. The projection $p_{\{s\}}$ projects then any vertex subset S containing s on the connected component of G_S containing s. The up set $F^{\{s\}}$ is then the set of vertex subsets inducing connected subgraphs containing s and the union of all these $F^{\{s\}}$ represents the whole set of connected subgraphs of G. For sake of simplicity we will further write singletons $\{s\}$ as s and subsets as words, as for instance 123. The subset $F^{1+3} = F^1 \cup F^3$ representing vertex subsets inducing connected subgraphs containing vertices 1 or 3 is also a confluence. Figure 1 displays the diagram of F^{1+3}. □

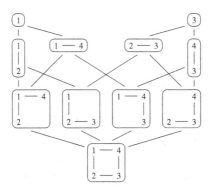

Fig. 1. The Hass diagram of a family F of connected subgraphs each generated by a vertex subset of the original graph whose vertex subset is $\{1, 2, 3, 4\}$ and whose edges form the square $\{12, 23, 34, 14\}$. We only display here the part $F^{1+3} = F^1 \cup F^3$ of F, which also has the confluence structure. Note that the pair $(123, 143)$ has two lower bounds, one in F^1 (namely 1) and the other in F^3 (namely 3). F^1 and F^3 both are lattices but F^{1+3} is a confluence.

3   Local Closures

In this section we will extend the concept lattice to the notion of local concept pre-confluence by restricting the extensional space 2^O to a pre-confluence of 2^O. First we recall a general result from [2, 9] that states that applying an interior operator to a lattice involved in a Galois connection preserves the connection:

Proposition 3. *Let X and L be two lattices, (int, ext) be a Galois connection on (X, L) and p be an interior operator on X, and $A = p[X]$, we have that $(\text{int}, p \circ \text{ext})$ is a Galois connection on (A, L).*

Corollary 1. *1. $f = \text{int} \circ p \circ \text{ext}$ is a closure operator on L and the set of closed elements $f[L] = \text{int}[A]$ is a topped $\wedge$-subsemilattice of L.*

2. $h = p \circ \text{ext} \circ \text{int}$ is a closure operator on A and the set of closed elements and $h[A] = p \circ \text{ext}[L]$ is a topped $\wedge$-subsemilattice of A.

3. $h[A]$ and $f[L]$ are two anti-isomorphic lattices and the lattice of the (e, c) pairs where $c = \text{int}(e)$ and $e = p \circ \text{ext}(c)$ form a Galois lattice, ordered following A, isomorphic to $h[A]$.

Note that the roles of X and L can be exchanged and, as a consequence, we can as well rephrase Proposition 3 and Corollary 1 using an interior operator on L. Regarding extensional abstractions, the extensional lattice X, on which the interior operator is applied, is a powerset 2^O of objects, the intensional map int is an intersection operator on the pattern language L, and the extensional map ext returns the whole set of objects whose description in L is greater than or equal to its argument:

Proposition 4. *Let $X = 2^O$ be a powerset of objects, L be a lattice and (int, ext) be the Galois connection on (X, L) associated to the description function $d : O \rightarrow L$:*

$$\text{int}(e) = \bigwedge_{o \in e} d(o)$$

$$\text{ext}(q) = \{o \in O \mid q \subseteq d(o)\}$$

then, let p be an interior operator on X, we have that $(\text{int}, p \circ \text{ext})$ is a Galois connection on $(p[X], L)$ and $\text{int} \circ p \circ \text{ext}$ is a closure operator on L.

This proposition allows to define abstract concept lattices, where the interior operator applies to the extensional space [2,9]. By rather applying the interior operator to the pattern lattice, we obtain projected pattern structures as proposed in [10].

3.1   Local Concept Pre-confluence

As we are interested in extensional confluences, we need to consider up sets of the confluence, which are lattices, and restrict our interest to the relevant parts of the pattern language. We first note hereunder that Proposition 4 holds when replacing $X = 2^O$ by any of its up set X^e as far as we consider only elements t of L such that $e \subseteq \text{ext}(t)$. In what follows we still note int and ext their restrictions to part of their domain.

Proposition 5. *Let e be an element of X, $L_{int(e)}$ be the down set of L whose maximum is $\text{int}(e)$ and X^e be the upset of X whose minimum is e, then the restrictions of int and ext to respectively X^e and $L_{int(e)}$ are such that*

$$(\text{int}, \text{ext}) \text{ define a Galois connection on } (X^e, L_{int(e)})$$

Proof. As int and ext define a Galois connection on X, L we know that they are anti monotonic and therefore:

- For any $y \in X^e$, $int(y) \leq int(e)$ and as a consequence $int(y) \in L_{int(e)}$, i.e. $int[X^e] \subseteq L_{int(e)}$
- For any $t \in L_{int(e)}$, $ext(t) \supseteq ext(int(e)) \supseteq e$ (as ext $\circ$ int is a closure operator) and as a consequence $ext(t) \in X^e$, i.e. $ext[L_{int(e)}] \subseteq X^e$.

This means that the restrictions of these functions have domains and co-domains as follows: $int : X^e \rightarrow L_{int(e)}$ and $ext : L_{int(e)} \rightarrow X^e$, and as they inherit from the properties of (int, ext) on (X, L), they also define a Galois connection on $(X^e, L_{int(e)})$. □

Now, recall that in a confluence F of a lattice X, to each element e in F is associated an interior operator p_e on X^e such that $p_e[X^e] = F^e$ (see Lemma 3 and Proposition 2). This leads to the following corollary of Proposition 5:

Corollary 2. *Let $X = 2^O$ be a powerset of objects, F be a confluence on X, e an element of F and p_e the corresponding interior operator on X^e, then*

1. $(int, p_e \circ ext)$ *define a Galois connection on* $(F^e, L_{int(e)})$.
2. $f_e = int \circ p_e \circ ext$ *is a closure operator on* $L_{int(e)}$ *and* $f_e[L_{int(e)}] = int[F^e]$ *is a topped $\wedge$-subsemilattice of* $L_{int(e)}$.
3. $h_e = p_e \circ ext \circ int$ *is a closure operator on* F^e *and* $h_e[F^e] = p_e \circ ext[L_{int(e)}]$ *is a topped $\wedge$-subsemilattice of* F^e.

Proof. From Proposition 5 we know that the restrictions of int and ext define a Galois connection on $(X^e, L_{int(e)})$. From Lemma 3 and Proposition 2 we know that for any element x of X^e and element e of F, the map p_e defined as $p_e(x) = \vee_{q \in F^e \cap X_x} q$ is an interior operator on X^e. To conclude we just need the general result presented in Proposition 3. □

f_e is a called a *local closure operator* with respect to e.

Example 2. We consider here the confluence F^{1+3} described in Example 1 and whose diagram is displayed Fig. 1. The pattern language will be $L = 2^{\{a,b,c,d,e\}}$ that we will simply write $L = 2^{abcde}$ again representing subsets as words. The corresponding context is described Table 1.

Consider the local closure operator $f_1 = int \circ p_1 \circ ext$ associated to the vertex subset $\{1\}$ simply noted 1. We have $X = O^{1234}$ and X^1 is $\{1\} \times 2^{234}$. $F^1 = p_1[X^1]$

Table 1. The context representing the descriptions in 2^{abcde} of the objects in confluence F^{1+3} described Example 1.

O/K	a	b	c	d	e
1	1	1	1	1	0
2	1	1	0	0	0
3	1	1	1	0	0
4	1	0	1	0	1

is then the abstraction $\{1, 12, 14, 123, 124, 134, 1234\}$. We have $\text{int}(1) = abcd$ and therefore the pattern language is restricted to 2^{abcd}, ensuring that $p_1 : X^1 \to X^1$ will be applied to some $\text{ext}(q)$ where q is less than or equal to $abcd$ and therefore such that $\text{ext}(q) \supseteq \{1\}$, i.e. does belong to X^1. Let us then consider the pattern abc, we have $\text{ext}(abc) = 13$ and $p_1(13) = 1$, i.e. the greatest element of F^1 included in 13. As a consequence $f_1(abc) = \text{int}(1) = abcd$. □

In the following result we show that the set $\text{int}[F]$ of intensions of an extensional confluence F is obtained by joining the ranges of these local closure operators:

Theorem 2. *Let F be a confluence of $X = 2^O$, L be a lattice and (int, ext) be the Galois connection on (X, L), then we have that:*

$$\text{int}[F] = \bigcup_{e \in F} f_e[L_{int(e)}] = \bigcup_{m \in \min(F)} f_m[L_{int(m)}]$$

Proof. Regarding the first equality, from Proposition 3 and considering p as the identity function, we deduce that the right part of the equality rewrites as $\bigcup_{e \in F} \text{int}[F^e]$. Furthermore, as F may be rewritten as the union of its up sets, we have that $F = \bigcup_{e \in F} F^e$. By applying the intensional function to both sides we obtain $\text{int}[F] = \text{int}[\bigcup_{e \in F} F^e]$. As $\text{int}[F]$ is the image of F by int, it is straightforward that the image of the union of subsets is the union of the image of these subsets, and therefore that $\text{int}[F] = \bigcup_{e \in F} \text{int}[F^e]$.

The second equality states that we only need the closure operators associated to the minimal elements of F. From Theorem 2 we have $\text{int}[F] = C = \bigcup_{e \in F)} f_e[L_{int(e)}]$. Let $C' = \bigcup_{m \in \min(F)} f_m[L_{int}(m)]$, as $\min[F] \subseteq F$ we clearly have $C' \subseteq C$. We have then to show that any element in C may be rewritten as an element of C'. Let then c be an element of C, this means that there exists $e \in F$ and $l \in L_{int(e)}$ such that $c = f_e(l)$. First we note that there necessarily exists $m \in \min[F]$ such that $e \geq m$ and also that because int is anti monotonic we have $\text{int}(m) \geq \text{int}(e)$ and therefore, whenever l belongs to $L_{int}(e)$ it also belongs to $L_{int}(m)$. Now, let $z = \text{ext}(l)$, recall that $z \in X^e$, since we have to apply p_e to z to build $f_e(l)$) and we have seen that $e \geq m$. From Lemma 4, we can then deduce that $p_e(\text{ext}(l)) = p_m(\text{ext}(l))$ and therefore $c = \text{int} \circ p_m \circ \text{ext}(l)$ with $l \in L_{int}(m)$. This means that c belongs to C'. Overall we have shown that $C' = C = \text{int}[F]$ □

This generalizes Proposition 3: we now have that the union of local closed elements of L with respect to a confluence F of X is the range of F under the intensional operator int. Again we only need the set of minimal elements $\min(F)$. Moreover, we have a stronger structure on the extensional space on which we may define a unique closure operator:

Theorem 3. *Let F be a confluence on $X = 2^O$, L be a lattice and (int, ext) be the Galois connection on (X, L), and $h : F \to F$ defined as*

$$\forall x \in F, h(x) = h_x(x)$$

is a closure operator on F and $E = h[F]$ is a pre-confluence.

Proof. We first show that h is a closure operator, i.e. is extensive, monotone and idempotent. Let x, y be elements of F.

- $h(x) \geq x$?

$h(x) = h_x(x)$ and as h_x is a closure operator we have that $h(x) = h_x(x) \geq x$.
- $x \geq y \Rightarrow h(x) \geq h(y)$?

Let $z = \text{ext} \circ \text{int}(x)$, then z belongs to the upset X^x and as $x \geq y$ this means following Lemma 4 that $p_x(z) = p_y(z)$ and therefore $h(x) = p_y(\text{ext} \circ \text{int}(x))$. Now as $\text{ext} \circ \text{int}$ is a closure operator it is monotone and therefore $\text{ext} \circ \text{int}(x) \geq \text{ext} \circ \text{int}(y)$, and as p_y is an interior operator it is also monotone and we have that $p_y \circ \text{ext} \circ \text{int}(x) \geq p_y \circ \text{ext} \circ \text{int}(y)$. To summarize, we have shown that $h(x) = p_y(\text{ext} \circ \text{int}(x)) \geq p_y(\text{ext} \circ \text{int}(y)) = h(y)$.
- $h(y) = h \circ h(y)$?

Let $x = h_y(y)$, we have $x \geq y$ and following the definition of h that $h \circ h(y) = h_x \circ h_y(y) = h(x)$. Again, let $z = \text{ext} \circ \text{int}(x)$, then z belongs to the upset X^x and as $x \geq y$ this means following Lemma 4 that $p_x(z) = p_y(z)$ and therefore $h \circ h(y) = h(x) = p_y(\text{ext} \circ \text{int}(x)) = h_y(h_y(y))$. Furthermore, as h_y is a closure operator, we have that $h_y(h_y(y)) = h_y(y)$ and therefore, $h \circ h(y) = h_y(y) = h(y)$.

Now, as h is a closure operator on the confluence F which is also a pre-confluence, this means following Theorem 1 that $h[F]$ is closed under local meet and therefore also is a pre-confluence. □

The closure subset $h[F]$ is isomorphic with the set of pairs $P = \{(e, l) | e \in F, e = h(e), l = \text{int}(e)\} = \{(e, l) | e \in F, e = p_e \circ \text{ext}(l), l = \text{int}(e)\}$, and as a consequence P, ordered following F, also is a pre-confluence. This leads to generalize Galois or Concept lattices and define (extensional) Galois pre-confluences we also call local concept pre-confluences. Each pair (e, l) is called a *local concept*, e is the local extension (or *local extent*) and l is the local intension (or *local intent*).

Definition 5. *The set* $P = \{(e, l) | e \in F, l \in L, e = p_e \circ \text{ext}(l), l = \text{int}(e)\}$ *is a pre-confluence isomorphic with* $h[F]$ *and is called the extensional Galois pre-confluence defined on the confluence* F *of* X *and the lattice* L *by the maps* $\text{int} : X \rightarrow L$, *and* $\text{ext} : L \rightarrow X$

Example 3. We consider here the confluence F^{1+3} and the pattern language and context described Example 2.

We report hereunder the local concepts (e, l) of the associated Galois pre-confluence. The corresponding local extents are obtained using the range closure operator $h[F^{1+3}]$ as stated in Theorem 3. We have $e = h_x(x) = p_i \circ \text{ext} \circ \text{int}(x)$ where $i = 1$ (resp. $i = 3$) whenever $\{1\} \subseteq x$ (resp. $\{3\} \subseteq x$), and the corresponding local intents are obtained as $\text{int}(e)$.

- $h(1) = p_1 \circ \text{ext} \circ \text{int}(1) = p_1 \circ \text{ext}(abcd) = 1$ and $l = \text{int}(1) = abcd$

- $h(3) = p_3 \circ \text{ext} \circ \text{int}(3) = p_3 \circ \text{ext}(abc) = 3$ and $l = \text{int}(3) = abc$
- $h(12) = p_1 \circ \text{ext} \circ \text{int}(12) = p_1 \circ \text{ext}(ab) = 123$ and $l = \text{int}(123) = ab$
- $h(14) = p_1 \circ \text{ext} \circ \text{int}(14) = p_3 \circ \text{ext}(ac) = 134$ and $l = \text{int}(134) = ac$
- $h(23) = p_3 \circ \text{ext} \circ \text{int}(23) = p_1 \circ \text{ext}(ab) = 123$ and $l = ab$
- $h(43) = p_3 \circ \text{ext} \circ \text{int}(43) = p_3 \circ \text{ext}(ac) = 134$ and $l = ac$
- $h(124) = p_1 \circ \text{ext} \circ \text{int}(124) = p_1 \circ \text{ext}(a) = 1234$ and $l = \text{int}(1234) = a$
- $h(234) = p_3 \circ \text{ext} \circ \text{int}(234) = p_3 \circ \text{ext}(a) = 1234$ and $l = a$
- $h(123) = p_1 \circ \text{ext} \circ \text{int}(123) = p_1 \circ \text{ext}(ab) = 123$ and $l = ab$
- $h(134) = p_1 \circ \text{ext} \circ \text{int}(134) = p_1 \circ \text{ext}(ac) = 134$ and $l = ac$
- $h(1234) = p_1 \circ \text{ext} \circ \text{int}(1234) = p_1 \circ \text{ext}(a) = 1234$ and $l = a$

Overall we have only five local concepts, namely $(1234, a),(123, ab)$, $(134, ac)$, $(1, abcd)$, $(3, abc)$. Ordered following the order on F, they form a Galois pre-confluence.

The same local concepts may be obtained starting from the pattern language using Theorem 2 by applying the local closure operators $f_1 = \text{int} \circ p_1 \circ \text{ext}$ and $f_3 = \text{int} \circ p_3 \circ \text{ext}$ to their respective sublanguages $L_{\text{int}(1)} = 2^{abcd}$ and $L_{\text{int}(3)} = 2^{abc}$. We obtain then $\text{int}[F]$ as $f_1[2^{abcd}] \cup f_3[2^{abc}]$ and therefore the set of local concepts as the union of the set of pairs $(p_1 \circ \text{ext}(q), q)$ where $q = f_1(q)$ and of the set of pairs $(p_3 \circ \text{ext}(q), q)$ where $q = f_3(q)$. Note that whenever we are interested by frequent local concepts, for which the local extent has to be larger than some threshold, then the latter way has to be preferred.   □

3.2   Local Implications

As we have seen, let q be a pattern and $m \in F$ be such that $m \subseteq \text{ext}(q)$, $p_m \circ \text{ext}(q)$ is defined and represent the local support of q in F which is greater than m. Whenever we have $p_m \circ \text{ext}(q) \subseteq p_m \circ \text{ext}(w)$ we rewrite this as the local implication $\square_m q \to \square_m w$ stating that if q has a local support set containing m, then w has a larger or equal local extension.

When considering a given closed pattern c with respect to 2^O, whose local extension e contains m, and whose corresponding local closed pattern in F is l, we have then that the implication rule $\square_m c \to \square_m l$ holds.

The set of such $\square_m c \to \square_m l$ local implications, with $c \neq l$, represents (a basis for) the local knowledge deriving from the reduction of the extensional space from 2^O to the confluence F.

3.3   Some Bounds on the Size of an Extensional Galois Pre-confluence

When considering the number of local closed patterns as defined above when considering a confluence of $X = 2^O$ as the extensional space, i.e. the size of $\text{int}([F]$, we may bound it by the number of closed patterns when considering $X = 2^O$ as the extensional space. However, if we are interested in the total number of pairs (e, l) such that $e = p_m \circ \text{ext}(l)$ and $l = \text{int}(e)$, we have a larger bound, because a given local closed pattern may have as many local extensions as minimal elements in F. We have then to consider the set of closed elements $h[F]$ each element e of which is the local extension of exactly one (e, l) pair: whenever e is given, l is defined as $\text{int}[e]$.

Proposition 6. *Let $f[L] = \text{int}[X]$ be the set of closed patterns on L with respect to the lattice $X = 2^O$, $\text{int}[F]$ be the set of local closed patterns with respect to the confluence F, and $P_F = \{(e, l)|e = h(e), l = \text{int}(e)\}$, be the set of local concepts, we have that:*

1. $\text{int}[F] \subseteq \text{int}[X]$
2. $|P_F| \leq |\min(F)| * |\text{int}[F]|.$

Proof. First, as $X \subseteq F$ it is straightforward that $\text{int}[F] \subseteq \text{int}[X]$. Then, each pair (e, l) belongs to at least one Galois lattice of a Galois connection on $(F^m, L_{\text{int}(m)})$, as there are $\min[F]$ such Galois connections and as the size $|\text{int}[F^m]|$ of each corresponding Galois lattice is smaller than $\text{int}[F]$, the total number of (e, l) pairs is smaller than or equal to $|\min[F]| * |\text{int}[F]|$.   □

As we will see in the simple example that follows, the bound is tight.

Example 4. Let $O = \{1, 2\}$ and the elements of O be described as $d(1) = \text{int}(\{1\}) = ab$, and $d(2) = \text{int}(\{2\}) = ab$, with $L = 2^{\{a,b\}}$. There is then only one concept $(\{1, 2\}, ab)$ and $|\text{int}[2^O]| = 1$.

When considering the confluence $F = \{\{1\}, \{2\}\}$, we have as the set of local closed patterns $\text{int}[\{\{1\}, \{2\}\}] = \{ab\}$ and we have as expected that the number (1) of local closed patterns with respect to F is smaller than or equal to the number (1) of closed patterns with respect to 2^O.

Now, when considering the number of local closed patterns we have now

- $L_{\text{int}(\{1\})} = L_{ab} = 2^{\{a,b\}} = L$ and therefore, $f_{\{1\}}[L] = \text{int}[F^{\{1\}}] = \{\text{int}(\{1\})\} = \{ab\}$ and $p_{\{1\}} \circ \text{ext}(ab) = \{1\}$, which gives as (local extension, local closed pattern) pairs the unique pair $(\{1\}, ab)$
- $L_{\text{int}(\{2\})} = L_{ab} = 2^{\{a,b\}} = L$ and therefore, $f_{\{2\}}[L] = \text{int}[F^{\{2\}}] = \{\text{int}(\{2\})\} = \{ab\}$ and $p_{\{2\}} \circ \text{ext}(ab) = \{2\}$, which gives as (local extension, local closed pattern) pairs the unique pair $(\{2\}, ab)$.

As a result we have 2 (e, l) pairs, $|\min[F]| = 2$, $|\text{int}[2^O]| = 1$, $\text{int}[F] = \text{int}[X] = \{ab\}$, and $|P_F| = |\min[F]| * |\text{int}[X]|$.   □

4   Graph Confluence

Whenever the objects are the vertices of some undirected graph, graph abstractions, as defined in [1], lead to abstract closed patterns. Here we discuss the graph confluence F, representing vertex subsets inducing connected subgraphs of a graph, which has been presented in Example 1. However, a large family of confluences may be obtained starting from some graph $(G = O, E)$, by simply deriving from G a new graph $G_T = (T, E_T)$ whose vertices are subsets of O, and whose edges E_T are deduced from G. Our motivating example considers T as the set of k-cliques of G and states that there is an edge in G_T between two k-cliques whenever they share $k-1$ vertices. In the field of social network analysis, the vertex subset in G corresponding to a connected component of G_T is known as a k-community [7]. A local concept pre-confluence in this particular case will be made of local concepts each associated to a k-community. Each k-community corresponds, in G_T, to a connected component of the subgraph of G_T induced by the extension of some pattern.

Consider a graph $G = (E_T, T)$ whose vertices are subsets of some object set O, i.e. $T \subseteq 2^O$ and let F_T be the confluence of subsets of 2^T inducing connected subgraphs of G_T. We also consider a pattern lattice L and two maps ext $: L \rightarrow 2^O$ and int $: 2^O \rightarrow L$ such that (int, ext) defines a Galois connection on $(L, 2^O)$.

Let $u : 2^T \rightarrow 2^O$ be such that $u(e_T) = \cup_{t \in e_T} t$. $u(e_T)$ is called the *flattening* of e_T. We then consider the two maps ext_T and int_T defined as follows:

- $\text{ext}_T : L \rightarrow 2^T$ with $\text{ext}_T(p) = \{t | t \subseteq \text{ext}(p)\}$
- $\text{int}_T : 2^T \rightarrow L$ with $\text{int}_T(e_T) = \text{int} \circ u(e_T)$.

It is straightforward that $(\text{int}_T, \text{ext}_T)$ defines a Galois connection on $(2^T, L)$. Therefore, given a graph confluence F of 2^T, we obtain a local concept pre-confluence defined on (F, L). We then have the following result when flattening the (local) extensions so found in F:

Proposition 7. *Let F be a confluence of 2^T and $U = u[F]$, where u is the flattening operator on O, then*

- $\text{int}_T[F] = \text{int}[U]$
- *Let (e_T, l) be a local concept in the Galois pre-confluence defined on (F, L) and $e_T \geq m \in \min[F]$, then $u(e_T)$ is the greatest element of $u[F^m]$ among elements e such that $\text{int}(e) = l$.*

Proof

- The first item is straightforward as for any element x of F, we have $\text{int}_T(x) = \text{int} \circ u(x)$.
- Let e be an element of $u[F^m]$, and $\text{int}(e) = l$. This means that there exists $e'_T \in u[F^m]$ such that $e = u(e'_T)$ and as a consequence we have $\text{int} \circ u(e'_T) = \text{int}_T(e'_T) = l$. But as e_T is the greatest element along the elements x of F^m such that $\text{int}_T(x) = l$ then necessarily $e'_T \subseteq e_T$ and as u is monotonic, $e = u(e'_T) \subseteq u(e_T)$. □

This means that the local closed patterns with respect to the confluence F are the same as the patterns support closed with respect to the extensional space $U = u[F]$, that we call the flattening of F[3]. This means that we may associate to each pair (e_T, l) of the Galois pre-confluence defined on F, the pair $(u(e_T), l)$

---

[3] A pattern is said support-closed whenever specializing the pattern decreases its extension [6].

with the same intension $l = \text{int}_T(e_T) = \text{int} \circ u(e_T)$ and such that $u(e_T)$ is the greatest element in $u[F^m] \subseteq 2^O$, whose intension is l. However, in general U is not a confluence of 2^O because there may be two elements e_T and e_T' with the same image $u(e_T) = u(e_T')$ in 2^O. In the following example we also discuss the corresponding local implications:

Example 5. Let $G = (O, E)$ be the graph displayed on the left part of Fig. 2. Each vertex has an itemset included in $\{a, b, c\}$ as a label. The set of triangles is $T = \{t_0, t_1, t_2, t_3, t_4, t_5, t_6, t_7\}$ and form a triangle graph G^T displayed on the right part of Fig. 2. An edge relates any pair of triangles sharing two vertices in G, as for instance (t_0, t_1). Each triangle in G^T has as its itemset the intersection of the itemsets of its three vertices in G. For instance, the description of t_1 in G^T is $ab = abc \cap ab \cap ab$. The vertex subsets inducing connected subgraphs of G^T form the confluence $F^T = \{\{t_0\}, \{t_1\}, \{t_0, t_1\}, \{t_2\}, \{t_3\}, \{t_2, t_3\}, \{t_4\}, \{t_5\}, \{t_4, t_5\}, \{t_6\}, \{t_7\}, \{t_6, t_7\}\}$. We do not consider in this example the empty pattern.

The support set of the pattern a is $\text{ext}_T(a) = \{t_0, t_1, t_2, t_3, t_6, t_7\}$. The local support with respect to t_0 is $p_{t_0}(\{t_0, t_1, t_2, t_3, t_6, t_7\}) = \{t_0, t_1\}$, i.e. the connected component containing the vertex t_0 of the subgraph induced by $\text{ext}_T(a)$. Hereunder, we note f_i the local closure operator $f_{\{t_i\}}$. The corresponding local closed patterns are as follows:

– $f_0(a) = f_1(a) = ac$, $f_2(a) = f_3(a) = ab$, $f_6(a) = f_7(a) = ab$.

In the same way, the pattern b whose support set is $\text{ext}(b) = \{t_2, t_3, t_4, t_5, t_6, t_7\}$ leads to the following local closed patterns:

– $f_2(b) = f_3(b) = ab$, $f_4(b) = f_5(b) = bc$, $f_6(b) = f_7(b) = ab$.

Note that ab appears both as a local closed pattern resulting from a with respect to f_0, f_1 and to f_6, f_7 and as a local closed pattern resulting from b with respect to f_2, f_3 and again to f_6, f_7. Now, as both a and b are closed patterns with respect to 2^T, we obtain various triples in the form (a, e_T, l) and (b, e_T, l) and corresponding local implications in the form $\square_{e_T} a \rightarrow \square_{e_T} l$ and $\square_{e_T} a \rightarrow \square_{e_T} l$. The former, for instance, also rewrites as $\square_{\{t_i\}} a \rightarrow \square_{\{t_i\}} l$ where t_i is any element of e_T. This leads to the following sets of local implications:

– $\square_{\{t_2\}} a \rightarrow \square_{\{t_2\}} ab$, $\square_{\{t_3\}} a \rightarrow \square_{\{t_3\}} ab$, equivalent to $\square_{\{t_2, t_3\}} a \rightarrow \square_{\{t_2, t_3\}} ab$
– $\square_{\{t_6\}} a \rightarrow \square_{\{t_6\}} ab$, $\square_{\{t_7\}} a \rightarrow \square_{\{t_7\}} ab$,
– $\square_{\{t_2\}} b \rightarrow \square_{\{t_2\}} ab$, $\square_{\{t_3\}} b \rightarrow \square_{\{t_3\}} ab$.

The 4 local concepts found in G_T are $(\{t_0, t_1\}, ac)$, $(\{t_2, t_3\}, ab)$, $(\{t_4, t_5\}, bc)$, $(\{t_6, t_7\}, ab)$, and correspond in the original graph to 4 3-communities of size 4.   □

In such a framework, a local closed pattern l is stated as *frequent* whenever its local extension in O, $u \circ \text{ext}(l)$, exceeds some threshold s. We consider then the following mining problem (Problem I):

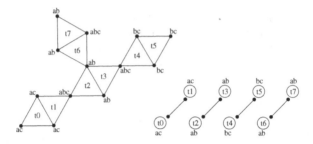

Fig. 2. On the left we have a graph of objects each described as an itemset included in $\{a, b, c\}$. This graph represents the triangle abstraction of some input graph. On the right, the graph G^T whose vertices are the triangles of G. The itemset describing a vertex in G^T is the intersection of the itemsets describing the elements of the corresponding triangle in G.

- Given an attributed graph $G_T = (T, E_T)$ where vertices are subsets of some object set O, and whose labels belong to an attribute lattice 2^K, find all the triples (c, e, l) where c is a closed pattern w.r.t. 2^T, l is a frequent local closed pattern w.r.t the connected subgraphs induced by 2^T, and e the associated local extension.

It should be clear that any algorithm solving Problem I allows to enumerate the frequent local implication rules $\square_e c \rightarrow \square_e l$ associated to an attributed graph G_T derived from some object set O. Such an algorithm is briefly suggested in the next section.

5   Algorithmics

An algorithmic way to solve Problem I is a top-down search in the pattern space and consists in adapting a separate and conquer algorithm enumerating frequent closed itemsets as PARAMINER [11]. Enumerating local closed patterns, in addition to each closed pattern c, means that when c is computed, the connected components of the subgraph of G_T induced by the extension of c are computed and the closure operator associated to each connected component is applied to c. All this results in an algorithm that outputs with no repetition triples (c, e, l) corresponding to local implication rules $\square_e c \rightarrow \square_e l$ where c is a closed pattern, e a local extension and l the local closed pattern associated to e. Such an algorithm searches patterns in a general to specify way and therefore allows to prune the search to only consider frequent local closed patterns and, therefore, solves Problem I. Note that in this approach we output several times each (e, l) local concept, because several closed patterns c may lead to triples with the same (e, l) part, but as each closed pattern c is output once, each (c, e, l) triple is also output once.

6   Conclusion

The present work follows various investigations in Formal Concept Analysis whose purpose is to extend the original FCA methodology, as presented in [12], in order to address various knowledge discovery problems. Most of these investigations concern how to extend FCA in order to address problems in which the pattern language is more sophisticated than the standard powerset of binary attributes. In particular *pattern structures* [10] have been recently introduced to represent complex data, associating such a *pattern structure* to each object and using interior operators on the pattern language in order to reduce the resulting concept lattice [13]. In the present article we are interested in addressing problems in which the extensional space is constrained, as it is the case when considering connected subgraphs.

We have addressed the question of the structure of the set of intensions $int[F]$ when F is a confluence of 2^O, i.e. a family of subsets of O with various minimal elements and such that for any pair of elements both greater than or equal to a same minimal element in F, their union belongs to F. We have shown that $int[F]$ is obtained as the set of closed elements from a family $\{f_m | m \in \min[F]\}$ of closure operators and that the extensions in F of these elements are obtained as the range $h[F]$ of a closure operator h on F. Overall we obtain a set of local concepts (e, l) where e is a closed element with respect to h and l is a closed element with respect to f_m with $m \leq e$. When ordered following $h[F]$, the set of local concepts is a pre-confluence, called an extensional Galois pre-confluence. These results also answer, when inverting the extensional and extensional spaces and corresponding maps ext and int, to the question of the structure of $ext[F]$ in the original presentation of intensional Galois pre-confluences in [5]. The practical motivation of this work was knowledge discovery in attributed graphs. From this point of view, the set of vertex subsets inducing connected subgraphs of an attributed graph is an extensional confluence, and this leads to, first, define local concepts (e, l) where e is a connected component of the subgraph induced by the support set of some pattern in the vertex set of an attributed graph, and l is the corresponding local closed pattern, and, second, to define local implications which are valid in the neighborhood of some vertex. Interestingly, this may be applied to graphs $G_T = (T, E_T)$ where T is a family of subsets of an original object set O. A direct application is the investigation of k-communities in attributed graphs.

Further work should consider both the extensional and extensional spaces as confluences, and develop definitions and interpretation of local implication bases.

Acknowledgments. Many thanks to Sylvie Borne for her help in drawing the figures.

References

1. Soldano, H., Santini, G.: Graph abstraction for closed pattern mining in attributed network. In: Schaub, T., Friedrich, G., O'Sullivan, B. (eds.) European Conference in Artificial Intelligence (ECAI). Frontiers in Artificial Intelligence and Applications, vol. 263, pp. 849–854. IOS Press, Amsterdam (2014)
2. Soldano, H., Ventos, V.: Abstract concept lattices. In: Jäschke, R. (ed.) ICFCA 2011. LNCS, vol. 6628, pp. 235–250. Springer, Heidelberg (2011)

3. Mougel, P.-N., Rigotti, C., Gandrillon, O.: Finding collections of k-clique percolated components in attributed graphs. In: Tan, P.-N., Chawla, S., Ho, C.K., Bailey, J. (eds.) PAKDD 2012, Part II. LNCS, vol. 7302, pp. 181–192. Springer, Heidelberg (2012)

4. Silva, A., Meira Jr., W., Zaki, M.J.: Mining attribute-structure correlated patterns in large attributed graphs. Proc. VLDB Endow. **5**(5), 466–477 (2012)

5. Soldano, H.: Closed patterns and abstraction beyond lattices. In: Glodeanu, C.V., Kaytoue, M., Sacarea, C. (eds.) ICFCA 2014. LNCS, vol. 8478, pp. 203–218. Springer, Heidelberg (2014)

6. Boley, M., Horváth, T., Poigné, A., Wrobel, S.: Listing closed sets of strongly accessible set systems with applications to data mining. Theor. Comput. Sci. **411**(3), 691–700 (2010)

7. Palla, G., Derenyi, I., Farkas, I., Vicsek, T.: Uncovering the overlapping community structure of complex networks in nature and society. Nature **435**(7043), 814–818 (2005)

8. Blyth, T.S.: Lattices and Ordered Algebraic Structures. Universitext. Springer, London (2005)

9. Pernelle, N., Rousset, M.C., Soldano, H., Ventos, V.: Zoom: a nested Galois lattices-based system for conceptual clustering. J. Exp. Theor. Artif. Intell. **2/3**(14), 157–187 (2002)

10. Ganter, B., Kuznetsov, S.O.: Pattern structures and their projections. In: Delugach, H.S., Stumme, G. (eds.) ICCS 2001. LNCS (LNAI), vol. 2120, pp. 129–142. Springer, Heidelberg (2001)

11. Negrevergne, B., Termier, A., Rousset, M.C., Méhaut, J.F.: Paraminer: a generic pattern mining algorithm for multi-core architectures. Data Min. Knowl. Disc. **28**(3), 593–633 (2013)

12. Ganter, B., Wille, R.: Formal Concept Analysis: Mathematical Foundations. Springer, Heidelberg (1999)

13. Kuznetsov, S.O., Samokhin, M.V.: Learning closed sets of labeled graphs for chemical applications. In: Kramer, S., Pfahringer, B. (eds.) ILP 2005. LNCS (LNAI), vol. 3625, pp. 190–208. Springer, Heidelberg (2005)

A Note on Pattern Structures and Their Projections

Lars Lumpe$^{(\boxtimes)}$ and Stefan E. Schmidt

Institut Für Algebra, Technische Universität Dresden, Dresden, Germany
larslumpe@gmail.com

Abstract. Literature on pattern structures suggests that projections lead again to pattern structures. To clarify the situation, we provide a counterexample. However, we also show that residual projections on pattern structures do indeed induce again pattern structures.

1 Introduction

Pattern structures within the framework of formal concept analysis have been introduced in [1]. Since then they have turned out to be a useful tool for analysing various real-world applications (cf. [1–5]). In our note we want to point out that the theoretical foundations of pattern structures encourage still some fruitful discussions. In particular, the role projections play within pattern structures for information reduction still needs some further investigation.

The goal of our paper is to look under which circumstances pattern structures can or cannot be replaced by simpler (meaningful) ones. Here it turns out that projections not always give rise to new pattern structures, however, residual projections do.

2 Preliminaries

For the counterexample we are going to construct, we need the following preparations:

(1) If $\mathbb{P}_1 = (P_1, R_1)$ and $\mathbb{P}_2 = (P_2, R_2)$ are posets (partially ordered sets) with $P_1 \cap P_2 = \emptyset$ then the vertical sum of $\mathbb{P}_1$ with $\mathbb{P}_2$ is defined as $\mathbb{P} := (P, R)$ with $P := P_1 \cup P_2$ and

$$R := R_1 \cup R_2 \cup (P_1 \times P_2);$$

we set $\mathbb{P}_1 +_{vert} \mathbb{P}_2 := \mathbb{P}$.

(2) For a poset $\mathbb{P} := (P, R)$ and a subset T of P, the restriction of $\mathbb{P}$ onto T is given by

$$\mathbb{P} \mid T := (T, R \cap (T \times T)).$$

© Springer International Publishing Switzerland 2015
J. Baixeries et al. (Eds.): ICFCA 2015, LNAI 9113, pp. 145–150, 2015.
DOI: 10.1007/978-3-319-19545-2_9

(3) For a poset $\mathbb{P} := (P, R)$ let $\mathbb{P}^d := (P, R^d)$ with

$$R^d := \{(y, x) \mid (x, y) \in R\}$$

denote the dual of $\mathbb{P}$.

(4) A pattern structure is defined as a triple $(G, \mathbb{D}, \delta)$ where G is a set of socalled *objects*, $\mathbb{D} := (D, \sqsubseteq)$ forms a meet-semilattice of socalled *patterns*, and $\delta : G \longrightarrow D$ is a map such that every subset X of $\delta G := \{\delta g \mid g \in G\}$ has an infimum (greatest lower bound) in $\mathbb{D}$, denoted by $\sqcap X$. The set D_δ of all infima of subsets of δG forms a complete subsemilattice of $\mathbb{D}$.

(5) A kernel operator on a poset $\mathbb{P} := (P, \leq)$ is a map $\psi : P \longrightarrow P$ such that $t \leq x$ is equivalent to $t \leq \psi x$, for all $x \in P$ and $t \in \psi P$. A subset K of P is called a kernel system in $\mathbb{P}$ if for every $x \in P$ the restriction of $\mathbb{P}$ onto $\{t \in K \mid t \leq x\}$ has a greatest element.

If $KO\mathbb{P}$ denotes the set of all kernel operators on $\mathbb{P}$ and $KS\mathbb{P}$ denotes the set of all kernel systems in $\mathbb{P}$, then the map

$$KO\mathbb{P} \longrightarrow KS\mathbb{P}, \psi \mapsto \psi P$$

is a bijection.

Dually, a closure operator on $\mathbb{P}$ is defined as a kernel operator on $\mathbb{P}^d$, and a closure system in $\mathbb{P}$ is defined as a kernel system in $\mathbb{P}^d$. Also, $CO\mathbb{P}$ denotes the set of all closure operators on $\mathbb{P}$ and $CS\mathbb{P}$ denotes the set of all closure systems in $\mathbb{P}$. Then

$$CO\mathbb{P} \longrightarrow CS\mathbb{P}, \psi \mapsto \psi P$$

is a bijection.

(6) If $(G, \mathbb{D}, \delta)$ is a pattern structure, then a kernel operator ψ on $\mathbb{D}$ will also be called a projection. Literature, compare for example [1,2], suggests that any such projection ψ on a pattern structure $(G, \mathbb{D}, \delta)$ induces via $(G, \mathbb{D}, \psi \circ \delta)$ a pattern structure. However, the following example shows that this is not always the case.

3 Construction of the Counterexample

Now we are prepared to construct a pattern structure on a lattice such that there exists a projection which does not induce a pattern structure.

Consider the chain $\mathbb{P}_0 := (\mathbb{N}, \leq)$ and the complete chain $\mathbb{P} := (\bar{\mathbb{N}}, \leq)$ where $\bar{\mathbb{N}} := \mathbb{N} \cup \{\infty\}$ and $x \leq \infty$ for all $x \in \bar{\mathbb{N}}$; then

$$\mathbb{D} := \mathbb{P}_0 +_{vert} (\mathbb{P}^d \times \mathbb{P}_0^d)$$

is a lattice, and for $G := \mathbb{N}$, the map $\delta : G \longrightarrow D, n \mapsto (n, 0)$ gives rise to a pattern structure $(G, \mathbb{D}, \delta)$, where $D_\delta = \bar{\mathbb{N}} \times \{0\}$. For $\Delta_\mathbb{N} := \{(n, n) \mid n \in \mathbb{N}\}$ the set $K := \Delta_\mathbb{N} \cup \{0\}$ forms a kernel system, the associated kernel operator ψ of which has the property that $(G, \mathbb{D}, \psi \circ \delta)$ is **not** a pattern structure, because Δ_N has no infimum in $\mathbb{D}$ (See also Fig. 1).

This may be even more surprising since ψ preserves finite meets in $\mathbb{D}$ and K forms a sublattice of $\mathbb{D}$.

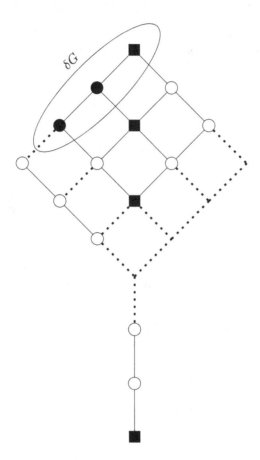

Fig. 1. Counterexample

4   Bipolar Systems

On a positive note, we are going to show that every residual projection on a pattern structure induces a pattern structure. For this we need some preparation:

For a poset $\mathbb{P} = (P, \leq)$, a pair (f, g) is an adjunction on $\mathbb{P}$ if $fx \leq y$ is equivalent to $x \leq gy$ for all $x, y \in P$. Here, f and g mutually determine each other; f is called *residuated* and g is called *residual* on $\mathbb{P}$. It easily follows that f and g are isotone maps on $\mathbb{P}$.

In the following theorem we will clarify the connection between residual kernel operators and socalled *bipolar systems* in posets. Here, a *bipolar system* in a poset $\mathbb{P}$ is defined as a kernel system in $\mathbb{P}$ which also is a closure system in $\mathbb{P}$. The set of all bipolar systems in $\mathbb{P}$ will be denoted by $BS\mathbb{P}$.

For a poset $\mathbb{P}$, let $KO^r\mathbb{P}$ denote the set of all residual kernel operators on $\mathbb{P}$, and dually, let $CO_r\mathbb{P}$ denote the set of all residuated closure operators on $\mathbb{P}$.

Lemma 1. *For a poset $\mathbb{P} = (P, \leq)$, a map $\psi : P \longrightarrow P$ is a kernel operator on $\mathbb{P}$ if and only if the following holds for all $x, y \in P$:*

$$x \leq \psi y \Leftrightarrow \exists t \in \psi P (x \leq t \leq y) \qquad (*)$$

Dually, a map $\varphi : P \longrightarrow P$ is a closure operator on $\mathbb{P}$ if and only if the following holds for all $x, y \in P$:

$$\varphi x \leq y \Leftrightarrow \exists t \in \varphi P (x \leq t \leq y) \qquad (**)$$

Proof. First, assume that ψ is a kernel operator on $\mathbb{P}$. Then $x \leq \psi y$ implies $x \leq t \leq y$ for $t := \psi y$ (since ψ is contractive, that is, $t \leq y$). On the other hand, if $x \leq t \leq y$ holds for $t \in \psi P$ then $t \leq \psi y$ (since ψ is a kernel operator), which immediately implies $x \leq \psi y$.

Second, assume that $(*)$ holds. Then $\psi y \leq \psi y$ implies $\psi y \leq t \leq y$ for some $t \in \psi P$; therefore, $\psi y \leq y$ holds. That is, ψ is contractive. Also, for $t \in \psi P$ with $t \leq y$ we get $t \leq t \leq y$, which by $(*)$ implies $t \leq \psi y$. It follows that ψ is a kernel operator on $\mathbb{P}$.

Theorem 1. *If $\mathbb{P}$ is a poset, then the map*

$$KO^r\mathbb{P} \longrightarrow BS\mathbb{P}, \ \psi \mapsto \psi P$$

is a bijection, and dually, the map

$$CO_r\mathbb{P} \longrightarrow BS\mathbb{P}, \ \psi \mapsto \psi P$$

is a bijection too.

Proof. Since $KO\mathbb{P} \longrightarrow KS\mathbb{P}, \psi \mapsto \psi P$ is a bijection, we have to show that $\psi \in KO^r\mathbb{P}$ always implies that ψP is a bipolar system in $\mathbb{P}$. On the other hand, we have to verify that the kernel operator associated with a bipolar system in $\mathbb{P}$ is always residual.

First, let $\psi \in KO^r\mathbb{P}$. Then ψP is a kernel system in $\mathbb{P}$. Since ψ is residual on $\mathbb{P}$, there exists φ such that (φ, ψ) is an adjunction on $\mathbb{P}$. For $x \in P$, the element $u := \psi(\varphi x) \in \psi P$ satisfies $x \leq u$ (since $\varphi x \leq \varphi x$ and the pair (φ, ψ) is an adjunction). Also, for $y \in P$, the element $w := \psi y \in \psi P$ with $x \leq w$ satisfies $\varphi x \leq y$. Thus $u \leq w$ (since ψ is isotone on $\mathbb{P}$), which yields that u is the least element in ψP with $x \leq u$. It follows that ψP is also a closure system and therefore a bipolar system in $\mathbb{P}$.

Second, let $B \in BS\mathbb{P}$. The associated kernel operator of B in $\mathbb{P}$ will be denoted by ψ and the associated closure operator of B in $\mathbb{P}$ will be denoted by φ. Since $\varphi P = B = \psi P$, the conditions $(**)$ and $(*)$ of *Lemma 1* yield the following for all $x, y \in P$:

$$\varphi x \leq y \Leftrightarrow \exists t \in B \, (x \leq t \leq y) \Leftrightarrow x \leq \psi y$$

Therefore, (φ, ψ) is an adjunction on $\mathbb{P}$, from which we derive that φ is a residuated closure operator and ψ is a residual kernel operator with $\varphi P = B = \psi P$.

5   Residual Projections

Our final result will state that residual projections on pattern structures induce again pattern structures.

Theorem 2. *Let* $\mathbb{S} := (G, \mathbb{D}, \delta)$ *be a pattern structure with* $\mathbb{D} := (D, \sqsubseteq)$, *and let* ψ *be a residual projection on* $\mathbb{S}$, *that is,* ψ *be a residual kernel operator on* $\mathbb{D}$. *Then* $\mathbb{S}_\psi := (G, \mathbb{D}, \psi \circ \delta)$ *is a pattern structure, which satisfies* $D_{\psi \circ \delta} = \psi(D_\delta)$. *Furthermore,* ψD *is a bipolar system in* $\mathbb{D}$ *satisfying that* $(G, \mathbb{D} \,|\, \psi D, \psi \circ \delta)$ *is a pattern structure.*

Remark 1. It is easy to observe that $(G, \mathbb{D} \,|\, \psi D, \psi \circ \delta)$ forms a pattern structure for any projection ψ on $\mathbb{S}$.

Proof. Let Y be a subset of $\psi(\delta G)$. Then there exists a subset X of δG with $Y = \psi X$. Since $\mathbb{S}$ is a pattern structure, $\sqcap X$ exists in $\mathbb{D}$.

Claim. The infimum of Y in $\mathbb{D}$ is given by $\psi(\sqcap X)$, thus $\psi(\sqcap X) = \sqcap Y$.
Since $\sqcap X$ is a lower bound of X in $\mathbb{D}$, the element $\psi(\sqcap X)$ is a lower bound of $Y = \psi X$ in $\mathbb{D}$ (because ψ is isotone). Let now t' be a lower bound of Y in $\mathbb{D}$. By our assumption on ψ there exists φ such that (φ, ψ) is an adjunction on $\mathbb{P}$. Thus for every $x \in X$, we have $t' \leq \psi x$ and therefore $\varphi t' \leq x$, which means that $\varphi t'$ is a lower bound of X. This implies $\varphi t' \leq \sqcap X$, and this yields $t' \leq \psi(\sqcap X)$. We conclude that $\mathbb{S}_\psi$ is a pattern structure; together with theorem 1 this completes the proof.

References

1. Ganter, B., Kuznetsov, S.O.: Pattern structures and their projections. In: Delugach, H.S., Stumme, G. (eds.) ICCS 2001. LNCS (LNAI), vol. 2120, pp. 129–142. Springer, Heidelberg (2001)
2. Kuznetsov, S.O.: Pattern structures for analyzing complex data. In: Sakai, H., Chakraborty, M.K., Hassanien, A.E., Ślęzak, D., Zhu, W. (eds.) RSFDGrC 2009. LNCS, vol. 5908, pp. 33–44. Springer, Heidelberg (2009)

3. Kuznetsov, S.O.: Scalable knowledge discovery in complex data with pattern structures. In: Maji, P., Ghosh, A., Murty, M.N., Ghosh, K., Pal, S.K. (eds.) PReMI 2013. LNCS, vol. 8251, pp. 30–39. Springer, Heidelberg (2013)
4. Kaytoue, M., Kuznetsov, S.O., Napoli, A., Duplessis, S.: Mining gene expression data with pattern structures in formal concept analysis. Inf. Sci. **181**, 1989–2001 (2011)
5. Kaiser, T.B., Schmidt, S.E.: Some remarks on the relation between annotated ordered sets and pattern structures. In: Kuznetsov, S.O., Mandal, D.P., Kundu, M.K., Pal, S.K. (eds.) PReMI 2011. LNCS, vol. 6744, pp. 43–48. Springer, Heidelberg (2011)

Enhanced FCA

Exploring Pattern Structures of Syntactic Trees for Relation Extraction

Artuur Leeuwenberg[(✉)], Aleksey Buzmakov, Yannick Toussaint,
and Amedeo Napoli

LORIA (CNRS – INRIA Nancy Grand Est – Université de Lorraine),
Équipe Orpailleur, Bâtiment B, BP 239, 54506 Vandœuvre-lès-Nancy Cedex, France
t.leeuwenberg@gmail.com,
{aleksey.buzmakov,yannick.toussaint,amedeo.napoli}@loria.fr

Abstract. In this paper we explore the possibility of defining an original pattern structure for managing syntactic trees. More precisely, we are interested in the extraction of relations such as drug-drug interactions (DDIs) in medical texts where sentences are represented as syntactic trees. In this specific pattern structure, called STPS, the similarity operator is based on rooted tree intersection. Moreover, we introduce "Lazy Pattern Structure Classification" (LPSC), which is a symbolic method able to extract and classify DDI sentences w.r.t. STPS. To decrease computation time, a projection and a set of tree-simplification operations are proposed. We evaluated the method by means of a 10-fold cross validation on the corpus of the DDI extraction challenge 2011, and we obtained very encouraging results that are reported at the end of the paper.

Keywords: Pattern structures · Relation extraction · Formal concept analysis · DDI extraction

1 Introduction

When a doctor wants to prescribe a drug to a patient, he/she would like to know when this drug interacts with other drugs that the patient may already take. A lot of research has been done on each drug, resulting in a lot of articles (often more than 1000 articles per drug). It would not be feasible for a human agent to read all these articles. For this reason it could be interesting to automatically find which drugs are interacting in these articles. Accordingly, in the extraction of drug-drug interactions –DDIs in the following– the task is to find pairs of drugs that are described as interacting in a sentence or a text.

In 2011, for the first time, a challenge on this task was initiated [12]. Several methods were proposed to perform this task [2–4,7,11,14]. The best performing system, i.e. the system with the highest F_1-measure on the given test set, combined several different subsystems in which information from different feature spaces was exploited [14]. Their highest F_1 on the test set was 65.7, and their F_1 for a document-wise 10-fold cross validation on the training data was 60.6. Linguistics features were used, such as part-of-speech, together with different tree

© Springer International Publishing Switzerland 2015
J. Baixeries et al. (Eds.): ICFCA 2015, LNAI 9113, pp. 153–168, 2015.
DOI: 10.1007/978-3-319-19545-2_10

kernels of the dependency parses, i.e. trees describing the grammatical dependencies between words, of the sentences. Another system that was successful in the challenge was based on a union of two different machine learning techniques [3]. The first machine learning technique is a feature-based SVM using different words, morphosyntactic features (internal structural features of words, like number and case) and contextual features (words in between the two considered drugs). The second machine learning technique is a kernel-based method combining three different kernels, namely "shallow linguistic information" (like part-of-speech and word-inflection information), "mildly extended dependency trees" and "phrase structure".

It appears that the most successful systems combine both deep linguistic information, such as dependency trees or phrase structures, and shallow linguistic features, such as word features and morphological information. Thus, we propose to apply a symbolic method, based on pattern structures [6] and Formal Concept Analysis (FCA), to deal with the phrase structure, i.e. the syntactic level, in a different way. A pattern structure can manage a complex data type, such as a tree, and allows one to build a hierarchy of elements of this data type, in the present case a hierarchy of trees. Such a pattern structure comes with a classification technique, called "Lazy Pattern Structure Classification" (LPSC) [9], which classifies the syntactic trees containing drug-drug interactions. This is one original application of pattern structures to syntactic trees and to the task of text-mining (here the mining of DDIs). The method is novel and deserves more research work but we already obtained substantial results showing that the current approach is suitable and valuable.

The rest of the paper is organized as follows. Firstly we explain the pipeline on which relies the proposed system. Then we define the pattern structure for syntactic trees, namely STPS, and as well Lazy Pattern Structure Classification (LPSC). After that, we introduce a projection related to STPS and a set of tree-simplification operations to reduce computational time. Finally we evaluate the method on the corpus of the DDI challenge 2011 [12].

2   The Data and the Pipeline

Our data consists of medical texts containing potential drug-drug interactions, i.e. the training corpus of the DDI extraction challenge 2011 [12]. This corpus consists of around 4200 sentences containing around 23000 potential interactions of which a small portion ($\sim$10 %) is annotated as positive and the rest as negative. In these data drugs in the sentences are already tagged (see Example 1).

If we take a sentence from the data containing n drugs, there are $\binom{n}{2}$ pairs of drugs in the sentence that can potentially interact. Each such pair is represented by a separate sentence, where the two potentially interacting drugs in the sentence are replaced with a drug_tag_r tag, and all other drugs by a drug_tag tag (see Examples 2 and 3 where the corresponding tags are following the name of the tagged drug).

Example 1. Antihistamines (**drug**) may enhance the effects of tricyclic antide-
pressants (**drug**), barbiturates (**drug**), alcohol (**drug**), and other CNS depres-
sants (**drug**).

Example 2. **drug_tag_r** may enhance the effects of **drug_tag**, **drug_tag**, **drug_
tag**, and other **drug_tag_r**.

Example 3. **drug_tag** may enhance the effects of **drug_tag**, **drug_tag_r**, **drug_
tag**, and other **drug_tag_r**.

Each such tagged sentence, representing a possible drug-drug interaction, is
parsed by the Stanford constituency parser v3.4 [8,13]. The resulting trees are
simplified by means of operations that preserve the parts of the tree describing the potential interaction as much as possible. Trees, representing drug-drug
pairs, can be considered as positive or negative. Trees are "positive" when an
interaction is described between the two drugs replaced by the **drug_tag_r** tag
(Example 2). Trees are "negative" when no such interaction is present (Example 3). The positive simplified tree of Example 2 is shown in Fig. 1.

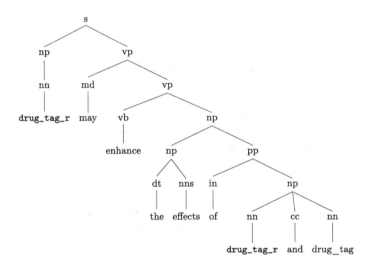

Fig. 1. The simplified syntax tree from Example 2.

A pattern structure is defined on such syntactic trees, whose similarity operator is based on unordered rooted tree intersection. The trees are interpreted
as unordered w.r.t the constituent order in the sentence in order to be able to
generalize over some grammatical structures (eg. conjunctions or enumerations)
without losing important grammatical relations (eg. verb argument relations,
prepositions) as they are also encoded in the hierarchy of the tree. To improve
the computational time of similarity, a projection is introduced, which can be
considered as a simplification of the similarity operator. Pattern structures, similarity and projections are introduced and discussed here after.

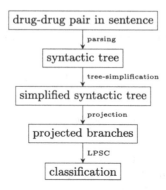

Fig. 2. A schematic view of the pipeline.

The set of trees, obtained from parsing the tagged sentences, is split into a "training set" and a "testing set" and LPSC is used to classify the trees. In the experiments, different settings based on tree simplifications are evaluated. Finally, a schematic view of the pipeline, starting from DDIs and going to LPSC classification, is shown in Fig. 2.

3   A Pattern Structure for Syntactic Trees

Pattern Structures were introduced in [6] and are a generalization of Formal Concept Analysis (FCA) [15]. From the pattern structure perspective, data can be thought of as a set of objects (G) with corresponding records. Each record is an object description, also called a *pattern*, in contrast to a set of binary attributes, as in standard FCA. On the set of potential descriptions (D) a similarity operator should be defined, which should be idempotent, associative and commutative. In this way, the partial ordering on object descriptions is a semilattice and can be used in a similar way as in Formal Concept Analysis to extract (meaningful) concepts from an unstructured data set, in an unsupervised way. More precisely, we have the following definitions.

Definition 1. *Let G be a set of objects, let $(D, \sqcap)$ be a meet-semilattice of potential object descriptions, and let $\delta : G \to D$ be a mapping. Then $(G, (D, \sqcap), \delta)$ is called a* pattern structure, *provided that the set*

$$\delta(G) := \{\delta(g) | g \in G\}$$

generates a complete sub-semilattice $(D_\delta, \sqcap)$ of $(D, \sqcap)$, i.e. every subset X of $\delta(G)$ has an infimum $\sqcap X$ in $(D, \sqcap)$ and D_δ is the set of these infima [6].

On a pattern structure $(G, (D, \sqcap), \delta)$ a Galois connection can be defined, linking sets of objects with set of descriptions.

$$A^\diamond := \prod_{g \in A} \delta(g) \quad \text{for } A \subseteq G$$

$$d^\diamond := \{g \in G | d \sqsubseteq \delta(g)\} \quad \text{for } d \in D$$

A concept (A, d) in $(G, (D, \sqcap), \delta)$ verifies $A^\diamond = d$ and $d^\diamond = A$ where A is the extent and d the intent of the concept. The subsumption order ($\sqsubseteq$) between descriptions c and d is defined as follows:

$$c \sqsubseteq d \Leftrightarrow c \sqcap d = c$$

Concepts are maximal closed sets of objects and their corresponding descriptions. Formal concepts form a concept lattice where the ordering between concepts is given as usual by inclusion of concept extents. The meaning of such a lattice depends on the similarity operator. The same data can be associated with different pattern structures. We look further into defining our pattern structure on trees, in particular syntax trees based on natural language sentences.

3.1   Objects and Object Descriptions

In the current case, the set of objects G in the considered pattern structure $(G, (D, \sqcap), \delta)$ consists of drug-drug pairs, i.e. DDIs, extracted from the collection of sentences. Then the set of object descriptions D is composed of "unordered labeled trees". The resulting pattern structure will be called "Syntactic Tree Pattern Structure" or STPS for short.

Definition 2. *An unordered labeled rooted tree t is a simple connected graph $t = \langle N, E \rangle$, where N is a set of nodes, and E a set of ordered pairs from $N \times N$, called edges. It should satisfy two conditions:*

- *t does not contain any cycle (it is a tree)*
- *t has one distinguished node $r \in N$, called the root node, that is an ancestor of every node $n \in N$.*

In unordered labeled rooted trees, nodes carry a label while there exists no order between the children of each node. This means that the trees in Fig. 3 are considered to be equivalent.

The mapping δ gives for each potential drug-drug pair the corresponding unordered syntactic tree of the sentence in which it occurs, where the drugs are replaced by the tags. Intuitively, one could think of δ as the function that parses the sentence and simplifies the resulting tree.

Fig. 3. Two equivalent unordered labeled rooted trees.

3.2 Similarity Operators

A similarity operator $\sqcap_t$ is defined on the set of object descriptions D. This operator is based on rooted tree intersection. In [1], rooted tree intersection is defined for unordered unlabeled trees and a corresponding algorithm is given. Our definition and implementation follow those in [1], except that we consider trees with labeled nodes.

To define our rooted tree intersection for unordered labeled trees we need to define the notion of rooted subtree first.

Definition 3. *Rooted tree $t_1 = \langle N_1, E_1 \rangle$ is a* rooted subtree *of rooted tree $t_2 = \langle N_2, E_2 \rangle$ (from now written as $t_1 \subseteq_t t_2$) iff the following conditions hold:*

- $N_1 \subseteq N_2$
- $E_1 \subseteq E_2$
- t_1 *and* t_2 *have the same root.*

Using this notion of subtree, we can define a rooted intersection operator on trees.

Definition 4. *The* rooted tree intersection *between tree t_1 and t_2, from now written as $t_1 \sqcap_t t_2$, is the set containing all maximal trees[1] from*

$$\{t \mid t \subseteq_t t_1\} \cap \{t \mid t \subseteq_t t_2\}$$

i.e. the intersection between all subtrees of t_1 and all subtrees of t_2.
An example of such intersection is shown in Fig. 4.

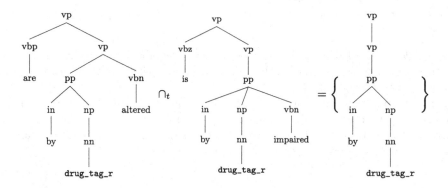

Fig. 4. An example of rooted unordered tree intersection ($\sqcap_t$) of two syntactic tree fragments. The tree on the right side is the maximal rooted subtree of both trees on the left side.

---

[1] The maximal trees from a set X are all trees of X that are not a rooted subtree of another tree in X.

With the notion of rooted tree intersection we can define the similarity operator of our pattern structure.

Definition 5. *The* similarity *between a set of trees A and a set of trees B, written as $A \sqcap_t B$, is the subset of maximal trees from*

$$\bigcup_{(a,b) \in A \times B} a \cap_t b$$

The corresponding subsumption operator is defined as mentioned previously.

$$A \sqsubseteq_t B \Leftrightarrow A \sqcap_t B = A$$

3.3 The Projections for the Syntactic Tree Pattern Structure

Projections for pattern structures were introduced in [6]. A projection is used for weakening the object descriptions and for allowing better performances in computation. Moreover, "good" projections always try to minimize the loss of information.

Definition 6. *A projection of a pattern structure $(G, (D, \sqcap), \delta)$ is a mapping $\psi : D \to D$ that replaces every object description $d \in D$ by $\psi(d)$, such that the original pattern structure is replaced by $(G, (D, \sqcap), \psi \circ \delta)$. It is required that ψ is a kernel operation, i.e. ψ is*

monotone: if $x \sqsubseteq y$, then $\psi(x) \sqsubseteq \psi(y)$,
contractive: $\psi(x) \sqsubseteq x$, and
idempotent: $\psi(\psi(x)) = \psi(x)$.

Projections can be used efficiently to reduce computation time of the similarity operator. For pattern structures of graphs several projections were already proposed and applied in the chemical domain [10]. Here we propose a projection that maps each tree description onto the set of its maximal branches.

Definition 7. *Rooted tree $t_1 = \langle N_1, E_1 \rangle$ is a branch of rooted tree $t_2 = \langle N_2, E_2 \rangle$ iff the following conditions hold:*

– *$t_1 \subseteq_t t_2$*
– *Each node $n_1 \in N_1$ has at most one outgoing edge.*

Definition 8. *The branch projection of a set of rooted trees T, from now written as $\psi_b(T)$, is the set of maximal trees from*

$$\bigcup_{t \in T} \{ b \longrightarrow b \text{ is a branch of } t \}$$

Thus, a tree t defined by a root with n leaves will be projected to a set of size n, containing its branches (see Fig. 5).

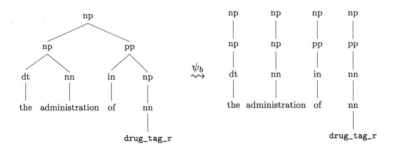

Fig. 5. A tree and its maximal branches.

4 Classification Based on Lazy Hypothesis Evaluation

Actually, the concept lattice resulting from the pattern structure which is defined above has not to be built. Instead, we follow a (kind of) supervised classification method for determining objects whose description includes a syntactic tree effectively representing a DDI, i.e. the drugs lying in the syntactic tree and marked with **drug_tag_r** tags are interacting. We follow a "Lazy Pattern Structure Classification" (LPSC) introduced in [9]. LPSC can classify objects from a given pattern structure in polynomial time w.r.t the cardinality of the set of objects G considered as training data. It is based on a set of positive examples G_+ and a set of negative examples G_-. In the current experiment, positive examples are sentences including interacting drug-drug pairs while negative examples are sentences which do not include interacting drug-drug pairs.

In [9], the classification of a new object o_n is performed w.r.t. two questions:

(1.) Is there a "positive hypothesis" for o_n?
(2.) Is there a "negative hypothesis" for o_n?

A positive hypothesis is defined as a pattern intent in the pattern structure $(G_+, (D, \sqcap), \delta)$ that does not subsume any pattern from $\delta(G_-)$, i.e. does not subsume any negative example. A positive hypothesis for o_n is found iff:

$$\exists g_+ \in G_+ \ \forall g_- \in G_- : (o_n \sqcap g_+^\diamond) \not\sqsubseteq g_-^\diamond$$

In other words, a positive hypothesis for o_n is found if and only if o_n is similar to a positive example g_+, i.e. the potential positive hypothesis, and o_n does not share this similarity with any negative example g_-. A negative hypothesis for o_n is defined symmetrically, by switching the negative and positive examples. How an object is classified depends on the answers for the questions (1.) and (2.), as shown in Table 1.

Our classification criteria differ from that in [9] as we are only looking for positive hypotheses and not for negative hypotheses. The underlying idea is that we assume that typical syntactic trees containing a DDI have some characteristic structures, while trees that do not contain any DDI do not have such characteristic structures. Thus, we discriminate positive and negative hypotheses w.r.t.

Table 1. Criteria in Lazy Pattern Structure Classification according to [9] are displayed on the left, and criteria in LPSC restricted to only positive hypotheses evaluation –used in our experiments– are displayed on the right.

1.	2.	Classification		1.	2.	Classification
yes	yes	undefined		yes	yes	positive
yes	no	positive		yes	no	positive
no	yes	negative		no	yes	negative
no	no	undefined		no	no	negative

the classification criteria, contrasting with [9] where there are also unclassified objects. In our experiment, an object is classified as positive when the first question is answered with "yes", and by complementarity, an object is classified as negative when this first question is answered with "no". This kind of classification was exclusively used in our experiments and is termed as "*Lazy Positive Hypothesis Classification*" (LPHC) (see Table 1).

An example of a positive hypothesis that was found in the experiments with LPHC is shown in Fig. 6. This positive hypothesis was created when classifying the tree corresponding to the potential DDI described in Example 4. Moreover, the positive example from the training set is the tree corresponding to Example 5.

Example 4. Antihistamines (**drug_tag_r**) may partially counteract the anticoagulation effects of heparin (**drug_tag_r**) or warfarin (**drug_tag**).

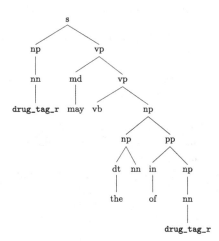

Fig. 6. A positive pattern found in the experiments, created from the two sentences in Example 4 and Example 5. It should be noticed that, for the sake of clarity, this pattern is represented as a tree respecting the word ordering, but actually it is an unordered set of branches.

Example 5. Tricyclic_antidepressants (drug_tag_r) may block the antihypertensive (drug_tag) action of guanethidine (drug_tag_r) and similarly acting compounds.

The tree in Fig. 6 materializes the similarity between Example 4 and Example 5, and is not subsumed by any negative example in the training data. For this reason it is classified as a positive hypothesis for Example 4.

5   The Simplification of Syntactic Trees

When we looked manually at the sentences in the dataset, we remarked that not all parts of some sentences seem to contain useful information about the described DDIs. When a syntactic tree is large, it often takes more time to compute similarity with other trees. Therefore, it is interesting to remove parts of the sentence that are not required to find a DDI. Accordingly, we introduce "tree simplification operations" which are described below.

Constituent Simplification. By means of manually checking the trees, we noticed that some of the constituents are not very informative for describing a DDI in a sentence. In Example 6, it can be seen that an interaction is described between two drug_tag_r tags.

Example 6. In diabetic patients, the metabolic effects of drug_tag_r may decrease blood glucose and therefore drug_tag_r requirements.

However, it can be seen in Example 7 that some parts of the sentence can be removed without altering the description of the interaction.

Example 7. The effects of drug_tag_r may decrease blood glucose and drug_tag_r requirements.

Usually, we can remove the constituents when the tree corresponding to the constituent does not contain any of the possibly interacting drugs, i.e.. any of the two drug_tag_r nodes.

The candidate constituent to be removed that we considered are: (i) adjectives (JJ), (ii) prepositional phrases (PP), (iii) declarative clauses and clauses introduced by a subordinate conjunction such as relative clauses (S, SBAR), (v) adverbial phrases (ADVP) and (vi) parenthetical expressions (PRN). Subtrees of all these six categories that do not contain any of the drug_tag_r nodes are removed from the initial tree. The simplification of the tree corresponding to Example 6 is given in Fig. 7.

NEGVP Renaming. To deal on a simple level with negation, each VP-node, i.e. representing a verb phrase, that directly contains a negating expression (not/no) is renamed as a NEGVP node. In this way a normal VP will not be matched with, or considered similar to, a negated VP.

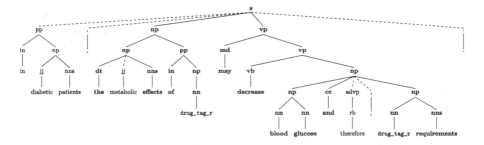

Fig. 7. The original syntactic tree associated with the sentence "In diabetic patients, the metabolic effects of **drug_tag_r** may decrease blood glucose and therefore **drug_tag_r** requirements." The subtrees that will fall off after simplification are indicated with dashed lines.

Lowest-S Simplification. Because relations can sometimes be described very deep in a subordinate clause, only the deepest S-node (i.e. declarative clause) containing both **drug_tag_r** tags is considered, as shown in Fig. 8. This makes sure that deeply nested interaction descriptions can be compared in an easier way to surface interaction descriptions. This way the lowest-S constituent in Example 8 (i.e. in the inner brackets) can be compared to the sentence in Example 9.

Example 8. [*S* **drug_tag**: Clinical studies, as well as post marketing observations, have shown that [*S* **drug_tag_r** can reduce the **drug_tag** effect of **drug_tag_r** and **drug_tag** in some patients].]

Example 9. [*S* **drug_tag_r** agents reduce the renal clearance of **drug_tag_r** and add a high risk of **drug_tag** toxicity.]

However, this rule does not always preserve all crucial information about the potential DDI. In some cases important information can be described at a meta level.

Example 10. [*S* It is not known if [*S* **drug_tag_r** differ in their effectiveness when used with **drug_tag_r**].]

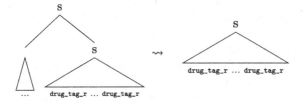

Fig. 8. Schematic view of lowest-S simplification.

In Example 10, both **drug_tag_r** tags occur in the S-constituent indicated by the inner brackets. Thus, when using lowest-S simplification, only the expression in the inner brackets is considered. However, the expression outside of the brackets, i.e. "It is not known if..." contains important information about the DDI description inside. It weakens or even nullifies the interaction that is described inside. For now, we do not have any clear solution to deal with such cases and we ignored them.

Link Contraction. After applying the constituent simplification operation, a resulting tree might contain branches that link nodes holding the same label with only one child. Such cases can be considered as redundant and can be simplified by removing the redundant non-branching duplicate nodes and linking the contracted new node with its single child node. An example is given in Fig. 9.

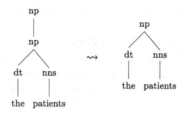

Fig. 9. A tree and its contracted version.

If we apply all these tree simplifications on the trees obtained after parsing the experiment dataset, the average number of nodes in each tree drops from 130 to 41 and the maximum number of nodes from 311 to 138. This shows that the application of these simplification operations have a substantial impact on the set of resulting syntactic trees.

6   Experiments and Discussion

6.1   The Experiment

In this experiment, different settings were evaluated. Each system classifies the potential DDIs by means of lazy pattern structure classification (actually positive hypothesis classification or LPHC). The underlying pattern structure is the one which is described in Sect. 3, using the branch projection. The settings are differing only in the tree simplifications that were applied.

For each setting, a 10-fold cross validation was performed on the data set. The corpus that is used is the training corpus of the DDI extraction challenge 2011 [12]. In this corpus, the drugs are annotated and the interactions are build using the DrugBank, and then manually checked by a domain specialist.

We ran the experiments on a laptop with an i7 Intel processor (using 4 of its 8 virtual cores). The algorithm was implemented in Python. On average, each

Table 2. Results from 10-fold cross validation on the DDI 2011 data set. performance is measured in precision (P), recall (R) and F_1-measure (F_1). In all conditions constituent simplification is applied.

Simplifications	P	R	F_1
1. NEGVP, lowest-S, contraction	0.29786	0.48900	0.37022
2. NEGVP, contraction	0.32261	0.39044	0.35330
3. lowest-S, contraction	0.27073	**0.49450**	0.34990
4. NEGVP, lowest-S	0.33598	0.44712	0.38367
5. NEGVP, lowest-S, vp-map	0.35216	0.44585	0.39350
6. NEGVP, lowest-S, vp-map, prep-map	**0.38556**	0.41328	**0.39894**

object classification took around 2 seconds. This long duration is primarily due to the search for positive hypotheses for each classification. It could be also possible to extract these positive hypotheses on a training set offline. Then they could be used as features in a different classification paradigm, maybe more optimized for a particular task. Here we did not do this as we were mostly interested in increasing the quality of the patterns.

The results from six settings we tested in the experiment are shown in Table 2. When we look at condition 1 and 2 in Table 2, we can see that applying lowest-S simplification strongly increases the recall, by 9.9 %, but also reduces precision by 2.5 %. Overall, F_1 increased by 1.69 %. The reduction in precision, is probably due to some cases where the interaction is not fully described in the lowest declarative clause (lowest S-node). The increase in recall is probably due to the fact that surface clauses can now be compared better to deeper ones.

Applying the link contraction seems to have a weaker but similar effect. However, it decreases the F_1-measure. This can be noticed if we compare setting 1 and 4. After applying link contraction, the precision reduces with 3.8 %, while the recall increases with 4,2 % and the F_1-measure decreases with 1.4 %. It appears that even if trees are non-branching, the hierarchy and its depth are important. Furthermore, if we compare setting 1 and 3, we can see that the NEGVP renaming has a positive effect on precision and only a minor negative effect on recall. It increases the F_1-measure with 2 %.

Settings 5 and 6 are discussed below, in the error analysis.

6.2   Error Analysis

We manually looked both at false positives (i.e. negative trees classified as positive) and false negatives (i.e. positive trees classified as negative). False positives can be analyzed very precisely, because for each positively classified tree, the positive hypothesis from the positive training examples can be examined as well. A few non-mutually exclusive error categories that we found are the following.

1. *Insufficient similarity:* Sometimes, the similarity between the to be classified drug-drug pair and the positive hypothesis is too small to make a proper

classification. This can be due to data sparseness or lack of information in the trees. Often in these cases the similarity between the to be classified tree and its corresponding positive hypothesis does not even contain a verb phrase node. Another frequent case is that the prepositions in the to be classified tree and its positive hypothesis do not match. An example of such poor similarity is given in Fig. 10.

2. *Non-sentences:* Some mistakes seem to occur in phrases that are not full sentences or that are not parsed as such. Often the parser considers these phrases as noun phrases or as "fragments" (i.e. the root node is NP or FRAG). A reason for errors to occur in this category can be that there is not enough training data for these cases, or the parser made a mistake. Again the pattern in Fig. 10 is an example of a non-sentence (an NP).

3. *Mistakes in annotation:* In some cases, a misclassification is due to errors in the drug annotations or in the interaction annotations. Examples of such cases can be found in [14].

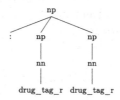

Fig. 10. An example for error category 1 and 2. This pattern is clearly not sufficient for classification. This is due to the lack of a negative example in the training data that subsumes this pattern.

It can be noticed that some patterns may cause false positives, but can at the same time be responsible for a lot of true positives. In our experiments, we did not do any filtering directly based on performance. When the interest is in pure performance, it could be interesting to filter patterns that do not cause any true positives or those that cause more false positives then true positives.

6.3   Similarity Mappings

In error category 1, the similarity between the to be classified tree and its positive hypothesis was too small to make a proper classification. To prevent insufficient similarity, one could manually introduce some linguistically based constraints on the hypotheses and exclude hypotheses that do not satisfy them. We do this by mapping outputs of the similarity operator that do not fulfill the constraints to the empty set, and therefore have no potential for being a hypotheses. Based on the found errors, we introduce two types of similarity mappings: (i) *VP-mapping,* which maps outputs of the similarity operator that do not contain either a VP-node or a NEGVP-node to the empty set, (ii) *Prep-mapping,* which maps outputs of the similarity operator that do contain a prepositional phrase (PP-node) but not the exact preposition to the empty set.

Their result on performances can be found in Table 2. When we compare settings 4 and 5, the "vp-mapping" seems to have a small positive effect on precision (+ 1.6 %), and hardly any effect on recall. When we compare settings 5 and 6, the "prep-mapping" also seems to have a positive effect on precision (+ 3.34 %). However, the recall seems to decrease as well (- 3.3 %). A reason for this could be that a side effect of the "prep-mapping" is that if two trees share a PP-node, but do not share the same preposition, this is considered the same as no PP-node match at all.

7   Conclusions and Future Work

In this paper we presented a new way of analyzing drug-drug interactions in sentences based on FCA. We defined a pattern structure and introduced a projection for syntactic trees. Lazy pattern structure classification was also used to discover informative syntactic patterns, i.e. including DDIs. Furthermore we introduced a set of tree-simplification operations to reduce the size of the syntactic trees. The whole method was evaluated on the training corpus of the DDI extraction challenge 2011.

At present, it can be concluded that in terms of performance the system in its current state does not achieve very high performance. This is probably due to the rigid way the system deals with the found patterns. Furthermore, it should be noticed that this is a single system, using only phrase structure information.

However, from a qualitative point of view, many extracted syntactic patterns seem quite promising. For example, it would be interesting to use these extracted patterns as features in other classification paradigms and this could be included in future research. Another important direction could be to apply parse thickets [5] for the task of DDI detection. A parse thicket is a graph built from the set of syntactic trees of a paragraph. This graph is enriched with the semantic links such as pronoun redirections. The work in [5] is based on pattern structures and, hence, can be adapted to our framework. Finally, other possible future research work could include the search for negative hypotheses, and to enrich the syntactic trees with semantic or morphological features.

References

1. Balcázar, J.L., Bifet, A., Lozano, A.: Intersection algorithms and a closure operator on unordered trees. In: MLG, p. 1 (2006)
2. Björne, J., Airola, A., Pahikkala, T., Salakoski, T.: Drug-drug interaction extraction from biomedical texts with svm and rls classifiers. In: Proceedings of DDIExtraction-2011 challenge task, pp. 35–42 (2011)
3. Chowdhury, F.M., Abacha, A.B., Lavelli, A., Zweigenbaum, P.: Two different machine learning techniques for drug-drug interaction extraction. In: Challenge Task on Drug-Drug Interaction Extraction, pp. 19–26 (2011)

4. Chowdhury, M.F.M., Lavelli, A.: Drug-drug interaction extraction using composite kernels. In: Challenge Task on Drug-Drug Interaction Extraction, pp. 27–33 (2011)
5. Galitsky, B.A., Ilvovsky, D., Kuznetsov, S.O., Strok, F.: Finding maximal common sub-parse thickets for multi-sentence search. In: Croitoru, M., Rudolph, S., Woltran, S., Gonzales, C. (eds.) GKR 2013. LNCS, vol. 8323, pp. 39–57. Springer, Heidelberg (2014)
6. Ganter, B., Kuznetsov, S.O.: Pattern structures and their projections. In: Delugach, H.S., Stumme, G. (eds.) ICCS 2001. LNCS (LNAI), vol. 2120, pp. 129–142. Springer, Heidelberg (2001)
7. Garcia-Blasco, S., Mola-Velasco, S.M., Danger, R., Rosso, P.: Automatic drug-drug interaction detection: A machine learning approach with maximal frequent sequence extraction. In: Challenge Task on Drug-Drug Interaction Extraction, pp. 51–58 (2011)
8. Klein, D., Manning, C.D.: Accurate unlexicalized parsing. In: Proceedings of the 41st Annual Meeting on Association for Computational Linguistics, vol. 1, pp. 423–430. Association for Computational Linguistics (2003)
9. Kuznetsov, S.O.: Fitting pattern structures to knowledge discovery in big data. In: Cellier, P., Distel, F., Ganter, B. (eds.) ICFCA 2013. LNCS, vol. 7880, pp. 254–266. Springer, Heidelberg (2013)
10. Kuznetsov, S.O., Samokhin, M.V.: Learning closed sets of labeled graphs for chemical applications. In: Kramer, S., Pfahringer, B. (eds.) ILP 2005. LNCS (LNAI), vol. 3625, pp. 190–208. Springer, Heidelberg (2005)
11. Minard, A.L., Makour, L., Ligozat, A.L., Grau, B.: Feature selection for drug-drug interaction detection using machine-learning based approaches. In: Challenge Task on Drug-Drug Interaction Extraction, pp. 43–50 (2011)
12. Segura-Bedmar, I., Martınez, P., Sánchez-Cisneros, D.: The 1st DDIExtraction-2011 challenge task: extraction of drug-drug interactions from biomedical texts. Challenge Task on Drug-Drug interaction extraction **2011**, 1–9 (2011)
13. Socher, R., Bauer, J., Manning, C.D., Ng, A.Y.: Parsing with compositional vector grammars. In: Proceedings of the ACL conference, Citeseer (2013)
14. Thomas, P., Neves, M., Solt, I., Tikk, D., Leser, U.: Relation extraction for drug-drug interactions using ensemble learning. In: Challenge Task on Drug-Drug Interaction Extraction, pp. 11–18 (2011)
15. Wille, R.: Restructuring lattice theory: an approach based on hierarchies of concepts. In: Ferré, S., Rudolph, S. (eds.) ICFCA 2009. LNCS, vol. 5548, pp. 314–339. Springer, Heidelberg (2009)

Totally Balanced Formal Context Representation

François Brucker$^{(\boxtimes)}$ and Pascal Préa

École Centrale Marseille – LIF, CNRS UMR 7279, Marseille, France
{francois.brucker,pascal.prea}@centrale-marseille.fr

Abstract. We show in this paper that doubly lexical orders of totally balanced formal context matrices yield a unique graphical representation binding formal contexts, associated concepts and underlying lattice directed cover graphs. Moreover this representation can be done linearly in the size of the formal context matrix.

Keywords: Formal concept analysis · Totally balanced matrices · Dismantlable lattices · Graphical representation

1 Introduction

We will focus in this paper on a special case of formal contexts, those associated with dismantlable lattices. These formal contexts can be helpful in practice because they only have a polynomial number of concepts that are easily computable and, as we shall show in this paper, admit a convenient graphical representation.

Indeed, given a formal context matrix $\mathcal{M}$ whose associated concept lattice is dismantlable, we present a procedure which associates each formal concept to an element of the formal context matrix. This allows to superpose the associated directed cover graph onto the data (Fig. 2). Moreover, since the procedure is linear in the size of the matrix and the graphical representation involves non-overlapping ordered 2-dimensional boxes (see Figs. 5 and 6), it can be used to locally see the interactions between concepts or to explore areas of interest in very large data-sets (through approximation of the original data, which can also be done linearly in the size of the context matrix).

Finally, because dismantlable lattices generalize several models used in clustering (hierarchical clustering and seriation clustering for instance) this procedure can be used in many fields of applications. For instance they are well suited for phylogenetic problems because co-atomic dismantlable lattices are in bijection with strongly chordal graphs (which are a sub-class of chordal graphs — graphs whose every cycle $(x_1, x_2, \ldots, x_n, x_1)$ with $n \geq 3$ contains an edge $x_i x_j$ with $i < j + 1$ — used in perfect phylogeny) and the clusters generated by phylogenetic trees (X-trees) form a dismantlable lattice (see [6] or [5]).

In a theoretical point of view, dismantlable lattices are the "trees" for lattices and there is an ongoing work to see if one can decompose a given lattice into a sum of dismantlable lattices.

© Springer International Publishing Switzerland 2015
J. Baixeries et al. (Eds.): ICFCA 2015, LNAI 9113, pp. 169–182, 2015.
DOI: 10.1007/978-3-319-19545-2_11

Formally speaking, a *formal context* [7] $K = (G, M, I)$ is a triple where G is a set of objects, M a set of attributes and $I \subseteq G \times M$ a binary relation. For subsets $A \subseteq G$ and $B \subseteq M$ two operators (so called *derivation operators*) are defined: $A^{\uparrow} = \{y \mid xIy, \forall x \in A\}$ and $B^{\downarrow} = \{x \mid xIy, \forall y \in B\}$. A *formal concept* associated with K is a pair (A, B) where:

- $A \subseteq G$ and $B \subseteq M$,
- $A = B^{\downarrow}$,
- $B = A^{\uparrow}$.

We shall assume that all sets in this paper are finite. Thus if we label the objects and the attributes of a formal context $K = (G, M, I)$ such that $G = \{l_1, \ldots, l_n\}$ and $M = \{c_1, \ldots, c_m\}$, K is equivalent to a $n \times m$ binary matrix $\mathcal{M}$ (called *formal context matrix*) such that $\mathcal{M}_{i,j} = 1$ if $l_i I c_j$ and 0 otherwise. Formal contexts or its associated formal context matrix can be represented by a cross table (see Table 1).

Table 1. Example of cross table (left) and its associated formal context matrix (right).

	c_1	c_2	c_3	c_4	c_5	c_6	c_7	c_8
l_1		×	×	×				
l_2					×	×	×	
l_3		×	×	×		×		×
l_4							×	×
l_5	×	×	×					

	c_1	c_2	c_3	c_4	c_5	c_6	c_7	c_8
l_1	0	0	1	1	1	0	0	0
l_2	0	0	0	0	0	1	1	1
l_3	0	1	1	1	0	1	0	1
l_4	0	0	0	0	0	0	1	1
l_5	1	1	1	0	0	0	0	0

The formal concepts associated with $K = (G, M, I)$ are exactly the pairs $(A, A^{\uparrow})$ where $A^{\uparrow\downarrow} = A$ (equivalently the pairs $(B^{\downarrow}, B)$ where $B^{\downarrow\uparrow} = B$). It is well known that the set $\mathcal{B}(G, M, I)$ of formal concepts associated with the order $\leq$ defined as $(A_1, B_1) \leq (A_2, B_2)$ whenever $A_1 \subseteq A_2$ (equivalently $B_2 \subseteq B_1$) forms a formal lattice that one can represent by its order diagram (see Fig. 1). The order diagram is a graphical representation of the *directed cover graph* associated with the formal lattice $(\mathcal{B}(G, M, I), \leq)$ which is the directed graph $\mathcal{G} = (\mathcal{B}(G, M, I), E)$ where $uv \in E$ whenever $u \prec v$.

Finally, for any $n \times m$ binary matrix $\mathcal{M}$, one can define the formal context $K_{\mathcal{M}} = (\{1, \ldots, n\}, \{1 \ldots, m\}, I_{\mathcal{M}})$ such that $iI_{\mathcal{M}}j$ whenever $\mathcal{M}_{i,j} = 1$. The column j of $\mathcal{M}$ is equal to $C_j = \{i \mid \mathcal{M}_{i,j} = 1\}$ and the *line* i of $\mathcal{M}$ is equal to $L_i = \{j \mid \mathcal{M}_{i,j} = 1\}$. According to these definitions, it is known [4] that:

- the closure under intersection of $\{C_1, \ldots, C_m\} \cup \{\{1, \ldots, n\}\}$, noted $\mathcal{C}$, is equal to the set $\{A \mid A^{\uparrow\downarrow} = A\}$,
- the closure under intersection of $\{L_1, \ldots, L_n\}$, noted $\mathcal{L}$, is equal to the set $\{B \mid B^{\downarrow\uparrow} = B\}$.

The above equivalence gives a way to link known *hypergraph* classes to known lattice models (an hypergraph is a pair $\mathcal{H} = (X, H)$ where $H \subseteq 2^X$). Indeed it is

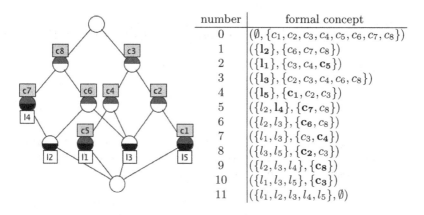

number	formal concept
0	$(\emptyset, \{c_1, c_2, c_3, c_4, c_5, c_6, c_7, c_8\})$
1	$(\{\mathbf{l_2}\}, \{c_6, c_7, c_8\})$
2	$(\{\mathbf{l_1}\}, \{c_3, c_4, \mathbf{c_5}\})$
3	$(\{\mathbf{l_3}\}, \{c_2, c_3, c_4, c_6, c_8\})$
4	$(\{\mathbf{l_5}\}, \{\mathbf{c_1}, c_2, c_3\})$
5	$(\{l_2, \mathbf{l_4}\}, \{\mathbf{c_7}, c_8\})$
6	$(\{l_2, l_3\}, \{\mathbf{c_6}, c_8\})$
7	$(\{l_1, l_3\}, \{c_3, \mathbf{c_4}\})$
8	$(\{l_3, l_5\}, \{\mathbf{c_2}, c_3\})$
9	$(\{l_2, l_3, l_4\}, \{\mathbf{c_8}\})$
10	$(\{l_1, l_3, l_5\}, \{\mathbf{c_3}\})$
11	$(\{l_1, l_2, l_3, l_4, l_5\}, \emptyset)$

Fig. 1. Order diagram (left) and formal concepts (right) associated with the formal context matrix of Table 1 (numbered from bottom to top and left to right).

clear that by labeling X as $\{x_1, \ldots, x_n\}$ and H as $\{h_1, \ldots, h_m\}$, each hypergraph $\mathcal{H}$ is equivalent to a $n \times m$ binary matrix $\mathcal{M}(\mathcal{H})$ where $\mathcal{M}(\mathcal{H})_{i,j} = 1$ whenever $x_i \in h_j$. Conversely, a $n \times m$ binary matrix $\mathcal{M}$ is equivalent to a hypergraph $\mathcal{H}(\mathcal{M}) = (\{1, \ldots, n\}, \{C_1, \ldots, C_m\})$.

We will in this paper focus on *totally balanced hypergraphs*. A hypergraph $\mathcal{H} = (X, H)$ is totally balanced [2] if there is no cycle $(v_1, e_1, \ldots, v_k, e_k)$ with $k \geq 3$ such that:

- $v_i \in e_i \cap e_{i-1}$ (for $i > 1$) and $v_1 \in e_k$,
- $v_i \notin e_j$ for $j \notin \{i, i-1\}$ and $(i, j) \neq (1, k)$.

This class has very nice combinatorial properties (several NP-hard problems become polynomial when focusing on this class) and a clear cluster interpretation when dealing with real data (clusters are connected parts of some tree and restrictions of a totally balanced hypergraph remains totally balanced [8]).

This class is in correspondence with *dismantlable lattices* [6]. Dismantlable lattices where defined recursively by Rival [9] as lattices L for which there is a *doubly irreducible element* x in L (a doubly irreducible element is such that there is at most one element x^- such that $x^- \prec x$ and at most one element x^+ such that $x \prec x^+$) such that $L \backslash \{x\}$ is also dismantlable. For instance, the order diagram of Fig. 1 represents a dismantlable lattice (elements 2, 4 and 5 are doubly irreducible).

Considering a formal context $K = (G, M, I)$, a formal concept (A, B) is doubly irreducible for the associated lattice $(\mathcal{B}(G, M, I), \leq)$ if there exists $(g, m) \in (A, B)$ such that for any formal concept $(U, V) \in \mathcal{B}(G, M, I)$, if $g \in U$, then $A \subseteq U$ and if $m \in V$ then $B \subseteq V$. Thus, a *formal context is totally balanced* (*i.e.* its associated lattice is dismantlable) if and only if there is a decomposition order such that [7]:

- $\mathcal{K}_0 = \mathcal{B}(G, M, I)$,
- there is $(A_i, B_i) \in \mathcal{K}_i$ such that $\exists (g, m) \in (A_i, B_i)$ for which $\forall (U, V) \in \mathcal{K}_i$, $g \in U \Rightarrow A_i \subseteq U$, and $m \in V \Rightarrow B_i \subseteq V$,
- $\mathcal{K}_{i+1} = \mathcal{K}_i \setminus \{(A_i, B_i)\}$
- $\mathcal{K}_{|\mathcal{H}|} = \emptyset$.

Looking at Fig. 1, the formal context admits $2, 7, 4, 8, 10, 5, 1, 9, 6, 3, 0, 11$ as a decomposition order (among many other).

The paper is organized as follows. We will first (Sect. 2) recall some properties of totally balanced matrices and use them in Sects. 3 and 4 to show that there is a one-to-one correspondence between proper formal concepts and some elements of the associated formal concept matrix.

2   Doubly Lexical Ordering of Totally Balanced Matrices

We recall here some properties of totally balanced hypergraphs and show the implication to totally balanced formal concepts.

2.1   Formal Concepts of Totally Balanced Formal Context Matrices

A $n \times m$ binary matrix $\mathcal{M}$ is said to be totally balanced if its associated hypergraph $\mathcal{H}(\mathcal{M})$ is totally balanced. The result from which all the results of this paper will follow is given by Theorem 1 and is linked with gamma-free matrices.

A $n \times m$ binary matrix $\mathcal{M}$ is said to be *gamma-free* whenever for any $1 \leq i < i' \leq n$ and any $1 \leq j < j' \leq m$: $\mathcal{M}_{i,j} = \mathcal{M}_{i,j'} = \mathcal{M}_{i',j} = 1$ implies $\mathcal{M}_{i',j'} = 1$.

For instance, the formal context matrix from Table 1 is not gamma-free because $\mathcal{M}_{1,3} = \mathcal{M}_{1,5} = \mathcal{M}_{3,3} = 1$ and $\mathcal{M}_{3,5} = 0$.

Theorem 1 ([1]). *Let $\mathcal{M}$ be a $n \times m$ binary matrix. $\mathcal{M}$ is totally balanced if and only if there is a gamma-free ordering of its line and columns.*

Even though the formal context matrix from Table 1 is not gamma-free, it admits a gamma-free reordering. See for instance Table 2 which is gamma-free.

Table 2. Cross table associated with a gamma-free ordering of the formal context matrix from Table 1.

	c_1	c_2	c_3	c_4	c_5	c_6	c_7	c_8
l_5	×	×	×					
l_4						×	×	
l_3		×	×	×		×		×
l_2						×	×	×
l_1			×	×	×			

We will use a special gamma-free ordering in Sect. 2.2 which also characterizes totally balanced matrices, but before that, we just state another main property of totally balanced hypergraphs and precise it for totally balanced matrices.

Hypergraphs for which the intersection of three clusters is the intersection of two of them are called *weak hierarchies* [3] and are a very popular model in classification theory because they generalize the well known hierarchical model (the intersection of two clusters is either empty or is one of them). Moreover Proposition 1 shows that the closure under intersection which is usually an expensive operation can be done easily for weak hierarchies.

Proposition 1 ([6]). *Let $\mathcal{H} = (X, H)$ be a totally balanced hypergraph and $A, B, C \in H$. We have: $A \cap B \cap C \in \{A \cap B, A \cap C, B \cap C\}$.*

Indeed, for a weak hierarchical hypergraph $\mathcal{H} = (X, H)$ its closure is simply equal to the clusters $\{A \cap B \mid A, B \in H\}$, which can be done in $\mathcal{O}(|X|^3)$ operations. Moreover, $|H|$ is bound by $|X|^2$ (it derives directly from the closure property), thus:

- the number of formal concepts associated with a totally balanced formal context $K = (G, M, I)$ is bound by $\min(|G|, |M|)^2$,
- the formal concepts associated with a totally balanced formal context $n \times m$ matrix $\mathcal{M}$ are $(C_i \cap C_j, (C_i \cap C_j)^\intercal)$ with $1 \leq i, j \leq m$.

The fact that the intersections of two columns are sufficient to compute all the formal concepts, combined with the doubly lexical ordering of Sect. 2.2 will be the key of all the demonstrations given in Sect. 3.

2.2   Doubly Lexical Ordering of Totally Balanced Matrices

A doubly lexical ordering of an $n \times m$ binary matrix is an ordering such that if the rows and columns are viewed as n or m digit numbers read from right to left for lines and from bottom to top for columns, both rows and columns occur in increasing order. Clearly:

Proposition 2. *A n by m binary matrix $\mathcal{M}$ is doubly lexically ordered if the two following assertions are satisfied:*

- *for any $1 \leq j \leq m$ if there exist $1 \leq i < i' \leq n$ such that $\mathcal{M}_{i,j} = 1$ and $\mathcal{M}_{i',j} = 0$ then there exists $j' > j$ with $\mathcal{M}_{i,j'} = 0$ and $\mathcal{M}_{i',j'} = 1$,*
- *for any $1 \leq i \leq n$ if there exist $1 \leq j < j' \leq m$ such that $\mathcal{M}_{i,j} = 1$ and $\mathcal{M}_{i,j'} = 0$ then there exists $i' > i$ with $\mathcal{M}_{i',j} = 0$ and $\mathcal{M}_{i',j'} = 1$.*

Any $n \times m$ binary matrix can be doubly lexically ordered. Several polynomial algorithms to perform such an ordering exist, see for instance Spinrad [10] for a $\mathcal{O}(nm)$ algorithm (linear in the size of $\mathcal{M}$). Table 3 is a doubly lexical ordering of Table 1.

This ordering allows another characterization of totally balanced matrices:

Table 3. Cross table associated with a doubly lexical ordering of the formal context matrix from Table 1.

	c_5	c_1	c_4	c_2	c_3	c_6	c_7	c_8
l_1	×		×		×			
l_5		×		×	×			
l_3			×	×	×	×		×
l_4						×	×	
l_2						×	×	×

Theorem 2 ([1]). *The three following assertions are equivalent:*

1. $\mathcal{M}$ *is a totally balanced binary matrix,*
2. *there is a doubly lexical ordering of* $\mathcal{M}$ *which is gamma-free,*
3. *every doubly lexical ordering of* $\mathcal{M}$ *is gamma-free.*

Since the formal context from Table 1 is totally balanced, its ordering given in Table 3 is gamma-free. Note that a gamma-free ordering is not necessarily doubly lexically ordered. For instance Table 2 is gamma-free but not doubly lexically ordered. In fact the doubly lexical order "packs" the formal concepts together as we shall show it hereafter.

3   Formal Concepts of Totally Balanced Formal Context Matrices

We shall prove here that, given a totally balanced formal context $K = (G, M, I)$, one can associate to each formal concept (A, B) an element (i, j) of its doubly lexically ordered formal context matrix $\mathcal{M}$.

We will assume without any loss of generality that the formal context K is such that:

- $G = \{1, \ldots, n\}$,
- $M = \{1, \ldots, m\}$,
- its associated formal context matrix $\mathcal{M}$ such that $\mathcal{M}_{i,j} = 1$ whenever iIj is doubly lexically ordered.

If $\mathcal{B}(G, M, I)$ is the set of all the formal concepts, we denote by $\mathring{\mathcal{B}}(G, M, I)$ the set of all the *proper formal concepts*: $\mathring{\mathcal{B}}(G, M, I) = \mathcal{B}(G, M, I) \backslash \{(\emptyset, G), (M, \emptyset)\}$. We define $f : \mathring{\mathcal{B}}(G, M, I) \to G \times M$ and $g : G \times M \to 2^G \times 2^M$ by:

$$f((A, B)) = (\min(A), \min(B))$$
$$g((i, j)) = (\{i' \mid \mathcal{M}_{i',j} = 1, i' \geq i\}, \{j' \mid \mathcal{M}_{i,j'} = 1, j' \geq j\})$$

Consider for instance the doubly lexically ordered formal context of Table 3. In order to fit the above definitions, we have to write its associated formal concepts (see Fig. 1) according to the lines and the columns indices, thus the formal concept $(\{l_2, l_3\}, \{c_6, c_8\})$ (number 6 in Fig. 1) is here equal to $((3, 5), (6, 8))$. We have then:

- $f(((3,5),(6,8))) = (3,6)$,
- $g((3,6)) = ((3,5),(6,8))$

Let $\mathcal{M}$ be an $n \times m$ doubly lexically ordered totally balanced formal context matrix. We denote by $\mathcal{B}(\mathcal{M})$ the set of all formal representative of $\mathcal{M}$. A pair (i,j) of $G \times M$ is a *formal representative* if:

- $\mathcal{M}_{i,j} = 1$,
- either $i = 1$ or there exists $j' \geq j$ such that $\mathcal{M}_{i,j'} = 1$ and $\mathcal{M}_{i-1,j'} = 0$,
- either $j = 1$ or there exists $i' \geq i$ such that $\mathcal{M}_{i',j} = 1$ $\mathcal{M}_{i',j-1} = 0$.

The goal of this section is to establish links between formal representatives and formal concepts. We first need some Lemmas. Note that Lemmas 1 and 2 only suppose that the formal context matrix is gamma-free, only Lemma 3 requires that $\mathcal{M}$ sould be doubly lexically ordered.

Lemma 1. *Let $\mathcal{M}$ be a $n \times m$ gamma-free ordered formal context matrix associated with a formal context $F = (G,M,I)$ and $\mathring{\mathcal{B}}(G,M,I)$ the set of its associated proper formal concepts. For all $(A,B) \in \mathring{\mathcal{B}}(G,M,I)$, $g(f((A,B))) = (A,B)$.*

Proof. We will first prove that there exist j_A and i_B such that $A = \{i \mid \mathcal{M}_{i,j_A} = 1, i \geq min(A)\}$ and $B = \{j \mid \mathcal{M}_{i_B j} = 1, j \geq min(B)\}$.

According to Proposition 1, A is either a column $C_j = \{\mathcal{M}_{i,j} = 1 \mid i \geq 1\}$ or an intersection of two columns $C_j \cap C_{j'}$ and one can assume that $j < j'$. If $A = C_j$ then $A = \{i' \mid \mathcal{M}_{i',j} = 1 i' \geq 1\} = \{i' \mid \mathcal{M}_{i',j} = 1, i' \geq min(A)\}$. If $A = C_j \cap C_{j'}$ then $min(A)$ is the smallest line number i for which $\mathcal{M}_{i,j} = \mathcal{M}_{i,j'} = 1$ and since $\mathcal{M}$ is gamma-free, $\mathcal{M}_{i',j} = 1$ implies $\mathcal{M}_{i',j'} = 1$ for all $i' \geq i$: $A = C_j \cap C_{j'} = \{i' \mid \mathcal{M}_{i',j} = 1, i' \geq min(A)\}$.

Since $^t\mathcal{M}$ is also gamma-free, the same proof leads to the fact that there exists i such that $B = \{j' \mid \mathcal{M}_{i,j'} = 1, j' \geq min(B)\}$.

Now, since $\mathcal{M}_{i,j_A} = 1$ for any $i \in A$ we have that $j_A \in B$ (because $A^\uparrow = B$) thus $min(B) \leq j_A$. Moreover, $\mathcal{M}_{i,min(B)} = 1$ for all $i \in A$ (because $B^\downarrow = A$) thus $\{i \mid \mathcal{M}_{i,min(B)} = 1, i \geq min(A)\} = \{i \mid \mathcal{M}_{i,j_A} = 1, i \geq min(A)\}$ (because $\mathcal{M}$ is gamma-free). This proves that $A = \{i \mid \mathcal{M}_{i,min(B)} = 1, i \geq min(A)\}$.

The same kind of proof can be done to prove that $B = \{j \mid \mathcal{M}_{min(A),j} = 1, j \geq min(B)\}$ which concludes the proof. $\square$

Lemma 1 shows that f is an injection and that $g = f^{-1}$ for its image. The following two Lemmas will characterize the image of f.

Lemma 2. *Let $\mathcal{M}$ be a $n \times m$ gamma-free ordered formal context matrix associated with a formal context $F = (G,M,I)$ and $\mathring{\mathcal{B}}(G,M,I)$ the set of its associated proper formal concepts. For any $(A,B) \in \mathring{\mathcal{B}}(G,M,I)$:*

- *either $min(B) = 1$ or there exists $j \geq min(B)$ such that $\mathcal{M}_{min(A),j-1} = 0$,*
- *either $min(A) = 1$ or there exists $i \geq min(A)$ such that $\mathcal{M}_{i-1,min(B)} = 0$.*

Proof. We will only prove the first assertion, the second one follows by using $^t\mathcal{M}$ instead of $\mathcal{M}$.

Lemma 1 shows that $A = \{i \mid \mathcal{M}_{i,\min(B)} = 1, i \geq \min(A)\}$ and according to Proposition 1, A is either a column $C_j = \{\mathcal{M}_{i,j} = 1 \mid i \geq 1\}$ or an intersection of two columns $C_j \cap C_{j'}$ and one can assume that $j < j'$.

Columns C_j are equal to $\{i \mid \mathcal{M}_{i,j} = 1, i \geq i(C_j)\}$ with $i(C_j) = \min\{i \mid \mathcal{M}_{i,j} = 1\}$. Thus either $i(C_j) = 1$ or $\mathcal{M}_{i(C_j)-1,j} = 0$.

Intersection $C_j \cap C_{j'}$ with $1 \leq j < j' \leq m$. Let $i(C_j \cap C_{j'}) = \min\{i \mid \mathcal{M}_{i,j} = \mathcal{M}_{i,j'} = 1\}$. Since $\mathcal{M}$ is gamma-free $\mathcal{M}_{i',j} = 1$ implies $\mathcal{M}_{i',j'} = 1$ for all $i' \geq i(C_j \cap C_{j'})$, thus $C_j \cap C_{j'} = \{i \mid \mathcal{M}_{i,j} = 1, i \geq i(C_j \cap C_{j'})\}$. Note that if $i(C_j \cap C_{j'}) > 1$ then either $\mathcal{M}_{i(C_j \cap C_{j'})-1,j} = 0$ or $\mathcal{M}_{i(C_j \cap C_{j'})-1,j'} = 0$. The above two cases show that if $\min(B) > 1$ and $\mathcal{M}_{\min(A),j-1} = 1$ for all $j \geq \min(B)$ then $A = \{i \mid \mathcal{M}_{i,\min(B)} = 1, i \geq \min(A)\}$ cannot be an intersection of columns thus $(A, A^\uparrow)$ is not a formal concept, which is impossible.   □

Lemma 3. *Let $\mathcal{M}$ be an $n \times m$ doubly lexically ordered totally balanced formal context matrix associated with a formal context $F = (G, M, I)$. Let (i, j) a pair such that:*

- $\mathcal{M}_{i,j} = 1$,
- *either $i = 1$ or there exists $j' \geq j$ such that $\mathcal{M}_{i,j'} = 1$ and $\mathcal{M}_{i-1,j'} = 0$,*
- *either $j = 1$ or there exists $i' \geq i$ such that $\mathcal{M}_{i',j} = 1$ $\mathcal{M}_{i',j-1} = 0$.*

We have $g((i,j)) \in \mathring{\mathcal{B}}(G, M, I)$.

Proof. Let $\mathcal{C}$ be the closure under intersection of the matrix columns $C_j = \{i \mid \mathcal{M}_{i,j} = 1\}$. We will prove by induction on the column number that if $\mathcal{M}_{i,j} = 1$ and $\mathcal{M}_{i-1,j} = 0$ then $\{i' \mid \mathcal{M}_{i',j} = 1, i' \geq i\} \in \mathcal{C}$. First consider the last column. Since $\mathcal{M}$ is doubly lexically ordered one cannot have $\mathcal{M}_{i,m} = 1$ and $\mathcal{M}_{i',m} = 0$ with $1 \leq i < i' \leq n$. The last column is then either empty or there exists i_m for which $\mathcal{M}_{i,m} = 1$ if and only if $i \geq i_m$ thus the only possible pair (i, m) for which $\mathcal{M}_{i,m} = 1$ and $\mathcal{M}_{i-1,m} = 0$ is (i_m, m) and $\{i' \mid \mathcal{M}_{i',m} = 1, i' \geq i_m\} = C_m \in \mathcal{C}$. Suppose the property true for any $j' > j_0$ and let j be the largest column smaller or equal to j_0 such that there exists i for which $\mathcal{M}_{i,j} = 1$ and $\mathcal{M}_{i-1,j} = 0$. two cases may occur:

- $\mathcal{M}_{i',j} = 0$ for all $i' < i$. In this case $\{i' \mid \mathcal{M}_{i',j} = 1, i' \geq i\} = C_j \in \mathcal{C}$,
- otherwise let $i_2 < i$ be the largest index such that $\mathcal{M}_{i_2,j} = 1$. Since $\mathcal{M}$ is doubly lexically ordered, there exists $j_2 > j$ such that $\mathcal{M}_{i_2,j_2} = 0$ and $\mathcal{M}_{i-1,j_2} = 1$. Two sub-cases are possible:
  - $\mathcal{M}_{i,j_2} = 0$. Since $\mathcal{M}$ is doubly lexically ordered there exists $j_3 > j_2$ such that $\mathcal{M}_{i-1,j_3} = 0$ and $\mathcal{M}_{i,j_3} = 1$. The induction hypothesis holds, so $\{i' \mid \mathcal{M}_{i',j_3} = 1, i' \geq i\} \in \mathcal{C}$. Because $\mathcal{M}$ is gamma-free, $\{i' \mid \mathcal{M}_{i',j} = 1, i' \geq i\} = C_j \cap \{i' \mid \mathcal{M}_{i',j_3} = 1, i' \geq i\} \in \mathcal{C}$.
  - $\mathcal{M}_{i,j_2} = 1$. There exists $i_3 < i$ for which $\mathcal{M}_{i',j_2} = 1$ for any $i_3 \leq i' \leq i$ and $\mathcal{M}_{i_3-1,j_2} = 0$. The induction hypothesis holds, so $\{i' \mid \mathcal{M}_{i',j_2} = 1, i' \geq i_3\} \in \mathcal{C}$. As $\mathcal{M}$ is gamma-free, $\{i' \mid \mathcal{M}_{i',j} = 1, i' \geq i\} = C_j \cap \{i' \mid \mathcal{M}_{i',j_2} = 1, i' \geq i_3\} \in \mathcal{C}$.

In all cases $\{i' \mid \mathcal{M}_{i',j} = 1, i' \geq i\} \in \mathcal{C}$. This concludes the proof by induction.

Let now (i,j) be a pair as stated in the Lemma. If $i = 1$ then $A = \{i' \mid \mathcal{M}_{i',j} = 1, i' \geq i\} \in \mathcal{C}$ and if $i > 1$ there exists $j' \geq j$ such that $\mathcal{M}_{i,j'} = 1$ and $\mathcal{M}_{i-1,j'} = 0$. The above induction demonstration states that $\{i' \mid \mathcal{M}_{i',j'} = 1, i' \geq i\} \in \mathcal{C}$ thus $A = \{i' \mid \mathcal{M}_{i',j} = 1, i' \geq i\} = C_j \cap \{i' \mid \mathcal{M}_{i',j'} = 1, i' \geq i\} \in \mathcal{C}$.

Since $\mathcal{M}$ is gamma-free, $\mathcal{M}_{i,j'} = 1$ with $j' \geq j$ implies that $\mathcal{M}_{i',j'} = 1$ for all $i' \in A$ thus $B = \{j' \mid \mathcal{M}_{i,j'} = 1, j' \geq j\} \subseteq A^\uparrow$. If $j = 1$ we clearly have equality. If $j > 1$ let i' be the largest line larger than i for which $\mathcal{M}_{i',j} = 1$ $\mathcal{M}_{i',j-1} = 0$. If there exists $j' < j$ such that $j' \in A^\uparrow$, then $\mathcal{M}_{i',j'} = 1$. Since $\mathcal{M}$ is doubly lexically ordered and $\mathcal{M}_{i',j'} = 1$ and $\mathcal{M}_{i',j-1} = 0$ there exists $i'' > i'$ for which $\mathcal{M}_{i'',j'} = 0$ and $\mathcal{M}_{i'',j-1} = 1$ which implies $\mathcal{M}_{i'',j} = 0$ because $j' \in A^\uparrow$. The doubly lexical ordering of $\mathcal{M}$ then states that there exists $i''' > i''$ for which $\mathcal{M}_{i''',j-1} = 0$ and $\mathcal{M}_{i''',j} = 1$ which is impossible by maximality of i' so such a j' does not exist.

We finally have that $A = \{i' \mid \mathcal{M}_{i',j} = 1, i' \geq i\} \in \mathcal{C}$ and $A^\uparrow = \{j' \mid \mathcal{M}_{i,j'} = 1, j' \geq j\}$ which concludes the proof.   □

The hereunder Proposition 3 directly follows from Lemmas 1, 2 and 3.

Proposition 3. *The sets $\mathring{\mathcal{B}}(G, M, I)$ and $\mathcal{B}(\mathcal{M})$ are in one-to-one correspondence:*

- *for all $(A, B) \in \mathring{\mathcal{B}}(G, M, I)$: $f((A, B)) = (\min(A), \min(B)) \in \mathcal{B}(\mathcal{M})$,*
- *for all $(i, j) \in \mathcal{B}(\mathcal{M})$: $g((i, j)) = (\{i' \mid \mathcal{M}_{i',j} = 1, i' \geq i\}, \{j' \mid \mathcal{M}_{i,j'} = 1, j' \geq j\}) \in \mathring{\mathcal{B}}(G, M, I)$,*
- *$g \circ f((A, B)) = (A, B)$,*
- *$f \circ g((i, j)) = (i, j)$.*

Table 4. Formal concepts of Fig. 1 represented by their associated pair of the doubly lexical ordering of the formal context from Table 3.

	c_5	c_1	c_4	c_2	c_3	c_6	c_7	c_8
l_1	(2)		(7)		(10)			
l_5		(4)		(8)	×			
l_3			(3)	×	×	(6)		(9)
l_4						(5)	×	
l_2						(1)	×	×

0	$(\emptyset, \{c_1, c_2, c_3, c_4, c_5, c_6, c_7, c_8\})$
1	$(\{\mathbf{l_2}\}, \{c_6, c_7, c_8\})$
2	$(\{\mathbf{l_1}\}, \{c_3, c_4, \mathbf{c_5}\})$
3	$(\{\mathbf{l_3}\}, \{c_2, c_3, c_4, c_6, c_8\})$
4	$(\{\mathbf{l_5}\}, \{\mathbf{c_1}, c_2, c_3\})$
5	$(\{l_2, \mathbf{l_4}\}, \{\mathbf{c_7}, c_8\})$
6	$(\{l_2, l_3\}, \{\mathbf{c_6}, c_8\})$
7	$(\{l_1, l_3\}, \{c_3, \mathbf{c_4}\})$
8	$(\{l_3, l_5\}, \{\mathbf{c_2}, c_3\})$
9	$(\{l_2, l_3, l_4\}, \{\mathbf{c_8}\})$
10	$(\{l_1, l_3, l_5\}, \{\mathbf{c_3}\})$
11	$(\{l_1, l_2, l_3, l_4, l_5\}, \emptyset)$

Table 4 shows the proper formal concepts of the formal context of Table 1 with the reordering of Table 3. Note that it is easy to find all the pairs of $\mathcal{B}(\mathcal{M})$ in $\mathcal{O}(nm)$ (the size of the formal context matrix) operations.

Proposition 3 and the fact that the formal context matrix is gamma-free allow us to state Proposition 4.

Proposition 4.

- $(A, B) \leq (A', B')$ *implies that* $\min(A) \leq \min(A')$ *and* $\min(B) \leq \min(B')$: *the formal concepts larger that a given* (A, B) *are associated with a pair in the top-right corner of* $f((A, B))$,
- *let* $(i, j), (i', j') \in \mathcal{B}(\mathcal{M})$. $g((i, j)) \leq g((i', j'))$ *is equivalent to* $\mathcal{M}_{i,j'} = 1$,
- *let* $(i, j), (i', j') \in \mathcal{B}(\mathcal{M})$. $g((i, j)) \geq g((i', j'))$ *is equivalent to* $\mathcal{M}_{i',j} = 1$,

Proposition 4 gives a way to show all the formal concepts of a given totally balanced formal context and their order relationship on a special ordering of its associated formal context matrix. We shall extend this property in the next section by showing that the whole formal concept lattice can be embedded into the matrix.

4   Cover Graphs and Graphical Representation

In this section, we show that the formal representatives can be associated to non-overlapping boxes of the formal context matrix. This also represents the directed cover graph.

Let $\mathcal{M}$ be a $n \times m$ doubly lexically ordered totally balanced formal context matrix associated with a formal context $F = (G, M, I)$, $\overset{\circ}{\mathcal{B}}(G, M, I)$ the set of all its proper formal concepts and $\mathcal{B}(\mathcal{M})$ the set of formal representatives.

We associate the pair $(n + 1, 1)$ to the formal concept $(\emptyset, \{1, \ldots, m\})$ and the pair $(1, m + 1)$ to the formal concept $(\{1, \ldots, n\}, \emptyset)$. So we extend maps f and g to all the formal concepts $\mathcal{B}(G, M, I)$. Thus we can draw the directed cover graph directly on $\mathcal{M}$ as shown in Fig. 2. For the rest of this section, we shall then consider that $g((n + 1, 1)) = (\emptyset, \{1, \ldots, m\})$ and that $g((1, m + 1)) = (\{1, \ldots, n\}, \emptyset)$.

By Proposition 3, for such a figure, the edges $((u_i, u_j), (v_i, v_j))$ of the directed cover graph are such that $u_i \geq v_i$ and $u_j \leq v_j$ and at least one of these inequalities is strict.

We shall precise this drawing by using all the 1 s of the matrix. Let $(i, j) \in \mathcal{B}(\mathcal{M})$, we define:

- $n(i, j)$: the smallest column such that $n(i, j) > j$ and for which there exists $i' \geq j$ with $\mathcal{M}_{i',n(i,j)} = 1$ and $\mathcal{M}_{i',n(i,j)-1} = 0$. If i' does not exist, $n(i, j) = m + 1$.
- $w(i, j)$: the largest column such that $j \leq w(i, j) < n(i, j)$ and $\mathcal{M}_{i,j''} = 1$ for all $j \leq j'' \leq w(i, j)$. Note that $w(i, j)$ always exists (it can be equal to j).

It is clear that, given two formal representatives (i, j) and (i', j'), the intersection of the two intervals $[j, w(i, j)]$ and $[j', w(i', j')]$ is either empty or one of them. The set of all the intervals is a hierarchy. Moreover, if $[j, w(i, j)] \not\subset [j', w(i', j')]$ then $i < i'$.

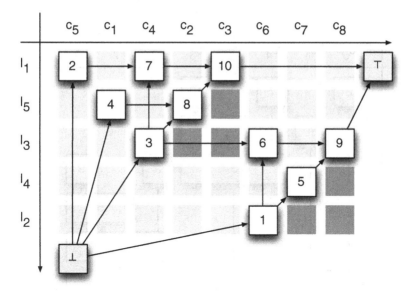

Fig. 2. Directed cover graph of the totally balanced formal context of Table 3. The unmarked 1 of the matrix are in dark grey (Color figure online).

Figure 3 shows the hierarchy formed by the formal concepts of the totally balanced formal context of Table 3. Note that this hierarchy can be directly drawn on the doubly lexically ordered context matrix and that it is a part of the associated directed cover graph.

Since ${}^t\mathcal{M}$ is also doubly lexically ordered and gamma-free, if we note ${}^t w(i,j)$ the element of $\mathcal{M}$ associated with $w(j,i)$ in ${}^t\mathcal{M}$, there is a also a hierarchy formed by the intervals $[i, {}^t w(i,j)]$. Figure 4 shows this hierarchy for the formal concepts of the totally balanced formal context of Table 3.

Note that the union of the two hierarchies forms the directed cover graph. Some of the 1 s of the matrix ${}^t\mathcal{M}$ are not "used". The hereunder procedure combines the two drawings into a unique one. Given a formal representative (i,j), the box formed by the two intervals $[i, {}^t w(i,j)] \times [j, w(i,j)]$ is full of 1 s in $\mathcal{M}$ and all the 1 s of $\mathcal{M}$ are at least in one box. If we do not want to allow boxes overlapping we have to favor one dimension upon the other. In the hereunder construction we favor $[j, w(i,j)]$ upon $[i, {}^t w(i,j)]$.

We associate to each $1 \leq i \leq n$, $1 \leq j \leq m$ such that $\mathcal{M}_{i,j} = 1$ a formal concept $g((i',j'))$:

- where $i' = i$ and j' is such that there exists $(i,j') \in \mathcal{B}(\mathcal{M})$ for which $j \leq j' < w(i,j')$,
- or where $i < i'$ and i' the largest element such that there exists $(i',j') \in \mathcal{B}(\mathcal{M})$ for which $j \leq j' < j + w(i',j')$.

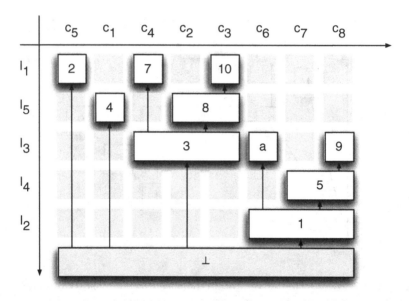

Fig. 3. Interval hierarchy of the columns of the formal concepts of the formal context of Table 3.

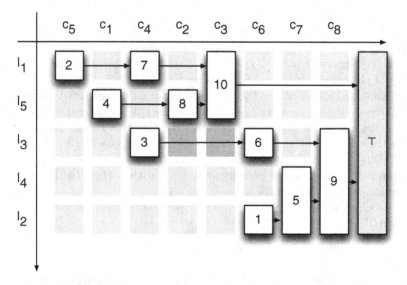

Fig. 4. Interval hierarchy of the lines of the formal concepts of the formal context of Table 3.

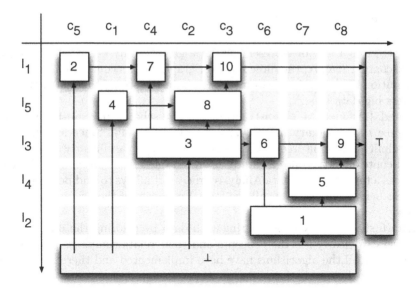

Fig. 5. Directed cover graph embedded into the formal context matrix of Table 4.

Fig. 6. Approximation of a 50×170 Matrix.

The above construction ensures that each formal concept is represented by two intervals forming a "box full of 1s" in $\mathcal{M}$ and that all these boxes never overlap. Note that the above association is unique and that each formal representative is self-associated. This association can be easily done in $\mathcal{O}(nm)$ operations.

In addition, one can prove that: $g((i,j)) \prec g((i',j'))$ if their associated boxes in $\mathcal{M}$ are neighbors. One can then draw the associated directed cover graph of the formal context on the matrix as shown in Fig. 5.

5   Conclusion

This paper gives a way of representing, for a given totally balanced formal context, the context matrix, the directed cover graph and its concepts into a unique representation. This can be performed linearly in the size of the matrix allowing to address big data sets.

Spinrad [11] gives a $\mathcal{O}(nm)$ approximation scheme to transform a non-gamma-free $n \times m$ binary matrix into a gamma-free one. Since a gamma-free matrix admits a gamma-free doubly lexical order, one can linearly approximate a formal context matrix into a totally balanced one. This can be done for large datasets as a first step of a Data Analysis process. It allows to find potential areas of interest before searching for all the formal concepts of the original context in this area.

Figure 6 shows a way of presenting such data by coloring the different concepts. One can clearly see the concepts and their relationships despite the size of the matrix. All the algorithms have been implemented and there is a ongoing study to apply the on big datasets.

References

1. Antsee, R., Farber, M.: Characterizations of totally balanced matrices. J. Algorithms **5**, 215–230 (1984)
2. Berge, C.: Sur certains hypergraphes généralisant les graphes bipartites. Comb. Theor. Appl. I, Colloquium Math. Soc. Janos Bolyai **4**, 119–133 (1970)
3. Bandelt, H.-J., Dress, A.: Weak hierarchies associated with similarity measures - an additive clustering technique. Bull. Math. Biol. **51**, 133–166 (1989)
4. Barbut, C., Monjardet, B.: Ordre et Classification, algèbre et Combinatoire. Hachette, Paris (1970)
5. Brucker, F., Gély, A.: Crown-free lattices and their related graphs. Order **28**, 443–454 (2009)
6. Brucker, F., Gély, A.: Parsimonious cluster systems. Adv. Data Anal. Classif. **3**, 189–204 (2009)
7. Ganter, B., Wille, R.: Formal Concept Analysis. Springer, Heidelberg (1999)
8. Lehel, J.: A characterization of totally balanced hypergraphs. Discrete Math. **57**, 59–65 (1985)
9. Rival, I.: Lattices with doubly irreducible elements. Can. Math. Bull. **17**(1), 91–95 (1974)
10. Spinrad, J.: Doubly lexical ordering of dense 0/1 matrices. Inf. Process. Lett. **45**, 229–235 (1993)
11. Spinrad, J.: Efficient Graph representations. American Mathematical Society, USA (2003)

Randomized Fuzzy Formal Contexts and Relevance of One-Sided Concepts

Lubomir Antoni$^{(\boxtimes)}$, Stanislav Krajči, and Ondrej Krídlo

Institute of Computer Science, Faculty of Science, Pavol Jozef Šafárik University
in Košice, Jesenná 5, 040 01 Košice, Slovakia
lubomir.antoni@student.upjs.sk,
{stanislav.krajci,ondrej.kridlo}@upjs.sk

Abstract. We define the randomized fuzzy formal context using the random variables with a normal distribution and explore the one-sided formal concept stability. Since the modified Rice-Siff algorithm aims at reducing the concept lattice and represents a crisp index in selecting the relevant clusters from the set of all one-sided formal concepts, we describe the probabilistic method and algorithm how to rank these clusters. Therefore, the proposed Gaussian probabilistic index in combination with the modified Rice-Siff algorithm gives the answer how to select top-k relevant one-sided formal concepts.

Keywords: One-sided formal concept · Randomized context · Gauss normal distribution · Stability

1 Introduction

The efficient selection of relevant formal concepts is an interesting and important issue for investigation and several studies have focused on this scalability question in formal concept analysis [21]. In this direction, the stability index of Kuznetsov [32] represents the proportion of subsets of attributes of a given concept whose closure is equal to the extent of this concept (in an extensional formulation). A high stability index signalizes that extent does not disappear if the extent of some of its attributes is modified. It helps to isolate concepts that appear because of noisy objects in [25] and the completely restoring of the original concept lattice is possible with combination of two other indices.

The phenomenon of the basic level of concepts is advocated to select important formal concepts in Bělohlávek et al. [7]. Five quantitative approaches on the basic level of concepts and their metrics are comparatively analyzed in [8]. The approaches on selecting of the formal concepts and simplifying the concept lattices are examined by [5,13,16,26,31,33], as well. The notes on the scalability and the applications of concept stability in the various fields from social networks to linguistics can reader find in thorough overview [37].

This work was supported by grant VEGA 1/0073/15 by the Ministry of Education, Science, Research and Sport of the Slovak republic.

J. Baixeries et al. (Eds.): ICFCA 2015, LNAI 9113, pp. 183–199, 2015.
DOI: 10.1007/978-3-319-19545-2_12

Regarding one-sided concept lattices [6,9,27], the antecedent method [40] of searching for significant concepts takes into account the idea of α-cuts from fuzzy logic. A modified Rice-Siff algorithm represents the crisp version of a method for selecting the significant concepts [28] on the set of one-sided fuzzy formal concepts, the application of this algorithm in the field of social network were presented by [27,29]. The generalized one-sided concept lattices [10,24] and the reduction of concepts from the set of generalized one-sided fuzzy concepts based on subset quality measure [11] were recently introduced in this area. Besides one-sided concept lattices and the selection of relevant concepts, the properties of other generalizations in formal concept analysis [1,2,10,34–36,38], intercontextual relationships [30], construction of adjoints in the unstructured codomain or the residuated operations on the multilattices [12,22] are intensively studied. In [41], Valverde-Albacete et al. explore the entropy conformation process with respect to the behaviour of incidences in L-valued contexts for a complete idempotent semifield L which generates different kinds of Galois connections. Possibility theory [17], as an alternative to probability theory, leads to partition a formal context into disjoint subsets of objects having distinct properties. Bartl [3] describes the simple probabilistic algorithm for finding all minimal solutions for generalized fuzzy relational equation. A formula for the mean of the size of the random Galois lattice and convergence of random empirical intents are established by Emillion and Lévy in [18,19].

In this paper, we propose a probabilistic approach to the one-sided formal concepts stability. We define the randomized formal contexts which are derived from the original fuzzy formal context by fluctuation of values using the random variables with a normal distribution. The proposed Gaussian probabilistic index represents the probability of some subset being the extent of the particular randomized formal context. We describe the algorithm and study the properties of the Gaussian probabilistic index with respect to the multivariate normal distribution, the variances of normal distributions and the boundary constraints of the incidence relation in the randomized formal contexts.

We remind that the idea of joining the probability and the stability of crisp formal concepts can be already found in [25], but the notion of concept probability reflects here the need to normalize the stability index in a crisp setting and the binomial distribution is sufficient. Our approach concerns with the probability of the extent stability focusing on the idea of random fluctuation of values in a fuzzy formal context and the Gaussian normal distribution is applied.

2   Preliminaries

The fuzzy formal context and the modified Rice-Siff algorithm are now recalled.

Definition 1. *Consider two nonempty sets A a B and a fuzzy relation R such that $R : A \times B \rightarrow [0,1]$. Then the triple $\langle A, B, R \rangle$ is called a fuzzy formal context, the elements of the sets A and B are called attributes and objects, respectively. The relation R is called an incidence fuzzy relation.*

We define the function which assigns the degree of each attribute for every crisp subset of objects X, whereby each object from X satisfies this degree.

Definition 2. *Let $X \subseteq B$ and $\uparrow: \mathcal{P}(B) \rightarrow [0,1]^A$. Then $\uparrow$ is a mapping that assigns to every crisp set X of objects a fuzzy set $X^\uparrow$ of attributes, such that a value in a point $a \in A$ is:*

$$X^\uparrow(a) = \inf\{R(a,b) : b \in X\}. \tag{1}$$

Conversely, we define the function which for fuzzy set of attributes assigns the set of objects, such that every object has all attributes at least in a value given by this fuzzy set.

Definition 3. *Let $f : A \rightarrow [0,1]$ and $\downarrow: [0,1]^A \rightarrow \mathcal{P}(B)$. Then $\downarrow$ is a mapping that assigns to every fuzzy set f of attributes a crisp set $\downarrow (f)$ of objects, such that:*

$$f^\downarrow = \{b \in B : (\forall a \in A)R(a,b) \geq f(a)\}. \tag{2}$$

The notion of one-sided fuzzy concept involves the formulas (1) and (2).

Definition 4. *The pair $\langle X, f \rangle$ is called a one-sided fuzzy concept, if $X^\uparrow = f$ and $f^\downarrow = X$. The set of objects X is called the extent and the fuzzy set $X^\uparrow$ is called the intent of this concept.*

The set of all one-sided fuzzy concepts ordered by inclusion is a complete lattice, called one-sided fuzzy concept lattice, as introduced in [27].

To reduce the number of one-sided formal concepts, the modified Rice-Siff algorithm was proposed in [27]. The method focuses on the distance function and its metric properties. The distance function $\rho : \mathcal{P}(B) \times \mathcal{P}(B) \rightarrow \mathbb{R}$ is defined for $X_1, X_2 \subseteq B$ by:

$$\rho(X_1, X_2) = 1 - \frac{\sum_{a \in A} \min\{\uparrow (X_1)(a), \uparrow (X_2)(a)\}}{\sum_{a \in A} \max\{\uparrow (X_1)(a), \uparrow (X_2)(a)\}}. \tag{3}$$

The formula (3) represents a metric on the set of all extents and comprises the cornerstone of a following clustering algorithm:

```
input: ⟨B, A, R⟩
    C ← D ← {{b}^↑↓ : b ∈ B};
    while (|D > 1|) do
        m ← min{ρ(X₁, X₂) : X₁, X₂ ∈ D, X₁ ≠ X₂}
        Ψ ← {⟨X₁, X₂⟩ ∈ D × D : ρ(X₁, X₂) = m}
        V ← {X ∈ D : (∃Y ∈ D)⟨X, Y⟩ ∈ Ψ}
        N ← {(X₁ ∪ X₂)^↑↓ : ⟨X₁, X₂⟩ ∈ Ψ}
        D ← (D \ V) ∪ N
        C ← C ∪ N
output: C
```

Two clusters with minimal distance are joined in each step of algorithm and the closure of their union is returned as the output. Such closures are gathered in a tree-based structure on the subset hierarchy with the cluster of all objects in the root. The zero iterations gather the closures of singletons, therefore the value of minimal distance function is not computed in the zero step. The more detailed properties of this clustering method with the special defined metric are described in [27,28]. The important results on the relationships between the values in L-fuzzy contexts due to replacing of the table entries are processed by Bĕlohlávek in [4].

3   Continuous Random Variables

The sample space Ω is a set of all possible finite or infinite outcomes of a random study. An event T is an arbitrary subset of Ω. The probability function p on a finite ($\{\omega_1, \ldots, \omega_n\}$) or infinite (e.g. interval of real numbers) sample space Ω assigns to each event $T \subseteq \Omega$ a number $p(T) \in [0, 1]$ such that $p(\Omega) = 1$ and $p(T_1 \cup T_2 \cup \ldots) = p(T_1) + p(T_2) + \ldots$ for $T_1, T_2, \ldots$ which are disjoint. From $T \cup T^c = \Omega$, we deduce that $p(T^c) = 1 - p(T)$. Events $T_1, T_2, \ldots T_m$ are called independent if $p(T_1 \cap T_2 \cap \ldots \cap T_m) = \prod_{i=1}^{m} p(T_i)$.

The real-valued function $\mathcal{E} : \Omega \to \mathbb{R}_1$, i. e. $\mathcal{E}(\omega) = \varepsilon \in \mathbb{R}_1$ for all $\omega \in \Omega$, defined on the sample space is called random variable. A random variable $\mathcal{E}$ is called continuous if exists function $f_{\mathcal{E}} : \mathbb{R}_1 \to \mathbb{R}$ integrable on $\mathbb{R}_1$ and satisfying $f_{\mathcal{E}}(\varepsilon) \geq 0$ for all $\varepsilon \in \mathbb{R}_1$ such that

$$\int_{-\infty}^{\infty} f_{\mathcal{E}}(\varepsilon)\mathrm{d}\varepsilon = 1 \quad \text{and} \quad p(\{\omega \in \Omega : \mathcal{E}(\omega) \leq c\}) = \int_{-\infty}^{c} f_{\mathcal{E}}(\varepsilon)\mathrm{d}\varepsilon. \tag{4}$$

The function $f_{\mathcal{E}}$ is called probability density function of $\mathcal{E}$. The value $p(\{\omega \in \Omega : \mathcal{E}(\omega) \leq c\})$ is written shortly by $p(\mathcal{E} \leq c)$ and it holds $p(\mathcal{E} \leq c) = p(\mathcal{E} < c)$.

A continuous random variable $\mathcal{E}$ has a normal distribution with parameters μ (the mean) and $\sigma^2 > 0$ (the variance) if its probability density function[1]

$$f_{\mathcal{E}}(\varepsilon) = \frac{1}{\sigma\sqrt{2\pi}} \cdot \exp\left(-\frac{1}{2}\left(\frac{\varepsilon - \mu}{\sigma}\right)^2\right) \tag{5}$$

for $-\infty < \varepsilon < \infty$. The parameter σ is called standard deviation. We denote the random variable having a normal distribution with parameters μ and σ^2 by $\mathcal{E} \sim \mathrm{N}(\mu, \sigma^2)$ and it is said to be normally distributed. Probability of $\mathcal{E} \sim \mathrm{N}(\mu, \sigma^2)$ being in the range $\langle -\infty, c \rangle$ for all $c \in \mathbb{R}_1$ is given by:

$$p(\mathcal{E} < c) = \frac{1}{\sigma\sqrt{2\pi}} \cdot \int_{-\infty}^{c} \exp\left(-\frac{1}{2}\left(\frac{\varepsilon - \mu}{\sigma}\right)^2\right)\mathrm{d}\varepsilon. \tag{6}$$

---

[1] To avoid the impractical superscript of the exponential function, the form $\exp(x)$ is written. We recall that the stability index is denoted by symbol σ in [25].

There is not an explicit solution for this integral, since the probability density function has no antiderivative. However, an arbitrary $\mathcal{E} \sim \mathrm{N}(\mu, \sigma^2)$ can be turned into $\mathcal{E}^* \sim \mathrm{N}(0, 1)$ by a substitution

$$\mathcal{E}^* = \frac{\mathcal{E} - \mu}{\sigma} \tag{7}$$

and random variable $\mathcal{E}^*$ is said to be standardizing normally distributed. In this manner, we have for all $c \in \mathbb{R}_1$

$$p(\mathcal{E} < c) = p\left(\mathcal{E}^* < \frac{c - \mu}{\sigma}\right) = \Phi\left(\frac{c - \mu}{\sigma}\right), \tag{8}$$

whereby the values of the function $\Phi(v)$ for all $v \in \mathbb{R}_1$ are available for the computations in the form of table. For a multivariate normal distribution of k-dimensional random vector, we denote $p(\mathcal{E}_1 < c_1, \mathcal{E}_2 < c_2, \ldots, \mathcal{E}_k < c_k) = \Phi(c_1, c_2, \ldots, c_k, \mathrm{corr}(\mathcal{E}_1, \mathcal{E}_2, \ldots, \mathcal{E}_k))$, where $\mathrm{corr}(\mathcal{E}_1, \mathcal{E}_2, \ldots, \mathcal{E}_k)$ is a correlation matrix of random vector $(\mathcal{E}_1, \mathcal{E}_2, \ldots, \mathcal{E}_k)$. Algorithms for the numerical computations of bivariate $(k = 2)$ and trivariate $(k = 3)$ normal probabilities are reviewed in [23].

The distributions of many natural phenomena are at least approximately normally distributed. One of the first applications of the normal distribution was devoted to the analysis of errors of measurement made in astronomical observations. Galileo noted the symmetricity of these errors and that small errors occured more frequently than large errors. The more detailed results can be found in [14, 15].

4   Gaussian Probabilistic Index in Framework of One-Sided Approach

4.1   Motivation and Definition

The motivation of our approach comes from the assumptions that students do not obtain the equal results in the exams performed continuously during the school year. The excellent student will obtain roughly 90 % in the most of exams, but once a time it can happened that he/she will pass 70 % for different reasons, otherwise will reach 98 %, pleasantly surprising.

R	a_1	a_2	a_3
b_1	1	0.9	0.8
b_2	**0.8**	**0.7**	**0.7**
b_3	0.3	0.3	0.3
b_4	0.8	0.6	0.9

In other way, consider four students $\{b_1, b_2, b_3, b_4\}$ and their results in the tests from three different subjects $\{a_1, a_2, a_3\}$. Take for example student b_2 and find the students with better results as b_2 in all subjects. From Sect. 2 we have that $\{b_2\}^{\uparrow\downarrow} = \{b_1, b_2\}$. Will it be valid after the repeated exams? We suppose that student b_3 will not be better than b_1 or b_2, but what about the student b_4? What is the probability of that some other student will join the group $\{b_1, b_2\}$ in other testing?

The notions of fuzzy formal contexts and the random variables provide a way to simulate the previous situation as follows.

Definition 5. *Let $\langle B, A, R \rangle$ be a fuzzy formal context and for $i \in \{1, \ldots, n\}$ consider the formal context $\langle B, A, R_i \rangle$ such that*

$$R_i(b, a) = \min\Big\{1, \max\{0, R(b, a) + \varepsilon_{b,a,i}\}\Big\}, \tag{9}$$

whereby $\varepsilon_{b,a,i}$ is a normally distributed value of random variable $\mathcal{E}_{b,a}$ with the mean 0 and variance σ^2, i. e. $\mathcal{E}_{b,a} \sim N(0, \sigma^2)$, for all $b \in B, a \in A$.

Let $X \subseteq B$. The Gaussian probability index $\mathrm{gpi} : \mathcal{P}(B) \times \mathbb{R}^+ \to [0, 1]$ is the function given by

$$\mathrm{gpi}(X, \sigma) = p(X \text{ is an extent of } \langle B, A, R_i \rangle) \tag{10}$$

for an arbitrary subset of objects X, an arbitrary standard deviation σ and mean 0. The fuzzy formal context $\langle B, A, R_i \rangle$ will be called the randomized (fuzzy) formal context for each $i \in \{1, \ldots, n\}$.

The values given by formula (10) express the probability of X being the extent of the arbitrary randomized formal context. We suppose the standard deviation σ to construct the incidence relation R_i from the original incidence relation R. Alternatively, the values of the Gaussian probability index one can compute by the following construction. Consider the randomized formal contexts $\langle B, A, R_1 \rangle, \langle B, A, R_2 \rangle \ldots, \langle B, A, R_n \rangle$ for a large positive integer n. Then by the classical definition of probabilistic function p one can write

$$\mathrm{gpi}(X, \sigma) = \frac{|i, i \in \{1, 2, \ldots, n\} : X \text{ is an extent of } \langle B, A, R_i \rangle|}{n}. \tag{11}$$

The algorithm for computing the formula (11) can be described as follows:

```
input: ⟨B, A, R⟩, X, σ, n (a large number)
    k ← 0;
    for i := 1 to n do
        {
        for all b ∈ B do
            for all a ∈ A do
                {
                ε_{b,a,i} ← Random.nextGaussian() * σ;
                R_i(b, a) ← min{1, max{0, R(b, a) + ε_{b,a,i}}};
                }
            if (X is an extent of ⟨B, A, R_i⟩) then
```

$$k \leftarrow k + 1;$$
$$\}$$
$$\text{gpi}(X, \sigma) \leftarrow \frac{k}{n};$$
output: $\text{gpi}(X, \sigma)$

The set of pairs $\{\langle X, \text{gpi}(X, \sigma)\rangle : X \subseteq B\}$ for some σ can be ordered by the second coordinate. The first remarks including the comparison between the Gaussian probabilistic index and the modified Rice-Siff algorithm is now briefly outlined.

Remark 1. Note that:

– every cluster $\mathcal{N}$ obtained by modified Rice-Siff is the extent of one-sided formal concept (because $\mathcal{N} = \{(X_1 \cup X_2)^{\uparrow\downarrow} : \langle X_1, X_2\rangle \in \Psi\}$),
– modified Rice-Siff algorithm represents the crisp method for selection of the one-sided concepts, the Gaussian probabilistic index is a fuzzy index,
– we can suppose that the clusters obtained by modified Rice-Siff have the higher $\text{gpi}(X, \sigma)$ as the other extents of one-sided formal concepts,
– the Gaussian probabilistic index gpi works with the relations of randomized formal contexts which can be investigated in connection with the ordinally equivalent relations defined in [4] with respect to Gödel logic connectives.

4.2   Observations on Gaussian Probabilistic Index

In effort to compute the Gaussian probabilistic index by formula (10), consider the special formal context $\langle\{b_1, b_2, b_3\}, \{a\}, R\rangle$ and $\mathcal{E}_{b,a} \sim N(0, \sigma^2)$ for all $b \in \{b_1, b_2, b_3\}$. For $\sigma = 0.1$, the sequence of randomized formal contexts $\langle\{b_1, b_2, b_3\}, \{a\}, R_1\rangle, \ldots, \langle\{b_1, b_2, b_3\}, \{a\}, R_n\rangle$ for $i \in \{1, \ldots, n\}$ is illustrated in Table 1.

Table 1. The randomized formal contexts of $\langle\{b_1, b_2, b_3\}, \{a\}, R\rangle$

R	a		R_1	a		R_2	a		R_3	a		R_n	a
b_1	0.8	$\rightarrow$	b_1	0.76		b_1	0.74		b_1	0.82	$\cdots$	b_1	0.82
b_2	0.6		b_2	0.58		b_2	0.56		b_2	0.52		b_2	0.57
b_3	0.5		b_3	0.52		b_3	0.46		b_3	0.54		b_3	0.48

First note, that in a special case, if $0 < R_i(b, a) < 1$ for all $a \in A$, $b \in B$, $i \in \{1, \ldots, n\}$, then the expression $R_i(b, a) = \min\{1, \max\{0, R(b, a) + \varepsilon_{b,a,i}\}\}$ can be reduced to $R_i(b, a) = R(b, a) + \varepsilon_{b,a,i}$. Later, we will solve the cases which require the boundaries, as well. Now, for $X = \{b_1, b_2\}$, we have that $X = X^{\uparrow\downarrow}$ in $\langle B, \{a\}, R\rangle$, hence $\{b_1, b_2\}$ is an extent of $\langle B, \{a\}, R\rangle$. To preserve this extent in $\langle B, \{a\}, R_i\rangle$ for some $i \in \{1, \ldots, n\}$, it has to hold

$$R_i(b_3, a) < R_i(b_1, a) \quad \wedge \quad R_i(b_3, a) < R_i(b_2, a),$$

which is in the special case of $0 < R_i(b, a) < 1$ equivalent to

$$\varepsilon_{b_3, a, i} - \varepsilon_{b_1, a, i} < R(b_1, a) - R(b_3, a) \quad \wedge \quad \varepsilon_{b_3, a, i} - \varepsilon_{b_2, a, i} < R(b_2, a) - R(b_3, a).$$

In general, to find the probability $p(\mathcal{E}_{b_3, a} - \mathcal{E}_{b_j, a} < R(b_j, a) - R(b_3, a))$ for $j \in \{1, 2\}$, we need to explore the linear combination of two independent random variables.

Lemma 1. *Let $\mathcal{E}_1$ and $\mathcal{E}_2$ be two mutually independent normal random variables with mean μ and variance σ^2 and let $k_1, k_2 \in \mathbb{R}$. Then the linear combination $k_1\mathcal{E}_1 + k_2\mathcal{E}_2$ follows the normal distribution with the mean $k_1\mu + k_2\mu$ and the variance $k_1^2\sigma^2 + k_2^2\sigma^2$, i.e. $(k_1\mathcal{E}_1 + k_2\mathcal{E}_2) \sim N(k_1\mu + k_2\mu, k_1^2\sigma^2 + k_2^2\sigma^2)$.*

Proof. For two random variables $\mathcal{E}_1$ and $\mathcal{E}_1$, in general, the expectation $\mathrm{E}(k_1\mathcal{E}_1 + k_2\mathcal{E}_2) = k_1\mathrm{E}(\mathcal{E}_1) + k_2\mathrm{E}(\mathcal{E}_2)$. The variance $\mathrm{Var}(\mathcal{E}_1)$ of a random variable $\mathcal{E}_1$ is the number $\mathrm{Var}(\mathcal{E}_1) = \mathrm{E}((\mathcal{E}_1 - \mathrm{E}(\mathcal{E}_1))^2)$. Using the expectation, we have that $\mathrm{Var}(k_1\mathcal{E}_1 + k_2\mathcal{E}_2) = k_1^2\mathrm{Var}(\mathcal{E}_1) + k_2^2\mathrm{Var}(\mathcal{E}_2) + 2k_1k_2E[(\mathcal{E}_1 - \mathrm{E}(\mathcal{E}_1))(\mathcal{E}_2 - \mathrm{E}(\mathcal{E}_2))]$. The number $E[(\mathcal{E}_1 - \mathrm{E}(\mathcal{E}_1))(\mathcal{E}_2 - \mathrm{E}(\mathcal{E}_2))]$ is called the covariance $\mathrm{cov}(\mathcal{E}_1, \mathcal{E}_2)$ and if $\mathcal{E}_1$ and $\mathcal{E}_2$ are independent, it holds that $\mathrm{cov}(\mathcal{E}_1, \mathcal{E}_2) = 0$.

Let $\mathcal{E}_1 \sim N(\mu, \sigma^2)$ and $\mathcal{E}_2 \sim N(\mu, \sigma^2)$, i.e. the expectation $\mathrm{E}(\mathcal{E}_1) = \mathrm{E}(\mathcal{E}_2) = \mu$ and the variance $\mathrm{Var}(\mathcal{E}_1) = \mathrm{Var}(\mathcal{E}_2) = \sigma^2$. Therefore, we have that

$$\mathrm{E}(k_1\mathcal{E}_1 + k_2\mathcal{E}_2) = k_1\mathrm{E}(\mathcal{E}_1) + k_2\mathrm{E}(\mathcal{E}_2) = k_1\mu + k_2\mu$$

and since $\mathcal{E}_1$ and $\mathcal{E}_2$ are independent

$$\mathrm{Var}(k_1\mathcal{E}_1 + k_2\mathcal{E}_2) = k_1^2\mathrm{Var}(\mathcal{E}_1) + k_2^2\mathrm{Var}(\mathcal{E}_2) = k_1^2\sigma^2 + k_2^2\sigma^2. \qquad \square$$

Lemma 2. *Let $\mathcal{E}_1$ and $\mathcal{E}_2$ be two mutually independent normal random variables with mean $\mu = 0$ and variance σ^2 and let $c \in \mathbb{R}$. Then*

$$p(\mathcal{E}_1 - \mathcal{E}_2 < c) = \Phi\left(\frac{c}{\sqrt{2}\sigma}\right).$$

Proof. From Lemma 1 and for $k_1 = 1$, $k_2 = -1$, we have that $\mathcal{E}_1 - \mathcal{E}_2 \sim N(0, 2\sigma^2)$. By transformation to a standard normal distribution one can obtain that

$$\frac{\mathcal{E}_1 - \mathcal{E}_2}{\sqrt{2\sigma^2}} \quad \sim \quad N(0, 1).$$

Now, we specify the probability of $\mathcal{E}_1 - \mathcal{E}_2$ being smaller than $c \in \mathbb{R}$ by

$$p(\mathcal{E}_1 - \mathcal{E}_2 < c) =$$

$$p\left(\frac{\mathcal{E}_1 - \mathcal{E}_2}{\sqrt{2\sigma^2}} < \frac{c}{\sqrt{2\sigma^2}}\right) = \Phi\left(\frac{c}{\sqrt{2\sigma^2}}\right) = \Phi\left(\frac{c}{\sqrt{2}\sigma}\right). \qquad \square$$

Example 1. Consider randomized formal contexts for $\langle\{b_1, b_2, b_3\}, \{a\}, R\rangle$ from Table 1 and for $\sigma = 0.1$.

For $X = \{b_1, b_2\}$, from Lemma 2 we have that $c_1 = R(b_1, a) - R(b_3, a) = 0.3$ and $c_2 = R(b_2, a) - R(b_3, a) = 0.1$ and it holds

$$p(\mathcal{E}_{b_3,a} - \mathcal{E}_{b_1,a} < c_1 \wedge \mathcal{E}_{b_3,a} - \mathcal{E}_{b_2,a} < c_2)$$
$$= p(\mathcal{E}_{b_3,a} - \mathcal{E}_{b_1,a} < 0.3 \wedge \mathcal{E}_{b_3,a} - \mathcal{E}_{b_2,a} < 0.1)$$
$$= \Phi(2.12, 0.71, \ \text{corr}(\mathcal{E}_{b_3,a} - \mathcal{E}_{b_1,a}, \mathcal{E}_{b_3,a} - \mathcal{E}_{b_2,a})).$$

We can either compute Φ mathematically (see Sect. 3) or we can run a large positive number n of iterations in our algorithm (see Subsect. 4.1). By an algorithm, the subset X does not correspond to the extent in 25.2 % of cases. Therefore, the Gaussian probabilistic index of the subset $\{b_1, b_2\}$ is 74.8 %.

Example 2. Consider the formal context $\langle\{b_1, b_2, b_3\}, \{a\}, R\rangle$, whereby $R(b_1, a) = 0.9$, $R(b_2, a) = 0.5$, $R(b_3, a) = 0.1$ and its randomized formal contexts based on $\mathcal{E}_{b_1,a}, \mathcal{E}_{b_2,a}, \mathcal{E}_{b_3,a} \sim N(0, \sigma^2)$ with $\sigma = 0.1$.

For $X = \{b_1, b_2\}$, from Lemma 2 we have that $c_1 = R(b_1, a) - R(b_3, a) = 0.8$ and $c_2 = R(b_2, a) - R(b_3, a) = 0.4$ and it holds

$$p(\mathcal{E}_{b_3,a} - \mathcal{E}_{b_1,a} < c_1 \wedge \mathcal{E}_{b_3,a} - \mathcal{E}_{b_2,a} < c_2)$$
$$= p(\mathcal{E}_{b_3,a} - \mathcal{E}_{b_1,a} < 0.8 \wedge \mathcal{E}_{b_3,a} - \mathcal{E}_{b_2,a} < 0.4)$$
$$= \Phi(5.66, 2.83, \ \text{corr}(\mathcal{E}_{b_3,a} - \mathcal{E}_{b_1,a}, \mathcal{E}_{b_3,a} - \mathcal{E}_{b_2,a})).$$

Instead of computing Φ, after a large positive number n of iterations of our algorithm, the subset X does not correspond to the extent in 0.2 % of cases. Therefore, the Gaussian probabilistic index of the subset $\{b_1, b_2\}$ is 99.8 %. In conclusion, we can see that the subset $\{b_1, b_2\}$ is more stable here in comparison with the input formal context from Example 1.

4.3   Formalization and Boundary Conditions

We can extend our examples for more attributes and for boundary cases by considering the set of events. Therefore, we formulate the condition for testing whether the particular object from $B \setminus X$ is included in the extent $X^{\uparrow\downarrow}$ of randomized formal context $\langle B, A, R_i \rangle$ for some particular $X \subseteq B$ and $i \in \{1, 2, \dots, n\}$.

Lemma 3. *Let $\langle B, A, R_i \rangle$ be a randomized formal context for $i \in \{1, 2, \ldots, n\}$. Consider $X \subseteq B$ and some particular $o \in B \setminus X$. Then*

$$o \in X^{\uparrow\downarrow} \quad iff \quad R_i(o, a) \geq \bigwedge_{x \in X} (R_i(x, a)) \ for \ all \ a \in A.$$

Proof. From Definitions 2 and 3 we have that

$$
\begin{aligned}
X^{\uparrow\downarrow} &= \{b \in B : (\forall a \in A) R_i(b, a) \geq X^{\uparrow}(a)\} \\
&= \{b \in B : (\forall a \in A) R_i(b, a) \geq \inf\{R_i(x, a) : x \in X\}\} \\
&= \{b \in (X \cup (B \setminus X)) : (\forall a \in A) R_i(b, a) \geq \inf\{R_i(x, a) : x \in X\}\} \\
&= X \cup \{o \in B \setminus X : (\forall a \in A) R_i(o, a) \geq \inf\{R_i(x, a) : x \in X\}\},
\end{aligned}
$$

since the inequality $R_i(b, a) \geq \inf\{R_i(x, a) : x \in X\}$ is satisfied for all $b \in X$ and $a \in A$. Due to finiteness of X, the infimum equals the minimum. □

The presence of the attribute with the following condition is sufficient to determine that some o from $B \setminus X$ is not included in the extent $X^{\uparrow\downarrow}$ of $\langle B, A, R_i \rangle$.

Corollary 1. *Let $\langle B, A, R_i \rangle$ be a randomized formal context for $i \in \{1, 2, \ldots, n\}$. Consider $X \subseteq B$ and some particular $a \in A$, $o \in B \setminus X$. If*

$$R_i(o, a) < R_i(x, a) \ for \ all \ x \in X,$$

then $o \notin X^{\uparrow\downarrow}$, i. e. o is not included in the extent of $\langle B, A, R_i \rangle$.

Proof. From assumption we have that exists $a \in A$ and $o \in B \setminus X$ such that $R_i(o, a) < R_i(x, a)$ for all $x \in X$, hence from Lemma 3 we have $o \notin X^{\uparrow\downarrow}$. □

Moreover, we can extend the results for the set of all objects from $B \setminus X$, to prove the following assertion.

Lemma 4. *Let $\langle B, A, R_i \rangle$ be a randomized formal context for $i \in \{1, 2, \ldots, n\}$. Consider $X \subseteq B$. Then*

$$X = X^{\uparrow\downarrow} \quad iff \quad (\forall o \in B \setminus X)(\exists a \in A) R_i(o, a) < \bigwedge_{x \in X} (R_i(x, a)).$$

Proof. Let $X = X^{\uparrow\downarrow}$. From the proof of Lemma 3 we have that $X^{\uparrow\downarrow} = X \cup \{o \in B \setminus X : (\forall a \in A) R_i(o, a) \geq \inf\{R_i(x, a) : x \in X\}\}$, therefore $\{o \in B \setminus X : (\forall a \in A) R_i(o, a) \geq \inf\{R_i(x, a) : x \in X\}\} = \emptyset$. But it means that $(\forall o \in B \setminus X)(\exists a \in A) R_i(o, a) < \inf\{R_i(x, a) : x \in X\}$.

Conversely, from our assumption we have by Corollary 1 that $o \notin X^{\uparrow\downarrow}$ for all $o \in B \setminus X$. Hence $X = X^{\uparrow\downarrow}$. □

The previous results help us to express the Gaussian probability index of the arbitrary subset of objects by using the values of incidence relation from the input fuzzy formal context.

Theorem 1. *Let $X \subseteq B$ and let $\langle B, A, R_i \rangle$ be a randomized formal context for some $i \in \{1, \ldots, n\}$, i. e.*

$$R_i(b, a) = \min\left\{1, \max\{0, R(b, a) + \varepsilon_{b,a,i}\}\right\}$$

for the fuzzy formal context $\langle B, A, R \rangle$ and normally distributed value $\varepsilon_{b,a,i}$ of random variable $\mathcal{E}_{b,a} \sim N(0, \sigma^2)$ for all $b \in B, a \in A$. Then the value of Gaussian probabilistic index for the subset $X \subseteq B$ and standard deviation σ is given by

$$\mathrm{gpi}(X, \sigma) = p\left(\bigcap_{o \in B \setminus X} \left(\bigcap_{a \in A} \left(\left(\bigcap_{x \in X} T_x \right)^c \right)^c \right) \right),$$

where T_x represents the event

$$\mathcal{E}_{o,a} - \mathcal{E}_{x,a} < R(x, a) - R(o, a) \quad \wedge$$
$$\mathcal{E}_{o,a} < 1 - R(o, a) \quad \wedge$$
$$\mathcal{E}_{x,a} > -R(o, a)$$

and $p(T_x)$ equals

$$\Phi\left(\frac{R(x, a) - R(o, a)}{\sqrt{2}\sigma}, \frac{1 - R(o, a)}{\sigma}, \frac{R(x, a)}{\sigma}, \ \mathrm{corr}(\mathcal{E}_{o,a} - \mathcal{E}_{x,a}, \mathcal{E}_{o,a}, \mathcal{E}_{x,a}) \right).$$

Proof. The subset $X \subseteq B$ is an extent of $\langle B, A, R_i \rangle$ iff $X = X^{\uparrow\downarrow}$. For an arbitrary $a \in A$ and $o \in B \setminus X$, we have from Corollary 1 that if $R_i(o, a) < R_i(x, a)$ for all $x \in X$, then $o \notin X^{\uparrow\downarrow}$. We need to determine the probability that for all $o \in B \setminus X$ exists $a \in A$ with this property which is equivalent (by Lemma 4) to probability that $X = X^{\uparrow\downarrow}$.

Take $o \in B \setminus X$ and $a \in A$. Then let T_x be the event that $R_i(o, a) < R_i(x, a)$ for an arbitrary $x \in X$. Furthermore, for $o \in B \setminus X$ and $a \in A$, the event $\bigcap_{x \in X} T_x$ represents that $R_i(o, a) < R_i(x, a)$ for all $x \in X$. Denote the event $(\bigcap_{x \in X} T_x)^c$ for an arbitrary $a \in A$ by S_a and the event $(\bigcap_{a \in A} S_a)^c$ for an arbitrary $o \in B \setminus X$ by V_o. Then the event $V = \bigcap_{o \in B \setminus X} V_o$ represents that $(\forall o \in B \setminus X)(\exists a \in A)(\exists x \in X) R_i(o, a) < R_i(x, a)$ which is by Lemma 4 equivalent to $X = X^{\uparrow\downarrow}$. In summary, it holds that

$$p(V) = \mathrm{gpi}(X, \sigma) = p\left(\bigcap_{o \in B \setminus X} \left(\bigcap_{a \in A} \left(\left(\bigcap_{x \in X} T_x \right)^c \right)^c \right) \right).$$

Regarding the assumption $R_i(b, a) = \min\{1, \max\{0, R(b, a) + \varepsilon_{b,a,i}\}\}$ which holds for all $b \in B$, $a \in A$ and some $i \in \{1, 2, \ldots, n\}$, we have that $R_i(o, a) < R_i(x, a)$ is equivalent to $0 \leq \min\{1, R(o, a) + \varepsilon_{o,a,i}\} < \max\{0, R(x, a) + \varepsilon_{x,a,i}\} \leq 1$, which is equivalent to $R(o, a) + \varepsilon_{o,a,i} < R(x, a) + \varepsilon_{x,a,i} \wedge 0 < R(x, a) + \varepsilon_{x,a,i} \wedge R(o, a) + \varepsilon_{o,a,i} < 1$.

It remains to prove the probability of the event T_x. First denote by T_x^1 the event $\mathcal{E}_{o,a} - \mathcal{E}_{x,a} < R(x, a) - R(o, a)$, the event T_x^2 is given by $\mathcal{E}_{o,a} < 1 - R(o, a)$ and the event T_x^3 is represented by $\mathcal{E}_{x,a} > -R(x,a)$. Then by Lemma 2 and using the Φ for the multivariate normal distribution of 3-dimensional random vector we have

$$p(T_x) = p\big(R_i(o,a) < R_i(x,a)\big) = p(T_x^1, T_x^2, T_x^3)$$

$$= \Phi\left(\frac{R(x,a) - R(o,a)}{\sqrt{2}\sigma}, \frac{1 - R(o,a)}{\sigma}, \frac{R(x,a)}{\sigma}, \text{ corr}(\mathcal{E}_{o,a} - \mathcal{E}_{x,a}, \mathcal{E}_{o,a}, \mathcal{E}_{x,a}) \right),$$

which completes the proof. (Note that $1 - \Phi(-c) = \Phi(c)$ for all $c \in \mathbb{R}$, which we applied to derive the third coordinate of Φ.) □

Note that with the growing σ, the probability of the boundary test events $\mathcal{E}_{o,a} < 1 - R(o,a)$ or $\mathcal{E}_{x,a} > -R(x,a)$ naturally decrease, since the probability of being $R_i(o,a) = 1$ or $R_i(x,a) = 0$ is increasing with higher σ.

Finally, Lemmas 5 and 6 give the answer about the condition in which is practically impossible to change the ordering of two object-attribute values in a formal context modified by a random variables with a normal distribution.

Lemma 5. *Let $\langle B, A, R_i \rangle$ be a randomized formal context of $\langle B, A, R \rangle$ for some $i \in \{1, 2, \ldots, n\}$. Consider $X \subseteq B$. If $R(x,a) - R(o,a) \geq 3\sqrt{2}\sigma$ for some $x \in X$, $o \in B \setminus X$, $a \in A$ and $0 < \sigma \leq \frac{\sqrt{2}}{6}$, then it holds $p(\mathcal{E}_{o,a} - \mathcal{E}_{x,a} < R(x,a) - R(o,a)) \doteq p(\Omega) = 1$.*

Proof. First note that since $0 \leq R(b,a) \leq 1$ for all $b \in B$ and $a \in A$ and $\sigma > 0$, it holds that $0 \leq 3\sqrt{2}\sigma \leq 1$ iff $0 < \sigma \leq \frac{\sqrt{2}}{6}$.

From Lemma 2 and monotonicity of Φ,

$$p(\mathcal{E}_{o,a} - \mathcal{E}_{x,a} < R(x,a) - R(o,a)) =$$

$$\Phi\left(\frac{R(x,a) - R(o,a)}{\sqrt{2}\sigma} \right) \geq \Phi(3) = 0.99865 \doteq p(\Omega). \qquad □$$

Lemma 6. *Let $\langle B, A, R_i \rangle$ be a randomized formal context of $\langle B, A, R \rangle$ for some $i \in \{1, 2, \ldots, n\}$. Consider $X \subseteq B$. If $R(o,a) \leq 1 - 4\sigma$ and simultaneously $R(x,a) \geq 4\sigma$ for some $x \in X$, $o \in B \setminus X$, $a \in A$ and $0 < \sigma \leq \frac{1}{4}$, then $p(R_i(o,a) = 1 \vee R_i(x,a) = 0) \doteq p(\emptyset)$.*

Proof. From the assumptions, the monotonicity of Φ and since the boundary conditions $R_i(o,a) < 1$ and $R_i(x,a) > 0$ are independent events, we obtain that

$$p\big(R_i(o,a) = 1 \vee R_i(x,a) = 0\big)$$

$$= 1 - p(R_i(o,a) < 1 \wedge R_i(x,a) > 0)$$

$$= 1 - p\big(\mathcal{E}_{o,a} < 1 - R(o,a)\big) \cdot p\big(\mathcal{E}_{x,a} > -R(x,a)\big)$$

$$= 1 - \Phi\left(\frac{1 - R(o,a)}{\sigma} \right) \cdot \Phi\left(\frac{R(x,a)}{\sigma} \right)$$

$$\leq 1 - \Phi(4) \cdot \Phi(4) = 1 - (0.99997)^2 \leq 0.0001 \doteq p(\emptyset). \qquad □$$

It can be seen from Lemmas 5 and 6 that if $R(x, a) - R(o, a) \geq 3\sqrt{2}\sigma$ and the probability of being $R_i(o, a) = 1$ or $R_i(x, a) = 0$ is too small, the inequality $R_i(o, a) \geq R_i(x, a)$ is practically an impossible event.

Moreover, Lemma 6 is the reason why we need not to test the boundaries in Examples 1 and 2 from Subsect. 4.2, since it holds that $R_{b_1} \geq 4\sigma$, $R_{b_2} \geq 4\sigma$ and $R_{b_3} \leq 1 - 4\sigma$ and therefore $p(T_{b_j}) = p(\mathcal{E}_{b_3,a} - \mathcal{E}_{b_j,a} < R(b_j, a) - R(b_3, a))$ for $j = \{1, 2\}$.

5   Illustrative Example

The Java method $\sigma * \text{Random.nextGaussian}() + \mu$ returns the random value of Gaussian normal distribution with the parameters μ and σ^2 and provides the way how to obtain the experimental results. In this section, the values of the Gaussian probabilistic index gpi are computed by the introduced algorithm for $\sigma = 0.1, \sigma = 0.2$, $n = 1000000$ and for the extents X of the particular fuzzy formal context from Table 2 with 5 objects and 5 attributes.

Table 2. The fuzzy formal context

R	a_1	a_2	a_3	a_4	a_5
b_1	1.0	0.8	0.2	0.3	0.5
b_2	0.8	1.0	0.2	0.6	0.9
b_3	0.2	0.3	0.2	0.3	0.4
b_4	0.4	0.7	0.1	0.2	0.3
b_5	1.0	0.9	0.3	0.2	0.4

Table 3 offers the clusters of objects which are computed by the classical one-sided concept lattice construction (first column), by the modified Rice-Siff algorithm (second column) and by the Gaussian probabilistic index. In parentheses of second column, we highlight the particular iteration of the modified Rice-Siff algorithm in which the cluster was selected. The values of gpi are ordered from the highest to the smallest for $\sigma = 0.2$. We omit empty and full set of objects in our computations.

The least stable extent of $\langle B, A, R \rangle$ is $\{b_1, b_2, b_3\}$. Really, if we take the object b_5, it is a big probability of joining b_5 with the group $\{b_1, b_2, b_3\}$ in a randomized formal context $\langle B, A, R_i \rangle$. Otherwise, the extent $\{b_2\}$ is the most stable because of its large value of attribute a_5 in comparison with other objects. The cluster $\{b_1, b_2\}$ was not selected by the modified Rice-Siff algorithm, since the objects b_1 and b_5 were joined in the first iteration. Therefore, the objects b_1 and b_2 are joined in the second iteration including the object b_5, as well. The same reasoning is able in the case of the clusters $\{b_2, b_5\}$ and $\{b_1, b_2, b_3, b_5\}$ from Table 3.

For a large number of objects, the Gaussian probabilistic index can be applied in combination with the modified Rice-Siff method, which returns about $2|B|$

Table 3. The relevant clusters of objects

Extent	The modified Rice-Siff	Gaussian probabilistic index	
		$\sigma = 0.1$	$\sigma = 0.2$
$\{b_2\}$	1 (0. it.)	0,999	0,998
$\{b_1, b_2, b_5\}$	1 (2. it.)	0,999	0,990
$\{b_1, b_2, b_4, b_5\}$	1 (3. it.)	0,999	0,983
$\{b_1, b_2\}$	0	0,959	0,927
$\{b_5\}$	1 (0. it.)	0,959	0,898
$\{b_2, b_5\}$	0	0,912	0,894
$\{b_1\}$	1 (0. it.)	0,934	0,873
$\{b_1, b_5\}$	1 (1. it.)	0,920	0,798
$\{b_1, b_2, b_3, b_5\}$	0	0,897	0,797
$\{b_1, b_2, b_3\}$	1 (0. it.)	0,771	0,650

Table 4. Simple comparison of approaches on concept stability

Name of the approach on concept stability	Type of relation in a formal context	Type of extent	Type of index on concept stability
Kuznetsov stability index [32]	Crisp	Crisp	Fuzzy
Probability and separation [25]	Crisp	Crisp	Fuzzy
Basic level of concepts [7]	Crisp	Crisp	Fuzzy
Modified Rice-Siff alg. [27]	Fuzzy	Crisp	Crisp
Gaussian probabilistic index	Fuzzy	Crisp	Fuzzy

clusters for the most real data. Then, the values of Gaussian probabilistic index are computed only for subsets which are given as clusters in the modified Rice-Siff algorithm in effort to provide their ordering.

6   Conclusion and Future Work

We provide the theoretical and experimental results on the issue of the fuzzy formal concepts relevance and stability. The properties of the randomized formal contexts and the boundary test conditions were analyzed. Simple comparison of approaches on concept stability is shown in Table 4.

The random Galois lattices [18,19] and the randomized formal contexts of a discrete random variable or in a generalized probability framework [20,39] will be the point of interest in our future work. Our aim is to verify this method in the applications from the educational area or in the area of social networks.

References

1. Alcalde, C., Burusco, A., Fuentes-González, R.: Application of the L-fuzzy concept analysis in the morphological image and signal processing. Ann. Math. Artif. Intell. **72**(1–2), 115–128 (2014)
2. Antoni, L., Krajči, S., Krídlo, O., Macek, B., Pisková, L.: On heterogeneous formal contexts. Fuzzy Sets Syst. **234**, 22–33 (2014)
3. Bartl, E.: Minimal solutions of generalized fuzzy relational equations: probabilistic algorithm based on greedy approach. Fuzzy Sets Syst. **260**, 25–42 (2015)
4. Bělohlávek, R.: Ordinally equivalent data: a measurement-theoretic look at formal concept analysis of fuzzy attributes. Int. J. Approx. Reason. **54**(9), 1496–1506 (2013)
5. Bělohlávek, R., De Baets, B., Konečný, J.: Granularity of attributes in formal concept analysis. Inf. Sci. **260**, 149–170 (2014)
6. Bělohlávek, R., Sklenář, V., Zacpal, J.: Crisply generated fuzzy concepts. In: Ganter, B., Godin, R. (eds.) ICFCA 2005. LNCS (LNAI), vol. 3403, pp. 269–284. Springer, Heidelberg (2005)
7. Belohlavek, R., Trnecka, M.: Basic level of concepts in formal concept analysis. In: Domenach, F., Ignatov, D.I., Poelmans, J. (eds.) ICFCA 2012. LNCS, vol. 7278, pp. 28–44. Springer, Heidelberg (2012)
8. Bělohlávek, R., Trnečka, M.: Basic Level in Formal Concept Analysis: Interesting Concepts and Psychological Ramifications. In: Rossi, F. (eds.) IJCAI 2013, pp. 1233–1239. AAAI Press (2013)
9. Ben Yahia, S., Jaoua, A.: Discovering knowledge from fuzzy concept lattice. In: Kandel, A., Last, M., Bunke, H. (eds.) Data Mining and Computational Intelligence. STUDFUZZ, vol. 68, pp. 167–190. Physica-Verlag, Heidelberg (2001)
10. Butka, P., Pócs, J.: Generalization of one-sided concept lattices. Comput. Inform. **32**(2), 355–370 (2013)
11. Butka, P., Pócs, J., Pócsová, J.: Reduction of Concepts from Generalized One-sided Concept Lattice Based on Subsets Quality Measure. In: Zgrzywa, A., Choroś, K., Siemiński, A. (eds.) New Research in Multimedia and Internet Systems. AISC, vol. 314, pp. 101–111. Springer, Heidelberg (2015)
12. Cabrera, I.P., Cordero, P., Gutiérez, G., Martinez, J., Ojeda-Aciego, M.: On residuation in multilattices: Filters, congruences, and homorphisms. Fuzzy Sets Syst. **234**, 1–21 (2014)
13. Cellier, P., Ferré, S., Ridoux, O., Ducassé, M.: A parameterized algorithm to explore formal contexts with a taxonomy. Int. J. Found. of Comput. Sci. **2**, 319–343 (2008)
14. Chung, K.L., AitSahlia, F.: Elementary Probability Theory. Springer (2006)
15. Dekking, F.M., Kraaikamp, C.: Lopuhaä, H.P., Meester, L.E.: A modern introduction to probability and statistics. Springer, Heidelberg (2005)
16. Dias, S.M., Vieira, N.J.: Reducing the size of concept lattices: The JBOS Approach. In: Kryszkiewicz, M., Obiedkov, S. (eds.) CLA 2010. CEUR Workshop Proceedings, vol. 672, pp. 80–91. CEUR-WS.org (2010)
17. Dubois, D., Prade, H.: Possibility theory and formal concept analysis: characterizing independent sub-contexts. Fuzzy Sets Syst. **196**, 4–16 (2012)
18. Emillion, R.: Concepts of a discrete random variable. In: Brito, P., Bertrand, P., Cucumel, G., De Carvalho, F. (eds.) Selected contributions in Data Analysis and Classifications, pp. 247–259. Springer, Heidelberg (2007)

19. Emillion, R., Lévy, G.: Size of random Galois lattices and number of closed frequent itemsets. Discrete Appl. Math. **157**, 2945–2957 (2009)
20. Frič, R., Papčo, M.: A categorical approach to probability theory. Stud. Logica **94**(2), 215–230 (2010)
21. Ganter, B., Wille, R.: Formal Concept Analysis: Mathematical Foundation. Springer, Heidelberg (1999)
22. Garciá-Pardo, F., Cabrera, I.P., Cordero, P., Ojeda-Aciego, M., Rodríguez-Sanchez, F.J.: On the definition of suitable orderings to generate adjunctions over an unstructured codomain. Inf. Sci. **286**, 173–187 (2014)
23. Genz, A.: Numerical computations of rectangular bivariate and trivariate normal and t probabilities. Stat. Comput. **14**(3), 251–260 (2004)
24. Halaš, R., Pócs, J.: Generalized one-sided concept lattices with attribute preferences. Inf. Sci. **303**, 50–60 (2015)
25. Klimushkin, M., Obiedkov, S., Roth, C.: Approaches to the selection of relevant concepts in the case of noisy data. In: Kwuida, L., Sertkaya, B. (eds.) ICFCA 2010. LNCS, vol. 5986, pp. 255–266. Springer, Heidelberg (2010)
26. Konečný, J., Medina, J., Ojeda-Aciego, M.: Multi-adjoint concept lattices with heterogeneous conjunctors and hedges. Ann. Math. Artif. Intell. **72**(1–2), 73–89 (2014)
27. Krajči, S.: Cluster based efficient generation of fuzzy concepts. Neural Netw. World **13**(5), 521–530 (2003)
28. Krajči, S.: Social network and formal concept analysis. In: Pedrycz, W., Chen, S.-M. (eds.) Social Networks: A Framework of Computational Intelligence. Studies in Computational Intelligence, vol. 526, pp. 41–61. Springer, Heidelberg (2014)
29. Krajči, S., Krajčiová, J.: Social network and one-sided fuzzy concept lattices. In: Spyropoulos, C. (eds.) Fuzz-IEEE 2007, Proceedings of the IEEE International Conference on Fuzzy Systems, pp. 1–6. IEEE Press (2007)
30. Krídlo, O., Krajči, S., Ojeda-Aciego, M.: The category of L-Chu correspondences and the structure of L-bonds. Fund. Inform. **115**(4), 297–325 (2012)
31. Krupka, M.: On complexity reduction of concept lattices: three counterexamples. Inf. Retr. **15**(2), 151–156 (2012)
32. Kuznetsov, S.O.: On stability of a formal concept. Ann. Math. Artif. Intell. **49**, 101–115 (2007)
33. Kwuida, L., Missaoui, R., Ben Amor, B., Boumedjout, L., Vaillancourt, J.: Restrictions on concept lattices for pattern management. In: Kryszkiewicz, M., Obiedkov, S. (eds.) CLA 2010, CEUR Workshop Proceedings, vol. 672, pp. 235–246. CEUR-WS.org (2010)
34. Medina, J., Ojeda-Aciego, M.: On multi-adjoint concept lattices based on heterogeneous conjunctors. Fuzzy Sets Syst. **208**, 95–110 (2012)
35. Medina, J., Ojeda-Aciego, M.: Dual multi-adjoint concept lattices. Inf. Sci. **225**, 47–54 (2013)
36. Medina, J., Ojeda-Aciego, M., Vojtáš, P.: Similarity-based unification: a multi-adjoint approach. Fuzzy Sets Syst. **146**, 43–62 (2004)
37. Poelmans, J., Kuznetsov, S.O., Ignatov, D.I., Dedene, G.: Formal concept analysis in knowledge processing: a survey on models and techniques. Expert Syst. Appl. **40**, 6601–6623 (2013)
38. Singh, P.K., Kumar, C.A.: Bipolar fuzzy graph representation of concept lattice. Inf. Sci. **288**, 437–448 (2014)
39. Skřivánek, V., Frič, R.: Generalized random events. Int. J. Theor. Phys. (2015, in press). doi:10.1007/s10773-015-2594-2

40. Snášel, V., Duráková, D., Krajči, S., Vojtáš, P.: Merging concept lattices of α-cuts of fuzzy contexts. Contrib. Gen. Algebra **14**, 155–166 (2004)
41. Valverde-Albacete, F.J., Peláez-Moreno, C., Peñas, A.: On concept lattices as information channels. In: Bertet, K., Rudolph, S. (eds.) CLA 2014, CEUR Workshop Proceedings, vol. 1252, pp. 119–131. CEUR-WS.org (2014)

Revisiting Pattern Structure Projections

Aleksey Buzmakov[1,2]([⊠]), Sergei O. Kuznetsov[2], and Amedeo Napoli[1]

[1] LORIA (CNRS – Inria NGE – U. de Lorraine), Vandœuvre-lès-Nancy, France
amedeo.napoli@loria.fr
[2] National Research University Higher School of Economics, Moscow, Russia
aleksey.buzmakov@inria.fr, skuznetsov@hse.ru

Abstract. Formal concept analysis (FCA) is a well-founded method for data analysis and has many applications in data mining. Pattern structures is an extension of FCA for dealing with complex data such as sequences or graphs. However the computational complexity of computing with pattern structures is high and projections of pattern structures were introduced for simplifying computation. In this paper we introduce o-projections of pattern structures, a generalization of projections which defines a wider class of projections preserving the properties of the original approach. Moreover, we show that o-projections form a semilattice and we discuss the correspondence between o-projections and the representation contexts of o-projected pattern structures.

Keywords: Formal concept analysis · Pattern structures · Representation contexts · Projections

1 Introduction

A significant part of recorded data represents phenomena in a structured way, e.g., a molecule is better represented as a labeled graph than as a set of attributes. Pattern structures are an extension of FCA for dealing with such kind of data [1–3]. Such a pattern structure is defined by a set of objects, a set of descriptions associated with the set of objects, and a similarity operation on descriptions, matching a pair of descriptions to their common part. For instance, the set of objects can contain molecule names, the set of descriptions contains fragments of molecules, and the similarity operation taking two sets of graphs to a set of maximal common subgraphs. The similarity operation is a semilattice operation on the set of descriptions. It allows one to deal with data (objects and their descriptions) in a similar way as one deals with objects and their intents in standard FCA. Such kind of formalization allows one to describe many types of data, however processing can be computationally very demanding. For example, pattern structures on sets of graphs [2–4] is based on the operation of finding maximal common subgraphs for a set of graphs, which is #P-hard.

To deal with this complexity and to have a possibility to process most of the data, projections of pattern structures were introduced [2]. Projections are special mathematical functions on the set of descriptions that simplify the descriptions of objects. This approach reduces the number of concepts in the pattern

© Springer International Publishing Switzerland 2015
J. Baixeries et al. (Eds.): ICFCA 2015, LNAI 9113, pp. 200–215, 2015.
DOI: 10.1007/978-3-319-19545-2_13

lattice corresponding to a pattern structure. However, it does not impact the computational worst-case complexity of the similarity operation. Moreover, it cannot remove concepts of special kinds from the "middle" of the semilattice which can be important in some practical cases, e.g., concepts containing too small graphs can be considered useless but they cannot be removed with projections. For example, in [5] concepts having intents that include short sequences of patient hospitalisations have little sense. Hence, short sequences could be "removed" from the intent, but the descriptions of objects, i.e., patients, usually include only one long sequence and should not be changed.

In this paper we introduce *o-projections* of pattern structures, a generalization of projections of pattern structures, that allow one to reduce the computational complexity of similarity operations. They also allow one to remove certain kinds of descriptions in the "middle" of the semilattice while the descriptions of the objects can be preserved. By introducing o-projections of pattern structures, we correct also some formal problems of projections of pattern structures, which will be discussed later.

The main difference between o-projections and projections is that in o-projected pattern structures we modify the semilattice of descriptions, while in the case of projected pattern structures we can modify only the descriptions of single objects. It should be noticed that most of the properties of projections are valid for o-projections. However, the relation between representation contexts, a reduction from pattern structures to FCA, and projections is quite different from the relation between representation contexts and o-projections. The introduction and study of this difference is one of the main contributions of this work. In addition we have discovered the fact that the set of o-projections of a pattern structure forms a semilattice. From a practical point of view it allows one to apply a set of independent o-projections, e.g., o-projections obtained from several experts, to a pattern structure.

This work further develops the methodology introduced in [5], where it was applied for the analysis of sequential pattern structures by introducing projections that remove irrelevant concepts.

The rest of the paper is organized as follows. In Sect. 2 we introduce the definitions of a pattern structure, representation context of a pattern structure, and discuss how one can compute with pattern structures along the lines of FCA. Section 3 introduces projections and o-projections of a pattern structure, defines the partial order on o-projections and shows that this order is a semilattice. At the end of this section the relation between o-projections and representation contexts of o-projected pattern structure is discussed. Finally, we conclude the paper and discuss furture work.

2   Pattern Structures

In FCA a formal context (G, M, I), where G is a set of objects, M is a set of attributes, and $I \subseteq G \times M$ is a binary relation between G and M, is taken to a concept lattice $\mathfrak{L}(G, M, I)$ [1]. For non-binary data, such as sequences or graphs, lattices can be constructed in the same way using pattern structures [2].

Definition 1. *A pattern structure* $\mathbb{P}$ *is a triple* $(G, (D, \sqcap), \delta)$, *where* G, D *are sets, called the set of objects and the set of descriptions, and* $\delta : G \to D$ *maps an object to a description. Respectively,* $(D, \sqcap)$ *is a meet-semilattice on* D *w.r.t.* $\sqcap$, *called similarity operation such that* $\delta(G) := \{\delta(g) \mid g \in G\}$ *generates a complete subsemilattice* $(D_\delta, \sqcap)$ *of* $(D, \sqcap)$.

For illustration, let us represent standard FCA in terms of pattern structures. The set of objects G is preserved, the semilattice of descriptions is $(\wp(M), \cap)$, where $\wp(M)$ denotes the powerset of the set of attributes M, a description is a subset of attributes and $\cap$ is the set-theoretic intersection. If $x = \{a, b, c\}$ and $y = \{a, c, d\}$ then $x \sqcap y = x \cap y = \{a, c\}$, and $\delta : G \to \wp(M)$ is given by $\delta(g) = \{m \in M \mid (g, m) \in I\}$.

Note that Definition 1 has an important partial case where $(D, \sqcap)$ is a complete meet-semilattice. In this case the semilattice $(D_\delta, \sqcap)$ is necessarily complete. First, in practical applications one often needs finite lattices, which are always complete. Second, in many practical cases one can easily extend an incomplete semilattice to a complete one by introducing some extra elements. For example, given an incomplete semilattice w.r.t containment order on the interval (a, b), one can add a and b to obtain the interval $[a, b]$, which is a complete semilattice. In this paper some of the statements hold only for the partial case of $(D, \sqcap)$ being a complete meet-semilattice.

The Galois connection for a pattern structure $(G, (D, \sqcap), \delta)$, relating sets of objects and descriptions, is defined as follows:

$$A^\diamond := \bigsqcap_{g \in A} \delta(g), \qquad\qquad \text{for } A \subseteq G$$

$$d^\diamond := \{g \in G \mid d \sqsubseteq \delta(g)\}, \qquad\qquad \text{for } d \in D$$

Given a subset of objects A, $A^\diamond$ returns the description which is common to all objects in A. Given a description d, $d^\diamond$ is the set of all objects whose description subsumes d. The natural partial order (or subsumption order between descriptions) $\sqsubseteq$ on D is defined w.r.t. the similarity operation $\sqcap$: $c \sqsubseteq d \Leftrightarrow c \sqcap d = c$ (in this case we say that c is subsumed by d). In the case of standard FCA the natural partial order corresponds to the set-theoretical inclusion order, i.e., for two sets of attributes x and y $x \sqsubseteq y \Leftrightarrow x \subseteq y$.

Definition 2. *A pattern concept of a pattern structure* $(G, (D, \sqcap), \delta)$ *is a pair* (A, d), *where* $A \subseteq G$ *and* $d \in D$ *such that* $A^\diamond = d$ *and* $d^\diamond = A$; A *is called the pattern extent and* d *is called the pattern intent.*

As in standard FCA, a pattern concept corresponds to the maximal set of objects A whose description subsumes the description d, where d is the maximal common description of objects in A. The set of all pattern concepts is partially ordered w.r.t. inclusion of extents or, dually, w.r.t. subsumption of pattern intents within a concept lattice, these two antiisomorphic orders making a lattice, called pattern lattice.

2.1   Running Example

The authors of [6] have used interval pattern structures for gene expression analysis. Let us consider an example of such pattern structures. In Fig. 1a an interval context is shown. It has three objects and two attributes. Every attribute shows the interval of values the attribute can have. If we have two objects, then a numerical attribute can have all values from the interval of this attribute in the first object and from the interval of this attribute of the second object. Consequently, the similarity between two intervals can be defined as a convex hull of the intervals, i.e. $[a, b] \sqcap [c, d] = [\min(a, c), \max(b, d)]$. Then, given two tuples of intervals, the similarity between these tuples is computed as a component-wise similarity between intervals.

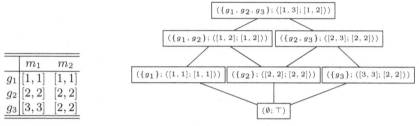

	m_1	m_2
g_1	[1, 1]	[1, 1]
g_2	[2, 2]	[2, 2]
g_3	[3, 3]	[2, 2]

(a) An interval context.   (b) An interval pattern lattice.

Fig. 1. An interval pattern structure and the corresponding lattice.

In this example, we have the pattern structure $(G, (D, \sqcap), \delta)$, where $G = \{g_1, g_2, g_3\}$, the set D is the set of all possible interval pairs with the similarity operation described above, and δ is given by the context in Fig. 1a, i.e., $\delta(g_1) = \langle [1, 1]; [1, 1] \rangle$ and $\delta(g_1) \sqcap \delta(g_2) = \langle [1, 2]; [1, 2] \rangle$.

Figure 1b shows the pattern lattice of the interval context in Fig. 1a. One can check that the extents and the intents in this lattice are connected by means of the Galois connection given above. The partial order in the semilattice of intervals is given by "the smaller the interval, the larger the description with this interval", i.e., the former description gives more certainty about the values than the latter.

2.2   Representation Context of a Pattern Structure

Note that any pattern structure can be represented by a formal context with the concept lattice isomorphic to the lattice of the pattern structure. Below we introduce a representation context of a pattern structure and its properties in the line of [2].

Given a pattern structure $(G, (D, \sqcap), \delta)$, we denote by $D_\delta \subseteq D$ the set of all intents of the concept lattice, i.e., $D_\delta = \{d \in D \mid (\exists X \subseteq G) \bigsqcap_{g \in X} \delta(x) = d\}$. Since

$(D_\delta, \sqcap)$ is a complete subsemilattice of $(D, \sqcap)$, for $X \subseteq D$ a join operation $\sqcup$ can be defined as follows:

$$\bigsqcup X = \bigsqcap \{d \in D_\delta \mid (\forall x \in X)x \sqsubseteq d\}.$$

Given this join operation, $(D_\delta, \sqcap, \sqcup)$ is a complete lattice. We say that a set $M \subseteq D$ is $\sqcup$-dense for $(D_\delta, \sqcap)$ if every element in D_δ is of the form $\sqcup X$ for some $X \subseteq M$. For example, $M = D_\delta$ is always $\sqcup$-dense for D_δ.

Definition 3. *Given a pattern structure* $\mathbb{P} = (G, (D, \sqcap), \delta)$ *and a set* $M \subseteq D$ $\sqcup$-*dense in* D_δ, *a formal context* (G, M, I) *is called the representation context of* $\mathbb{P}$, *if* I *is given by* $I = \{(g, m) \in G \times M \mid m \sqsubseteq \delta(g)\}$. *The representation context of* $\mathbb{P}$ *is denoted by* $\mathbb{R}(\mathbb{P})$.

The next theorem establishes a bijection between the pattern concepts in the lattice of pattern structure $\mathbb{P}$ and the concepts in the lattice of the representation context $\mathbb{R}(\mathbb{P})$. Here, the ideal of element $d \in D$ is denoted by $\downarrow d = \{e \in D \mid e \sqsubseteq d\}$.

Theorem 1 (Theorem 1 from [2]). *Let* $\mathbb{P} = (G, (D, \sqcap), \delta)$ *be a pattern structure and let* $\mathbb{R}(\mathbb{P}) = (G, M, I)$ *be a representation context of* $\mathbb{P}$. *Then for any* $A \subseteq G$, $B \subseteq M$, *and* $d \in D$ *the following conditions are equivalent:*

1. (A, d) *is a pattern concept of* $\mathbb{P}$ *and* $B = \downarrow d \cap M$.
2. (A, B) *is a formal concept of* $\mathbb{R}(\mathbb{P})$ *and* $d = \bigsqcup B$.

Example 1. A representation context for the pattern structure given in Fig. 1 can be given by the set M where every element $m \in M$ is of the form $\langle [-\infty, a]; [-\infty, +\infty] \rangle$ or $\langle [-\infty, +\infty]; [b, +\infty] \rangle$, and $a, b \in \{1, 2, 3\}$.

In fact, the element $\langle [-\infty, +\infty]; [a, +\infty] \rangle$ corresponds to the attribute '$m_2 \geq a$' in the case of the interordinal scaling [1] of numerical data. Another representation context can be constructed from the intents of join-irreducible concepts of the lattice in Fig. 1b. These two representation contexts of the pattern structure related to Fig. 1 are shown in Figs. 2a and b. It can be seen that the resulting lattices, e.g., the lattice in Fig. 2c, are isomorphic to the lattice in Fig. 1b.

It should be noticed that in some cases the representation context is hard to compute. For example, in case of numerical data with the set of all values W, to construct representation context, one needs to create $2|W| + 1$ binary attributes, which can be much more than the number of original real-valued attributes. The authors of [6] have shown that pattern structures provide more efficient computations than the equivalent approach based on FCA and scaling, which can be considered as a way to build representation context of interval pattern structures, e.g., see Fig. 2a.

In case of graph data the set of attributes of the representation context consists of all subgraphs of the original graph descriptions, which is hard to compute [4].

	$\langle[3,+\infty);[-\infty,+\infty)\rangle$ $m_1 \geq 3$	$\langle[2,+\infty);[-\infty,+\infty)\rangle$ $m_1 \geq 2$	$\langle[-\infty,1];[-\infty,+\infty)\rangle$ $m_1 \leq 1$	$\langle[-\infty,2];[-\infty,+\infty)\rangle$ $m_1 \leq 2$	$\langle[-\infty,+\infty);[2,\infty)\rangle$ $m_2 \geq 2$	$\langle[-\infty,+\infty);[-\infty,1]\rangle$ $m_2 \leq 1$
g_1			x	x		x
g_2		x		x	x	
g_3	x	x			x	

(a) Representation context corresponding to interordinal scaling.

	$\langle[1,1];[1,1]\rangle$ a_1	$\langle[3,3];[2,2]\rangle$ a_2	$\langle[1,2];[1,2]\rangle$ a_3	$\langle[2,3];[2,2]\rangle$ a_4	$\langle[1,3];[1,2]\rangle$ a_5
g_1	x		x		x
g_2			x	x	x
g_3		x		x	x

(b) Another possible representation context.

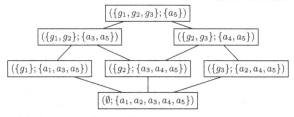

(c) A concept lattice for the context if Figure 2b.

Fig. 2. Possible representation contexts for the pattern structure in Fig. 1 and the concept lattice for the context in Fig. 2b.

2.3 Computation of Pattern Lattices

Nearly any algorithm for computing concept lattices from contexts can be adapted to compute pattern lattices from pattern structures. To adapt an algorithm, every set intersection operation on attributes is replaced by the semilattice operation $\sqcap$ on corresponding patterns, and every subset checking is replaced by the semilattice order $\sqsubseteq$ checking, in particular, all $(\cdot)'$ operations are replaced by $(\cdot)^\circ$. For example, let us consider a modified version of Close-by-One (CbO) algorithm [7].

Algorithm 1 shows the listing of the modified part of CbO. Here the canonical extension IsCanonicExtension and canonical order $\succ$ are defined on the set of objects and hence are the same as in [7]. We can see that only lines 3 and 4 are modified. In these lines the set intersection operation and the subset relation checking are replaced by the corresponding operators of a pattern structure.

3 Revised Projections of Pattern Structures

Pattern structures are hard to process due to the large number of pattern concepts in the pattern lattice and the algorithmic complexity of the similarity operation $\sqcap$. Projections of pattern structures "simplify" to some degree the

```
1 Function CloseByOne(Ext, Int)
       Data: P = (G, (D, ⊓), δ), the extent Ext and the intent Int of a concept.
       Result: All canonical ancestors of (Ext, Int) in the concept lattice.
2      foreach S ⊆ G, S ≻ Ext do
3          NewInt ⟵ ⊓ δ(g) ;                           /* ⊓ - the similarity */
                   g∈S
4          NewExt ⟵ {g ∈ G | NewInt ⊑ δ(g)} ;   /* ⊑ - the subsumption */
5          if IsCanonicExtension(Ext, NewExt) then
6              SaveConcept((NewExt, NewInt));
7              CloseByOne(NewExt, NewInt);
8 /* Looking for all concepts of the concept lattice                         */
9 CloseByOne(∅, ⊤);
```

Algorithm 1. The version of the Close-by-One algorithm computing the pattern lattice of a pattern structure $\mathbb{P}$.

computation and allow one to work with "simpler" descriptions. In fact, a projection can be considered as a mapping for pruning descriptions with certain mathematical properties. These properties ensure that a projection of a semilattice is a semilattice and that the concepts of a projected[1] pattern structure are related to the concepts of the original pattern structure [2].

In this section we introduce o-projected pattern structures ("o" coming from "order"), i.e. a revision of projected pattern structures in accordance with [5]. We discuss the properties of o-projected pattern structures and relate them to the projected pattern structures from [2]. The notion of (o-)projected pattern structure is based on a kernel operator (a projection).

Definition 4 ([2]). *A projection $\psi : D \rightarrow D$ is a kernel (interior) operator on the partial order $(D, \sqsubseteq)$, i.e. it is (1) monotone ($x \sqsubseteq y \Rightarrow \psi(x) \sqsubseteq \psi(y)$), (2) contractive ($\psi(x) \sqsubseteq x$) and (3) idempotent ($\psi(\psi(x)) = \psi(x)$).*

Given a projection ψ we say that the fixed point of ψ is the set of all elements from D such that they are mapped to themselves by ψ. The fixed point of ψ is denoted by $\psi(D) = \{d \in D \mid \psi(d) = d\}$. Note that, if $\psi(d) \neq d$, then there is no other $\tilde{d}$ such that $\psi(\tilde{d}) = d$ because of idempotency. Hence, any element outside the fixed point of the projection ψ is pruned.

3.1   Definition of Projected Pattern Structures

Let us first consider the projected pattern structure w.r.t. a projection ψ according to [2]. Given a pattern structure $\mathbb{P} = (G, (D, \sqcap), \delta)$ and a projection ψ on D, the projected pattern structure is defined as $(G, (D, \sqcap), \psi \circ \delta)$. As we can see, a projection only changes the descriptions of the objects but not the underlying

---

[1] We use the expression "a projected pattern structure" instead of "a projection of a pattern structure" to distinguish between projection as an operator ψ and as the result of applying the operator to a pattern structure.

semilattice $(D, \sqcap)$. There are two problems with this definition of the projected pattern structures. First, it is necessary to restrict the class of projections given by Definition 4 in order to ensure the property $\psi(x \sqcap y) = \psi(x) \sqcap \psi(y)$. Second, the complexity of computing $\sqcap$ can be very high, but with this kind of projected patten structures we cannot decrease the algorithmic complexity. Below we discuss these two points.

In [2] (Proposition 1) the following property of the projection operator is discussed: given a semilattice $(D, \sqcap)$ and a projection ψ on D, for any two elements x and y from D one has $\psi(x \sqcap y) = \psi(x) \sqcap \psi(y)$. Let us consider the example in Fig. 3 with the meet-semilattice $D = \{x, y, z, \perp\}$ given by its diagram and the projection ψ given by the dotted lines. It is easy to see that $\psi(x \sqcap y) = \perp \neq z = \psi(x) \sqcap \psi(y)$. One way of solving this problem is to give additional conditions on projection ψ that would imply the required property. An important example is the following condition: for all $x, y \in D$ if $x < y$ and $\psi(y) = y$, then $\psi(x) = x$. This kind of solution respects the intuition behind the definition of the projected pattern structure in [2], according to which the initial descriptions of objects are changed, but the similarity operation $\sqcap$ is not changed.

$$D = \{x, y, z, \perp\}$$

$$\psi : x \mapsto x, y \mapsto y,$$
$$z \mapsto \perp, \perp \mapsto \perp$$

$$\psi(x \sqcap y) = \psi(z) = \perp \neq$$
$$\neq z = \psi(x) \sqcap \psi(y)$$

Fig. 3. Contrexample to Proposition 1 from [2].

Another way of solving the problem above is to generalize the definition of the projected pattern structure, and we proceed in this way in the next section, by allowing to modify the similarity operation on descriptions.

3.2 Definition of O-Projected Pattern Structures

Below we propose a definition of o-projected pattern structures by means of a kernel operator ψ. The definition takes into account the problems discussed above. In the o-projected pattern structure we substitute the semilattice of descriptions by its suborder (the letter "o" comes from "order") with another similarity operation, which can be different from the initial one.

Let us first note that, given a meet-semilattice D and a kernel operator ψ, the fixed point $\psi(D)$ is a semilattice w.r.t. to the natural order on D.

Theorem 2. *Given a semilattice $(D, \sqcap)$ and a kernel operator ψ, the fixed point $(\psi(D), \sqcap_\psi)$ is a semilattice w.r.t. the natural order on $(D, \sqcap)$, i.e., $d_1 \sqsubseteq d_2 \Leftrightarrow d_1 \sqcap d_2 = d_1$. If $\bigsqcap X$ exists for a set $X \subseteq D$, then $\bigsqcap_\psi \psi(x)$ exists and is given by*

$$\bigsqcap_{x \in X} \psi(x) = \psi(\bigsqcap_{x \in X} x) \tag{1}$$

Proof. Let us denote $d = \bigsqcap_{x \in X} x$. Since $(\forall x \in X)d \sqsubseteq x$, one has $(\forall x \in X)\psi(d) \sqsubseteq \psi(x)$. Let us show that for any $p \in \psi(D)$, i.e. $\psi(p) = p$ such that $(\forall x \in X)p \sqsubseteq \psi(x)$, we have $p \sqsubseteq \psi(d)$, i.e., that $\psi(d) = \bigsqcap_{x \in X} \psi(x)$.

Since $(\forall x \in X)p \sqsubseteq \psi(x)$ then $(\forall x \in X)p \sqsubseteq x$. Since $d = \bigsqcap_{x \in X} x$, one has $p \sqsubseteq d$. Thus, $p = \psi(p) \sqsubseteq \psi(d)$ and $\psi(d)$ is the minimum of the set $\psi(X)$, i.e. $\psi(D)$ is a semilattice and the Eq. (1) holds.

Corollary 1. *Given a complete subsemilattice $\tilde{D}$ of $(D, \sqcap)$ and a kernel operator ψ on D, the image of $\tilde{D}$ is a complete subsemilattice $\psi(\tilde{D})$ of the fixed point $(\psi(D), \sqcap_\psi)$.*

Since according to Theorem 2 $\psi(D)$ is a semilattice and according to Corollary 1 $\psi(D_\delta)$ is a complete semilattice, we can define an o-projected pattern structure as a pattern structure with $\psi(D)$ as a semilattice.

Definition 5. *Given a pattern structure $\mathbb{P} = (G, (D, \sqcap), \delta)$ and a kernel operator ψ on D, the o-projected pattern structure $\psi(\mathbb{P})$ is a pattern structure $(G, (\psi(D), \sqcap_\psi), \psi \circ \delta)$, where $\psi(D) = \{d \in D \mid \psi(d) = d\}$ and $\forall x, y \in D, x \sqcap_\psi y := \psi(x \sqcap y)$.*

In the o-projected pattern structure the kernel operator ψ modifies not only the descriptions of the objects, but also the semilattice operation, i.e., the semilattice $(\psi(D), \sqcap_\psi)$ is not necessarily a subsemilattice of $(D, \sqcap)$ and so it is not always true that $x \sqcap y = x \sqcap_\psi y$ in D.

Example 2. Let us define an o-projection for the interval pattern structure from Subsect. 2.1. Let us suppose that the *aggregated size of a pattern*, i.e., the sum of the lengths of the intervals in the pattern, should be less than 2. First, we should define the corresponding kernel operator $\psi : D \to D$. Thus, if an aggregated length of a pattern p is less than 2, then $\psi(p) := p$, otherwise $\psi(p) := \bot = \langle[-\infty, +\infty]; [-\infty, +\infty]\rangle$. For instance, $\psi(\langle[1, 1]; [1, 1]\rangle) = \langle[1, 1]; [1, 1]\rangle$, while $\psi(\langle[1, 2]; [1, 2]\rangle) = \langle[-\infty, +\infty]; [-\infty, +\infty]\rangle$, because it has two intervals of length 1, i.e., the aggregated size is equal to 2.

Let us consider the o-projected interval pattern structure $(G, (\psi(D), \sqcap_\psi), \psi \circ \delta)$. It is clear that $\psi \circ \delta = \delta$, thus this o-projected interval pattern structure cannot be expressed as a projected pattern structure.

The concepts of a pattern structure and a projected pattern structure are connected through Proposition 1. This proposition can be found in [2], but thanks to Theorem 2, it is also valid in our case.

Proposition 1. *Given a pattern structure* $\mathbb{P} = (G, (D, \sqcap), \delta)$ *and a kernel operator* ψ *on* D:

1. *if* A *is an extent in* $\psi(\mathbb{P})$, *then* A *is also an extent in* $\mathbb{P}$.
2. *if* d *is an intent in* $\mathbb{P}$, *then* $\psi(d)$ *is also an intent in* $\psi(\mathbb{P})$.

It is easy to see that the other propositions from [2] concerning projected pattern structures hold for the o-projected pattern structures as well. Below we cite Proposition 3 from [2] that relates implications in a pattern structure and those in an o-projected pattern structure. We skip the propositions related to supervised classification with projected pattern structures by means of hypotheses, because it is out of the scope of this paper. However, they are valid in the case of o-projected pattern structures and can be proven with the help of Theorem 2.

Proposition 2 (Proposition 3 from [2]**).** *Let* $a, b \in D$. *If* $\psi(a) \to \psi(b)$ *and* $\psi(b) = b$ *then* $a \to b$, *where* $x \to y \Leftrightarrow$ *for all* $g \in G$ $(x \sqsubseteq \delta(g)$ *implies* $y \sqsubseteq \delta(g))$

3.3   Order of Projections

In this subsection we limit ourselves to the practically important case when a set of descriptions is a complete semilattice. We can consider projections as a means of description pruning in $(D, \sqcap)$. Indeed, given a semilattice $(D, \sqcap)$ and a projection ψ on this semilattice, the set D can be divided into two sets $D = \{d \in D \mid \psi(d) = d\} \cup \{d \in D \mid \psi(d) \neq d\}$, i.e., the fixed point of ψ and the rest. It can be seen that the intents of the o-projected pattern structure $\psi((G, (D, \sqcap), \delta))$ are in the fixed point of ψ, i.e., all elements of the form $\psi(d) \neq d$ are discarded. We recall that by $\psi(D) = \{d \in D \mid \psi(d) = d\}$ we denote the fixed point of ψ. But *under which condition do we have that for any* $D_1 \subset D_2$ *there is a projection* ψ *of* D_2 *such that* $\psi(D_2) = D_1$? The following theorem gives necessary and sufficient conditions for such a property.

Theorem 3. *Given a complete semilattice* $(D, \wedge)$, *with the natural order* $\leq$, *and* $D_s \subseteq D$, *there is a projection* $\psi : D \to D$ *such that* $\psi(D) = D_s$, *if and only if* $\perp \in D_s$ *and for any* $X \subseteq D_s \subseteq D$, *one has* $\bigvee X \in D_s$, *where* $\perp := \bigwedge D$ *and* $\bigvee X = \bigwedge \{d \in D \mid (\forall x \in X) d \geq x\}$.

Proof

1. Given a projection ψ such that $\psi(D) = D_s$, $\perp \in D_s$ because of contractivity of ψ, i.e., $\psi(\perp) = \perp$. Let us suppose that there is a set $X \subseteq D_s$, i.e., $(\forall x \in X) \psi(x) = x$ such that $\psi(\bigvee X) \neq \bigvee X$. Then, $(\forall x \in X)(x < \bigvee X \underset{\text{monotonicity}}{\Rightarrow} x \leq \psi(\bigvee X) \underset{\text{contractivity}}{<} \bigvee X)$. It is a contradiction, since $\bigvee X$ is the supremum of X. Hence for any $X \subseteq D_s$ we have $\psi(\bigvee X) = \bigvee X$.

2. Given $D_s \subseteq D$ such that $\perp \in D_s$ and for any $X \subseteq D_s$, one has $\bigvee X \in D_s$, let us construct the corresponding projection ψ. First, $\psi(d \in D_s) := d$ and for all $d \in D \setminus D_s$ we should have $\psi(d) \neq d$. For an element $d \in D \setminus D_s$, let us consider the set $S_d = \{x \in D_s \mid x < d\}$, which is not an empty set since $\perp \in D_s$. We know that $\bigvee S_d \in D_s$ and by definition of $\bigvee$ we have $\bigvee S_d < d$. Then we set $\psi(d) := \bigvee S_d$.

Let us show that the function ψ is a projection of D. Idempotency and contractivity are satisfied by the construction of ψ. Let us check monotonicity. Let us take any $a, b \in D$ such that $a > b$. Then, if $\psi(a) = a$, then $\psi(a) = a > b \geq \psi(b)$, i.e., the monotonicity holds. If $\psi(a) \neq a$, then $\psi(a) = \bigvee S_a$ by construction. Hence, if $\psi(b) = b$, then $b \in S_a$, i.e., $\psi(a) \geq \psi(b)$. Finally, if $\psi(b) \neq b$, then $S_b \subseteq S_a$, because if $d \in S_b$, i.e., $d < b$, then $d < b < a$, i.e. $d \in S_a$. In this case, $\psi(a) = \bigvee S_a \geq \bigvee S_b = \psi(b)$.

Corollary 2. *Given a complete semilattice* $(D, \wedge)$*, with the natural order* $\leq$*, and a subset* $D_s \subseteq D$ *such that* $\perp \in D_s$ *and for any* $X \subseteq D_s$*, one has* $\bigvee X \in D_s$*, the poset* $(D_s, \leq)$ *is a complete semilattice.*

Proof. According to Theorem 3 there is a projection $\psi : D \to D$ such that $\psi(D) = D_s$. Then, according to Theorem 2 D_s is a semilattice. $\qquad \blacksquare$

Since a projection of D can be considered as a mapping with the fixed point $\psi(D)$, we can introduce an order w.r.t. this fixed point.

Definition 6. *Given a complete semilattice* $(D, \sqcap)$ *and two projections* ψ_1 *and* ψ_2 *in* D*, we say that* $\psi_1 \leq \psi_2$ *if* $\psi_1(D) \subseteq \psi_2(D)$*.*

However in some cases, it is more convenient to order projections w.r.t. a superposition of projections or their "generality".

Definition 7. *Given a complete semilattice* $(D, \sqcap)$ *and two projections* ψ_1 *and* ψ_2 *in* D*, we say that* $\psi_1 \leq \psi_2$ *if there is a projection* $\psi : \psi_2(D) \to \psi_2(D)$ *such that* $\psi_1 = \psi \circ \psi_2$*.*

It can be seen that these two definitions yield the same ordering.

Proposition 3. *Definitions 6 and 7 are equivalent.*

Proof

1. Let $\psi_1 = \psi \circ \psi_2$. Since ψ is a projection in $\psi_2(D)$, then $\psi_1(D) = \psi(\psi_2(D)) \subseteq \psi_2(D)$.
2. Let $\psi_1(D) \subseteq \psi_2(D)$. Let us denote by $(\cdot)_1$ and $(\cdot)_2$ the operations in $(\psi_1(D), \sqcap_{\psi_1})$ and $(\psi_2(D), \sqcap_{\psi_2})$, respectively, and let us denote $D_i = \psi_i(D)$ the fixed points of ψ_i, where $i \in \{1, 2\}$.

   Let us build $\psi : D_2 \to D_1$ equal to ψ_1 in D_2, i.e., for all $d \in D_2$ we set $\psi(d) := \psi_1(d)$. Since ψ_1 is a projection in D, ψ is a projection in D_2 (the natural order is the same). Since D_1 is the fixed point of ψ_1 then $\psi_1(D_2) \subseteq D_1$. However, since $D_1 \subseteq D_2$ and $\psi_1(D_1) = D_1$ then $\psi_1(D_2) = D_1$, i.e., there is a projection ψ such that $\psi_1 = \psi \circ \psi_2$.

Example 3. Let us return to Example 2. We change the threshold for the aggregated size. In Example 2 it was set to 2 ($\psi_{al=2}$), but we can change it to 5 ($\psi_{al=5}$) or 10 ($\psi_{al=10}$). The higher the threshold, the more possible descriptions are projected to themselves, i.e., belong to the fixed point of the projection. Thus, we have $\psi_{al=2} \leq \psi_{al=5} \leq \psi_{al=10}$.

Thanks to Proposition 1 it can be seen that, given a pattern structure $\mathbb{P}$, if we have two projections $\psi_1 \leq \psi_2$, then the set of pattern extents of $\psi_1(\mathbb{P})$ is a subset of the set of pattern extents of $\psi_2(\mathbb{P})$, i.e., the smaller the projection, the smaller the number of concepts in the corresponding projected pattern structure.

Now it can be seen that projections actually form a semilattice with respect to the previously defined order.

Proposition 4. *Projections of a complete semilattice $(D, \sqcap)$ ordered by Definition 6 or 7 form a semilattice $(\mathbb{F}, \wedge)$, where the semilattice operation between $\psi_1, \psi_2 \in \mathbb{F}$ is given by $\psi_1 \wedge \psi_2 = \psi_3$ iff $\psi_3(D) = \psi_1(D) \cap \psi_2(D)$.*

Proof. It follows from the definitions that if for any ψ_1 and ψ_2 the projection ψ_3 exists, then projections of D form a semilattice. Let us describe the corresponding ψ_3.

Let us denote $D_1 = \psi_1(D)$ and $D_2 = \psi_2(D)$ and $D_3 = D_1 \cap D_2$. Let us suppose that there exist $x, y \in D_3$ such that $x \sqcup y \notin D_3$. But as $D_3 \subseteq D_1$ and $D_3 \subseteq D_2$, then, since ψ_1 is a projection of D and ψ_2 is a projection of D, we have $x \sqcup y \in D_1$ and $x \sqcup y \in D_2$, i.e., $x \sqcup y \in D_1 \cap D_2 = D_3$. Thus, $(\forall x, y \in D_3) x \sqcup y \in D_3$. Then, according to Theorem 3 there is a projection ψ_3 such that $\psi_3(D) = D_3$.

3.4   Analogue of Theorem II for Revised Projections

An important question is *how a projection changes the representation context of a pattern structure?* We limit the discussion of this question for the case when a set of description D is a complete semilattice. In [2] the authors describe this change by means of Theorem 2. The formulation of this theorem was corrected in [8]. Below we give the corrected version of the theorem.

Theorem 4 (Theorem 2 from [2]). *For two pattern structures $(G, (D, \sqcap), \delta_1)$ and $(G, (D, \sqcap), \delta_2)$ the following statements are equivalent:*

1. $\delta_2 = \psi \circ \delta_1$ for some ψ on $(D, \sqcap)$.
2. $(\forall g \in G)(\delta_2(g) \sqsubseteq \delta_1(g))$ and there is a representation context (G, M, I) of $(G, (D, \sqcap), \delta_1)$ and some $N \subseteq M$ such that $(G, N, I \cap (G \times N))$ is a representation context of $(G, (D, \sqcap), \delta_2)$.

In Theorem 4 one compares two pattern structures that differ in mapping functions. However, in the o-projected pattern structures we can modify the lattice structure itself. *How can we adjust the formulation of Theorem 4 in such a way that it can be applied to revised projections?* First, we should notice that in a pattern structure and in an o-projected pattern structure the set of objects is preserved. Second, the minimal representation context of a pattern structure can have less attributes than the minimal representation context of an o-projected pattern structure, as shown in Example 4.

Example 4. Let $M = \{a, b, c\}$ and the description semilattice be $D = (2^M, \cap)$. Let $\psi : 2^M \to 2^M$ be the following mapping: $\psi(\{a\}) = \emptyset$ and for any $A \neq \{a\}$ we put $\psi(A) = A$. This projection is visualised in Fig. 4a by dashed arrows. Let us consider the following pattern structure $(\{g_1, g_2, g_3\}, (2^M, \cap), \{g_1 \mapsto \{a, b\}, g_2 \mapsto \{a, c\}, g_3 \mapsto \{b, c\}\}$.

The minimal representation context of this pattern structure contains 3 attributes $M = \{a, b, c\}$, while the minimal representation context of the o-projected pattern structure contains 4 attributes $M_\psi = \{b, c, ab, ac\}$. The corresponding contexts are shown in Figs. 4b and c.

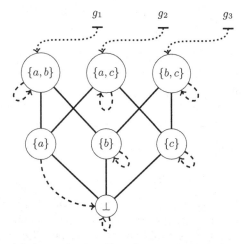

	a	b	c
g_1	x	x	
g_2	x		x
g_3		x	x

(b) Representation context of the pattern structure

	ab	ac	b	c
g_1	x		x	
g_2		x		x
g_3			x	x

(a) A semilattice D and its projection ψ.

(c) Representation context of the projected pattern structure

Fig. 4. An example of a projection that can increase the number of attributes in the minimal representation context.

We can see that to introduce the "revised Theorem 2" from [2] we have to define a special relation between contexts.

Definition 8. *Given two contexts* $\mathbb{K}_1 = (G, M_1, I_1)$ *and* $\mathbb{K}_2 = (G, M_2, I_2)$, $\mathbb{K}_1$ *is said to be simpler than* $\mathbb{K}_2$, *denoted by* $\mathbb{K}_1 \leq_S \mathbb{K}_2$, *if for any* $m_{1,i} \in M_1$ *there is a set* $B_2 \subseteq M_2$ *such that* $(\{m_{1,i}\})^1 = (B_2)^2$. *Here by* $(\cdot)^1$ *and* $(\cdot)^2$ *we denote the derivation operators in the contexts* $\mathbb{K}_1$ *and* $\mathbb{K}_2$, *respectively.*

Example 5. The context in Fig. 4c is smaller w.r.t. Definition 8 than the context in Fig. 4b because every column of the context in Fig. 4c is the intersection of a subset of columns of the context in Fig. 4b.

This relation between contexts is a preorder. Indeed, it is reflexive, transitive, but not necessarily antisymmetric: given two contexts $\mathbb{K}_1$ and $\mathbb{K}_2$, if $\mathbb{K}_1$ and $\mathbb{K}_2$ have the same closure system of attributes, i.e., the same set of intents in the concept lattice, then according to the definition $\mathbb{K}_1 \leq_S \mathbb{K}_2$ and $\mathbb{K}_1 \geq_S \mathbb{K}_2$. However, we can consider only the context with the minimal number of attributes in the class of equivalence, i.e., the attribute-reduced context. For simplicity in the rest of the paper we consider only attribute-reduced contexts.

This definition of the simplicity order on contexts can be related to context bonds [1] in the following way. Three formal contexts $\mathbb{K}_i = (G_i, M_i, I_i)$ form a bond if $\mathbb{K}_1 \leq_S \mathbb{K}_2$ and $\mathbb{K}_2^T \leq_S \mathbb{K}_3^T$, where $\mathbb{K}^T = (M, G, I^T)$. Simplicity order can also be considered as a generalization of "closed-relation-of" order between contexts:

Definition 9 (Definition 50 from [1]). *A binary relation $J \subseteq I$ is called a **closed relation** of the context (G, M, I) if every concept of the context (G, M, J) is also a concept of (G, M, I).*

From Definitions 8 and 9 it can be seen that if J is a closed relation of (G, M, I), then $(G, M, J) \leq_S (G, M, I)$, but not always in the other direction. The following theorem gives a relation between kernel operators of D and the change in the representation context of o-projected pattern structures.

Theorem 5. *Given a pattern structure $\mathbb{P} = (G, (D, \sqcap), \delta)$ such that $(D, \sqcap)$ is a complete semilattice the following holds:*

1. *for any projection ψ of D we have $\mathbb{R}(\psi(\mathbb{P})) \leq_S \mathbb{R}(\mathbb{P})$.*
2. *for any context $\mathbb{K} = (G, M, I)$ such that $\mathbb{K} \leq_S \mathbb{R}(\mathbb{P})$, there is a projection ψ of D such that $\mathbb{K}$ is a representation context of $\psi(\mathbb{P})$.*

Proof

1. The first statement follows from the fact that any extent of $\psi(\mathbb{P})$ is an extent of $\mathbb{P}$ (Proposition 1).
2. Given a pattern structure $\mathbb{P}$ and a context $\mathbb{K}$ such that $\mathbb{K} \leq_S \mathbb{R}(\mathbb{P})$, let us define the set $D_M = \{d \in D \mid (\exists m \in M)(m')^\diamond = d\}$ (notice that for $\mathbb{K}$ and $\mathbb{P}$ there is the same set G, thus, given $A \subseteq G$, both A' and $A^\diamond$ are defined in $\mathbb{K}$ and $\mathbb{P}$ correspondingly). Since $\mathbb{K} \leq_S \mathbb{R}(\mathbb{P})$, m' is an extent of $\mathbb{P}$. Thus, we can see that there is a bijection between D_M and M given by $m' = d^\diamond$. We denote this bijection by $f(m) = d$, i.e. $f(m) = d \Leftrightarrow m' = d^\diamond$. Correspondingly, given a subset $N \subseteq M$, we denote by $f(N) = \{d \in D_M \mid f^{-1}(d) \in N\}$, i.e., $f(M) = D_M$.

   Let us define $D_\psi = \{d \in D \mid (\exists X \subseteq D_M) \bigsqcup X = d\}$. According to Theorem 3 there is a projection ψ such that $D_\psi = \psi(D)$.

Let us consider the o-projected pattern structure $\psi(\mathbb{P})$. The set D_M is $\sqcup$-dense for $\psi(D)$, i.e., the context (G, D_M, I_{D_M}), where $(g, d) \in I_{D_M} \Leftrightarrow \psi \circ \delta(g) \sqsupseteq d$, is a representation context of $\psi(\mathbb{P})$. There is the bijection between D_M and M. Let us show that the relation I is similar to the relation I_M, i.e., $(g, m) \in I \Leftrightarrow (g, f(m)) \in I_{D_M}$.

It can be seen that for all $g \in G$ and all $d \in f(g')$, we get $\psi \circ \delta(g) \sqsupseteq d$, because for any $d \in f(g')$ we have $g \in d^\circ$. Moreover, for any $\tilde{d} \in D \setminus f(g')$ we have $d \not\sqsupseteq \psi \circ \delta(g)$. Thus, the context $\mathbb{K}$ and the context (G, D_M, I_{D_M}) are similar, and hence for any context $\mathbb{K} \leq_S \mathbb{R}(\mathbb{P})$ there is a projection such that $\mathbb{K}$ is a representation context of $\psi(\mathbb{P})$.

4   Conclusion

In this paper we have introduced o-projections of pattern structures that are based on kernel operators $\psi : D \to D$. O-projections are a generalization of projections of pattern structures and allow one to change the semilattice of descriptions in o-projected pattern structures. Thus, the complexity of similarity (semilattice) operation can be reduced. Moreover, O-projections also correct a formal problem of projections.

We have shown that o-projections form a semilattice. This can be important when several independent o-projections are applied to a pattern structure. For example, if projections are discussed with several experts it may happen that several types of projections should be combined. In the case of several independent projections we know that there is the only one o-projection w.r.t. the semilattice of o-projections that is a combination of these projections.

Finally, we have shown that the representation context of an o-projected pattern structure can have more attributes than the representation context of the pattern structure itself. To describe this change in the representation context after o-projection we have introduced a new order on contexts, with the use of which we have described the way the representation context can change.

An important direction of the future work is to formalize *transformations* of pattern structures, i.e., special homomorphisms between the semilattice of descriptions D and a different semilattice D_1. In particular, it allows one to formalize the mappings of the form $\psi : D \to \mathbb{R}$, an instance of which are kernel functions used in Support Vector Machines (SVM).

Acknowledgments. This research was supported by the Basic Research Program at the National Research University Higher School of Economics (Moscow, Russia) and by the BioIntelligence project (France). The second author was also supported by a grant from Russian Foundation for Basic Research, grant no. 13-0700504.

References

1. Ganter, B., Wille, R.: Formal Concept Analysis: Mathematical Foundations, 1st edn. Springer, Heidelberg (1999)

2. Ganter, B., Kuznetsov, S.O.: Pattern structures and their projections. In: Delugach, H.S., Stumme, G. (eds.) ICCS 2001. LNCS (LNAI), vol. 2120, pp. 129–142. Springer, Heidelberg (2001)

3. Ganter, B., Grigoriev, P.A., Kuznetsov, S.O., Samokhin, M.V.: Concept-based data mining with scaled labeled graphs. In: Wolff, K.E., Pfeiffer, H.D., Delugach, H.S. (eds.) ICCS 2004. LNCS (LNAI), vol. 3127, pp. 94–108. Springer, Heidelberg (2004)

4. Kuznetsov, S.O., Samokhin, M.V.: Learning closed sets of labeled graphs for chemical applications. In: Kramer, S., Pfahringer, B. (eds.) ILP 2005. LNCS (LNAI), vol. 3625, pp. 190–208. Springer, Heidelberg (2005)

5. Buzmakov, A., Egho, E., Jay, N., Kuznetsov, S.O., Napoli, A., Raïssi, C.: On projections of sequential pattern structures (with an application on care trajectories). In: Proceeding of the 10th International Conference on Concept Lattices Their Application, pp. 199–208 (2013)

6. Kaytoue, M., Kuznetsov, S.O., Napoli, A., Duplessis, S.: Mining gene expression data with pattern structures in formal concept analysis. Inf. Sci. (Ny) 181(10), 1989–2001 (2011)

7. Kuznetsov, S.O.: A fast algorithm for computing all intersections of objects from an arbitrary semilattice. Nauchno-Tekhnicheskaya Informatsiya Seriya 2 (Autom. Doc. Mathem. Ling.) (1), 17–20 (1993)

8. Kaiser, T.B., Schmidt, S.E.: Some remarks on the relation between annotated ordered sets and pattern structures. In: Kuznetsov, S.O., Mandal, D.P., Kundu, M.K., Pal, S.K. (eds.) PReMI 2011. LNCS, vol. 6744, pp. 43–48. Springer, Heidelberg (2011)

Methods and Applications

Exploring Faulty Data

Daniel Borchmann[1,2(✉)]

[1] Institute of Theoretical Computer Science, TU Dresden, Dresden, Germany
[2] Center for Advancing Electronics Dresden, Dresden, Germany
daniel.borchmann@tu-dresden.de

Abstract. Within formal concept analysis, attribute exploration is a powerful tool to semi-automatically check data for completeness with respect to a given domain. However, the classical formulation of attribute exploration does not take into account possible errors which are present in the initial data. To remedy this, we present in this work a generalization of attribute exploration based on the notion of *confidence*, that will allow for the exploration of implications which are not necessarily valid in the initial data, but instead enjoy a minimal confidence therein.

1 Introduction

Attribute exploration is one of the most important algorithms in the area of formal concept analysis [9], a branch of mathematical order theory with applications in artificial intelligence, machine learning and data mining. The main purpose of this algorithm is to check a given set of *initial data* for completeness, in the sense that this algorithm assists a domain expert in checking whether this initial data completely represents the particular domain the expert is interested in. In doing so, the algorithm presents *implications* to the expert, who has to either validate them or has to provide a counterexample from the domain of discourse. When the algorithm has finished, the initial data has been extended to a *complete* set of examples whose valid implications are exactly all implications valid in the domain.

However, this approach requires the initial data to be free of *errors* in the sense that all the data really stems from the domain. In practical applications, this may not be reasonable to assume, as it may likewise not be reasonable to check the data for correctness. However, the data itself may still be of "high quality" and could thus still be useful, yet only directly applying attribute exploration is not possible anymore.

One way to consider a data set to be of "high quality" is to say that errors occur only "rarely." To handle a scenario like this, an approach is proposed that is based on the notion of *confidence* from data mining [1]. The idea of this approach is not only to explore the implications which are valid in the initial data set, but also to explore those that satisfy a certain lower bound on their confidence. Of course, this will only provide us with a heuristic algorithm, but in a setting like this, where errors can occur randomly, this is the best we can expect to get. Moreover, an *exploration by confidence* has to be thought of as

© Springer International Publishing Switzerland 2015
J. Baixeries et al. (Eds.): ICFCA 2015, LNAI 9113, pp. 219–235, 2015.
DOI: 10.1007/978-3-319-19545-2_14

a first step in a completion process, where the resulting set of implications and set of data should be used further on. As an example, the implications obtained from the exploration by confidence could be used as a background knowledge for a classical attribute exploration which starts out with an empty data set.

Of course, this work is not the first to consider implications together with their confidence. The most notable previous work here is from Luxenburger [10], who considered implications together with their confidence and support in formal contexts. However, while he also considered bases of implications with confidence and support, he did not consider an attribute exploration of them.

On the other hand, there has also been previous research on making attribute exploration more suitable for practical applications. Notable works here are *exploration with incomplete knowledge* by Burmeister and Holzer [5], and *attribute exploration with background implications and exceptions* by Stumme [11]. The former extends attribute exploration to the setting of *incomplete data*, i.e., where the data-set in question may have unspecified entries. However, those entries specified must still be correct. The latter work allows *exceptions* in the exploration, by simply removing unwanted corner cases from the domain of discourse. But again, the data that is used for exploration must be free of errors. In this sense, the problem we want to consider in this paper, an exploration of data that may contain errors, is fundamentally different from previously considered extensions of attribute exploration.

The main contribution of this work is an algorithm for exploration by confidence, which shall be discussed in Sect. 5. This algorithm arises as an instance of a generalized formulation of attribute exploration, which shall be introduced in Sect. 4. A naive and direct application of this generalized algorithm will yield a first version of exploration by confidence, which however is only "approximative," in a sense that will be discussed in Sect. 5.1. A slight modification of this approximative version presented in Sect. 5.2 will then yield the desired algorithm for exploration by confidence.

The results of this work are taken from [4], which not only contains the proofs of the claims in this paper (which we omit due to space restrictions), but also an adaptation of exploration by confidence which also works with *general concept inclusions*, logical objects akin to implications used in the field of *description logics*. We shall give a very brief outlook about this adaptation results in Sect. 6.

2   Implications and Confidence

We assume the reader has some familiarity with the basic notions of formal concept analysis, as we will not repeat them here. However, we shall repeat some notions and fix some notations about implications that are crucial for the understanding of this paper.

Denote with $\mathrm{Imp}(M)$ the set of all implications on a set M. Recall that an implication $(A \to B) \in \mathrm{Imp}(M)$ is *valid* in a formal context $\mathbb{K} = (G, M, I)$ if and only if

$$A' \subseteq B'.$$

We shall denote with $\mathrm{Th}(\mathbb{K})$ the set of all valid implications on M that are valid in $\mathbb{K}$.

Let $\mathcal{L} \subseteq \mathrm{Imp}(M)$ be a set of implications, and let $(A \to B) \in \mathrm{Imp}(M)$. Recall that the set $\mathcal{L}$ *entails* $(A \to B)$ if and only if in all formal contexts $\mathbb{L}$ with attribute set M, it is true that if all implications from $\mathcal{L}$ are valid in $\mathbb{L}$, then $(A \to B)$ is valid in $\mathbb{L}$ as well. In other words,

$$\mathbb{L} \models \mathcal{L} \implies \mathbb{L} \models (A \to B),$$

where we write $\mathbb{L} \models \mathcal{L}$ to mean that all implications in $\mathcal{L}$ are valid in $\mathbb{L}$. If $\mathcal{L}$ entails $(A \to B)$ we shall also write $\mathcal{L} \models (A \to B)$. The subset of $\mathrm{Imp}(M)$ of all implications on M which is entailed by $\mathcal{L}$ is denoted by $\mathrm{Cn}_M(\mathcal{L})$. We shall drop the subscript if the set M is clear from the context.

Entailment between implications can be characterized in a different manner. For this we introduce the notion of *closure operators* induced by sets of implications. More precisely, we define for $A \subseteq M$ the operators

$$\mathcal{L}^1(A) := A \cup \bigcup \{ Y \mid (X \to Y) \in \mathcal{L}, X \subseteq A \},$$
$$\mathcal{L}^{i+1}(A) := \mathcal{L}^1(\mathcal{L}^i(A)) \quad (i \in \mathbb{N}_{>0}),$$
$$\mathcal{L}(A) := \bigcup_{i \in \mathbb{N}_{>0}} \mathcal{L}^i(A).$$

We shall call the mapping $A \mapsto \mathcal{L}(A)$ the *closure operator* induced by $\mathcal{L}$, and we shall call the set A to be *closed under* $\mathcal{L}$ if and only if $A = \mathcal{L}(A)$. The closure operator induced by $\mathcal{L}$ can now be used to characterize entailment of implications as follows:

$$\mathcal{L} \models (A \to B) \iff B \subseteq \mathcal{L}(A).$$

Let $\mathcal{K} \subseteq \mathrm{Imp}(M)$ be another set of implications. We shall call $\mathcal{L}$ a *base* of $\mathcal{K}$ if and only if $\mathrm{Cn}(\mathcal{L}) = \mathrm{Cn}(\mathcal{K})$. In other words, all implications in $\mathcal{K}$ are entailed by $\mathcal{L}$ and vice versa. If $\mathcal{K} = \mathrm{Th}(\mathbb{K})$, then we shall call $\mathcal{L}$ a *base of* $\mathbb{K}$. Note that a base of $\mathcal{K}$ is always a base of $\mathrm{Cn}(\mathcal{K})$, and vice versa.

Bases allow us to represent sets $\mathcal{K}$ of implications in different ways, without changing their behavior with respect to entailment. This fact is mostly exploited by searching for bases of $\mathcal{K}$ which are of considerably smaller size than $\mathcal{K}$ itself. Those bases are preferably *non-redundant* or even *minimal*. More precisely, if $\mathcal{L}$ is a base of $\mathcal{K}$, then $\mathcal{L}$ is called *non-redundant* if no proper subset of $\mathcal{L}$ is a base of $\mathcal{K}$ as well. Furthermore, $\mathcal{L}$ is called *minimal* if and only if there does not exist another base $\mathcal{L}'$ of $\mathcal{K}$ satisfying $|\mathcal{L}'| < |\mathcal{L}|$.

If we search for bases of $\mathcal{K}$, it might be the case that we do not want to include a certain set $\mathcal{L}_{\mathrm{back}}$ of implications which we already "know." We can think of these implications as given *a-priori*, or as *background knowledge*. If we are given such background knowledge, to find a base of $\mathcal{K}$ it only remains to find a base of all those implications in $\mathcal{K} \setminus \mathrm{Cn}(\mathcal{L}_{\mathrm{back}})$. We thus shall call a set $\mathcal{L} \subseteq \mathrm{Imp}(M)$ a *base of $\mathcal{K}$ relative to $\mathcal{L}_{\mathrm{back}}$* (or a *base of $\mathcal{K}$ with background knowledge $\mathcal{L}_{\mathrm{back}}$*) if and only if $\mathcal{L} \cup \mathcal{L}_{\mathrm{back}}$ is a base of $\mathcal{K}$. The notions of non-redundancy and minimality for relative bases are the same as in the case of bases. Note that if the background knowledge is empty, then relative bases are just bases.

A particular relative base that is known to have minimal cardinality is the *canonical base* $\mathrm{Can}(\mathcal{K}, \mathcal{L}_{\mathrm{back}})$. To define this base, we need to introduce the notion of $\mathcal{L}_{\mathrm{back}}$-*pseudo-closed sets of* $\mathcal{K}$ [11]. Let $P \subseteq M$. Then P is called an $\mathcal{L}_{\mathrm{back}}$-pseudo-closed set of $\mathcal{K}$ if and only if the following conditions hold.

i. $P = \mathcal{L}_{\mathrm{back}}(P)$;
ii. $P \neq \mathcal{K}(P)$;
iii. for all $Q \subsetneq P$ which are $\mathcal{L}_{\mathrm{back}}$-pseudo-closed sets of $\mathcal{K}$ it is true that $\mathcal{K}(Q) \subseteq P$.

Then

$$\mathrm{Can}(\mathcal{K}, \mathcal{L}_{\mathrm{back}}) := \{\, P \to \mathcal{K}(P) \mid P \subseteq M \text{ an } \mathcal{L}_{\mathrm{back}}\text{-pseudo-closed set of } \mathcal{K} \,\}.$$

It is well-known that $\mathrm{Can}(\mathcal{K}, \mathcal{L}_{\mathrm{back}})$ is a base of $\mathcal{K}$ with background-knowledge $\mathcal{L}_{\mathrm{back}}$ of minimal cardinality; see [6,9] for a proof on this.[1]

Let $\mathbb{K} = (G, M, I)$ be a formal context, and let $(A \to B) \in \mathrm{Imp}(M)$. A *counterexample (negative example)* for $(A \to B)$ in $\mathbb{K}$ is an object $g \in A' \setminus B'$. It is obvious that $A \to B$ is valid in $\mathbb{K}$ if and only if $\mathbb{K}$ does not contain counterexample for $A \to B$. Conversely, we call g a *model (positive example)* of $A \to B$ if and only if $g \notin A'$ or $g \in B'$.

Related to the notion of counterexamples we define the *confidence* of $A \to B$ in $\mathbb{K}$ as

$$\mathrm{conf}_{\mathbb{K}}(A \to B) := \begin{cases} 1 & \text{if } A' = \emptyset, \\ \frac{|(A \cup B)'|}{|A'|} & \text{otherwise .} \end{cases}$$

In other words, $\mathrm{conf}_{\mathbb{K}}(A \to B)$ is the conditional probability that a randomly chosen object $g \in G$ (in a uniform way), that has all the attributes from A also has all the attributes from B. It is clear that $A \to B$ holds in $\mathbb{K}$ if and only if its confidence in $\mathbb{K}$ is 1.

Let $c \in [0, 1]$. We shall denote with $\mathrm{Th}_c(\mathbb{K})$ the set of all implications in $\mathrm{Imp}(M)$ whose confidence is at least c. If c is chosen properly, we may think of $\mathrm{Th}_c(\mathbb{K})$ as the set of implications which are "almost valid" in $\mathbb{K}$; finding a base $\mathcal{L}$ for this set might therefore be desirable. However, the set $\mathrm{Th}_c(\mathbb{K})$ is not closed under entailment, and thus $\mathcal{L} \subseteq \mathrm{Th}_c(\mathbb{K})$ may not necessarily be true. However, a base of $\mathrm{Th}_c(\mathbb{K})$ might be of more use if the element of the base are also "almost valid," i.e., have a confidence in $\mathbb{K}$ which is at least c. We shall therefore call $\mathcal{L}$ a *confident base* of $\mathrm{Th}_c(\mathbb{K})$ (or just $\mathbb{K}$, if c is clear from the context) if and only if $\mathcal{L}$ is a base of $\mathrm{Th}_c(\mathbb{K})$ and $\mathcal{L} \subseteq \mathrm{Th}_c(\mathbb{K})$.

3   Classical Attribute Exploration

It is the purpose of this section to introduce attribute exploration as it is needed in the exposition of this paper. This includes a description of the classical

---

[1] This proof is only for the special case $\mathcal{K} = \mathrm{Th}(\mathbb{K})$, which however is easily generalized to our general case.

attribute exploration algorithm, which we shall give now. Thereafter, we shall discuss a generalized form of attribute exploration in Sect. 4, which uses similar ideas as but is different from the one given in [3].

We have already mentioned that attribute exploration is an algorithm which assists experts in completing implicational knowledge about a certain domain of interest. More specifically, let us suppose that we have fixed a finite set M of attributes which are relevant for our considerations. We then can understand the *domain of interest* as a collection $\mathcal{D}$ of objects where each object possesses some attributes from M. In other words, a domain $\mathcal{D}$ on a set M can be viewed as a formal context. Let us furthermore suppose that we are given a set $\mathcal{K}$ of implications from which we definitively know that they are valid in our domain $\mathcal{D}$. Finally, we assume that we have an initial collection of some *examples* from our domain, given again as a formal context.

We are now interested in finding all implications that hold in our domain $\mathcal{D}$, i.e., to find all implications which are not invalidated by objects from the domain $\mathcal{D}$. The difficulty of this problems stems from the fact that enumerating all these objects may be algorithmically infeasible. What we can assume, however, is that we are given an *expert* which is able to provide us with the information whether there *exists*, for a given implication $(A \to B) \in \mathrm{Imp}(M)$, an object in our domain $\mathcal{D}$ which is a counterexample for (i.e., not a model of) $A \to B$, and in that case, also provides such a counterexample.

Abstractly, attribute exploration now proceeds as follows. From all implications in $\mathrm{Cn}(\mathcal{K})$ we already known that they are valid in our domain $\mathcal{D}$. Furthermore, for all implications which are invalidated by objects from $\mathbb{K}$, we known that they are not valid in $\mathcal{D}$. For all other implications we do not know whether they hold in $\mathcal{D}$ or not, i.e., all implications in

$$U(\mathbb{K}, \mathcal{K}) := \mathrm{Th}(\mathbb{K}) \setminus \mathrm{Cn}(\mathcal{K})$$

are *undecided* in the sense that they could be valid in $\mathcal{D}$ or not. Then, for the implications in $U(\mathbb{K}, \mathcal{K})$ we have to consult the expert. Attribute exploration now does this in a systematic and somehow efficient way, provided that M is finite.

To make this more precise, we shall proceed by describing attribute exploration in a formal way. This description shall be much more formal than usual, to provide the necessary notions we need for our generalized attribute exploration. To this end, we shall first provide some necessary definitions. After that, we give a formal description of the algorithm. Finally, we shall note some well-known properties of attribute exploration.

We shall start by formalizing our initial, subjective notion of a *domain expert*. Intuitively, a domain expert for a domain $\mathcal{D}$ is just a "function" p that, given an implication $A \to B$, returns "true" if $A \to B$ is not invalidated in $\mathcal{D}$, or returns an object from $\mathcal{D}$ which is a counterexample for $A \to B$. We shall take this understanding as the motivation for the following definition. See also [3].

Definition 1. *Let M be a set. A* domain expert *on M is a function*

$$p \colon \mathrm{Imp}(M) \to \{\top\} \cup \mathfrak{P}(M),$$

where $\top \notin \mathfrak{P}(M)$, *such that the following conditions hold:*

i. *If* $(X \to Y) \in \text{Imp}(M)$ *such that* $p(X \to Y) = C \neq \top$, *then* $C \nvDash (X \to Y)$, *i.e.,* $X \subseteq C, Y \nsubseteq C$. (*p gives counterexamples for false implications*)
ii. *If* $(A \to B), (X \to Y) \in \text{Imp}(M)$ *such that* $p(A \to B) = \top, p(X \to Y) = C \neq \top$, *then* $C \vDash (A \to B)$. (*counterexamples do not invalidate correct implications*)

We say that p confirms an implication $A \to B$ if and only if $p(A \to B) = \top$. Otherwise, we say that p rejects $A \to B$ with counterexample $p(A \to B)$. The theory Th(p) of p is the set of all implications on M confirmed by p.

It is easy to see that every domain gives rise to a domain expert.

Lemma 1. *Let $\mathcal{D}$ be a domain (formal context) on a set M. For each $(A \to B) \in \text{Imp}(M)$ for which there exists a counterexample in $\mathcal{D}$, let $C_{A \to B}$ such a counterexample. Then the mapping*

$$p_{\mathcal{D}}(X \to Y) := \begin{cases} C_{X \to Y} & \text{if } C_{X \to Y} \text{ exists} \\ \top & \text{otherwise} \end{cases}$$

is a domain expert on M.

Note that the definition of $p_{\mathcal{D}}$ depends on the particular choice of the counterexamples, therefore $\mathcal{D}$ may give rise to more than one domain expert.

Let p be a domain expert on a set M, and define

$$\mathcal{D}_p := (\{ p(A \to B) | (A \to B) \in \text{Imp}(M) \} \setminus \{ \top \}, M, \ni).$$

Then clearly $\mathcal{D}_p$ is a domain, and it is easy to see that each domain expert p on M can be obtained as a domain expert of the form $p_{\mathcal{D}_p}$, and that for each domain $\mathcal{D}$ on M it is true that $\mathcal{D} = \mathcal{D}_{p_{\mathcal{D}}}$.

The crucial observation is now that domain experts can answer the question of *validity* in the domains they represent.

Lemma 2. *Let M be a set and let p be a domain expert on M. Then for each $(A \to B) \in \text{Imp}(M)$ it is true that*

$$(A \to B) \text{ is valid in } \mathcal{D}_p \iff p(A \to B) = \top.$$

We have now formally captured the notion of an expert, and we are ready to describe the algorithm of attribute exploration in a formal way, as presented in Algorithm 1. In this exposition, we assume that the set M is equipped with a strict linear order, which then gives rise to a lectic order as it is needed for applying the Next-Closure Algorithm [8]. Furthermore, for better readability, we denote a formal context that arises from another formal context $\mathbb{K}$ by adding a new object with attributes from C by $\mathbb{K} + C$.

Note that the computation of the set P_{i+1} from P_i, $\mathcal{K}_i$ and $\mathbb{K}_i$ and $\mathcal{L}_i$ can be done using the Next-Closure algorithm. As the details are not relevant for our further discussion, we refer the interested reader to the literature [4].

The following results are well known properties of Algorithm 1, and corresponding proofs can be found in [6,7,9,11].

Theorem 1. *Let p, $\mathcal{K}$ and $\mathbb{K}$ be valid input for Algorithm 1. Then Algorithm 1 terminates with input p, $\mathcal{K}$ and $\mathbb{K}$. If $\mathcal{K}'$ and $\mathbb{K}'$ are the corresponding values returned by the algorithm, then the following statements are true:*

i. $\mathcal{K} \subseteq \mathcal{K}' \subseteq \mathrm{Th}(\mathbb{K}') \subseteq \mathrm{Th}(\mathbb{K})$.
ii. $\mathrm{Th}(p) = \mathrm{Th}(\mathbb{K}') = \mathrm{Cn}(\mathcal{K}')$.
iii. The cardinality of $\mathcal{K}' \setminus \mathcal{K}$ is the smallest possible with respect to $\mathrm{Th}(p) = \mathrm{Cn}(\mathcal{K}')$. More specifically, $\mathcal{K}' \setminus \mathcal{K} = \mathrm{Can}(\mathbb{K}', \mathcal{K})$.

Algorithm 1

Input: A domain expert p on a finite set M, a set $\mathcal{K} \subseteq \mathrm{Imp}(M)$ and a formal context $\mathbb{K}$ with attribute set M such that $\mathcal{K} \subseteq \mathrm{Th}(p) \subseteq \mathrm{Th}(\mathbb{K})$.

Procedure

  i. Initialize $i := 0, P_i := \mathcal{K}(\emptyset), \mathcal{K}_i := \mathcal{K}, \mathbb{K}_i := \mathbb{K}$.
  ii. Let P_{i+1} be the smallest $\mathcal{K}_i$-closed set lectically larger or equal to P_i, which is not an intent of $\mathbb{K}_i$. If no such set exists, terminate.
  iii. If p confirms $P \to P''$, then
     – $\mathcal{K}_{i+1} := \mathcal{K}_i \cup \{P \to P''\}$,
     – $\mathbb{K}_{i+1} := \mathbb{K}_i$.
  iv. If p provides a counterexample C for $P \to P''$, then
     – $\mathcal{K}_{i+1} := \mathcal{K}_i$,
     – $\mathbb{K}_{i+1} := \mathbb{K}_i + C$.
  v. Set $i := i + 1$ and go to ii.

Output: Return $\mathcal{K}_i$ and $\mathbb{K}_i$.

4   Exploring Sets of Implications

We are given a precise formulation of attribute exploration in the previous section. However, this formulation is not applicable to our setting of exploring implications with a certain minimal confidence. To address this issue, we shall develop in this section a more general formulation of attribute exploration which goes beyond the classical one.

In the classical case, we are given a formal context $\mathbb{K}$ and a set of implications $\mathcal{K} \subseteq \mathrm{Th}(\mathbb{K})$, as well as a domain expert p, who confirms all implications in $\mathcal{K}$ and where all implications confirmed by p are contained in $\mathrm{Th}(\mathbb{K})$. The task attribute exploration then solves is to provide a method to guide the expert p through all implications in $\mathrm{Th}(\mathbb{K}) \setminus \mathrm{Cn}(\mathcal{K})$ for deciding whether these implications are valid in the domain or not. At the end, attribute exploration both provides a a relative base of all valid implications of the domain p represents, and a set of objects from the domain such that an implication is valid in the domain if and only if all these objects are models of this implication. This set of objects forms itself a domain, and it can be thought of as a sufficient excerpt of the domain represented by p.

We want to try to lift this description of attribute exploration to the case of exploration by confidence. There, our setting is a bit more involved. As in the case of classical attribute exploration, we are given a domain expert p, a formal context $\mathbb{K}$ and a set of implications $\mathcal{K}$. Additionally, we are given a $c\in[0,1]$, the *confidence threshold* for our exploration. Then, in contrast to the classical setting, exploration by confidence considers not only the implications $\mathrm{Th}(\mathbb{K}) \setminus \mathrm{Cn}(\mathcal{K})$, but also those in $\mathrm{Th}_c(\mathbb{K})\setminus\mathrm{Cn}(\mathcal{K})$. We assume that $\mathcal{K}$ is a set of implications with confidence at least c and that all implications in $\mathcal{K}$ are confirmed by p; in other words, $\mathcal{K}\subseteq\mathrm{Th}(p)$ and $\mathcal{K}\subseteq\mathrm{Th}_c(\mathbb{K})$. While the first condition is rather clear, the second is not strictly necessary, but adopted for simplicity.

An attribute exploration algorithm which then works in this setting should guide the expert through the implications in $\mathrm{Th}_c(\mathbb{K})\setminus\mathrm{Cn}(\mathcal{K})$, asking whether some implications are correct or not. The counterexamples provided by the expert are then used to falsify certain implications in $\mathrm{Th}_c(\mathbb{K})$. They are not used, however, for computing the confidence; this is solely done in the initial context $\mathbb{K}$, because we want to find a base of $\mathrm{Th}_c(\mathbb{K})$. At the end, the attribute exploration algorithm should both compute a set $\mathcal{L}$ of implications and a formal context $\mathbb{L}$ such that each implication in $\mathrm{Th}_c(\mathbb{K})$ is either not valid in $\mathbb{L}$ or follows from $\mathcal{L}\cup\mathcal{K}$.

What we now want to describe is a more general formulation of attribute exploration that is applicable to our setting of exploration by confidence. For this, we shall develop in the remainder of this section a general formulation of attribute exploration that works with a set of *certain implications* and a set of *interesting implications* and provides a method to guide an expert through the set of *undecided implications*, until no more are left. The properties this algorithm should have should be the same as in the classical case, as far as this is possible. Then later on, we shall apply this algorithm to our setting of exploration of confidence.

To this end, let us recapitulate our setting for the exploration algorithm, this time a bit more general: we are given a finite set M, a domain expert p on M, and two sets $\mathcal{K}, \mathcal{L}$ of implications. In our classical case, $\mathcal{L} = \mathrm{Th}(\mathbb{K})$ for some formal context $\mathbb{K}$; in our setting of exploration by confidence, we would have $\mathcal{L} = \mathrm{Th}_c(\mathbb{K})$, again for some formal context $\mathbb{K}$ and some $c\in[0,1]$. We assume that $\mathcal{K}\subseteq\mathrm{Th}(p)$ and $\mathcal{K}\subseteq\mathcal{L}$. We then consider the set $\mathcal{K}$ as the (initial) set of *certain implications*. During our exploration we only consider implications in $\mathcal{L}$, wherefore we shall call this set the set of *interesting implications*. Finally, for each implication in $\mathcal{L} \setminus \mathrm{Cn}(\mathcal{K})$ it is not clear yet whether p confirms it or not. Therefore, we call this set the (current) set of *undecided implications*.

An exploration for this abstract setting now should compute a relative base of $\mathcal{L} \cap \mathrm{Th}(p)$ with background knowledge $\mathcal{K}$ by interacting with the expert p. At best, this interaction is kept at a minimum (i.e., the number of times the expert is invoked is as small as possible), as expert interaction is assumed to be expensive.

Considering the classical attribute exploration algorithm, it is not very difficult to come up with a reformulation which is reasonably applicable to this general setting. To this end, let us fix a finite set M and a lectic order $\preceq$ on $\mathfrak{P}(M)$. Then such a reformulation is given in Algorithm 2.

The problem this algorithm has is that it does not ensure that the implications asked to the expert are elements of $\mathcal{L}_i$, the current set of all interesting implications. Because of this, we cannot expect this algorithm to actually compute a relative base of $\mathcal{L} \cap \mathrm{Th}(p)$. However, what this algorithm achieves is to compute an "approximation" of a relative base of $\mathcal{L} \cap \mathrm{Th}(p)$ in the sense that if n is the index of the last iteration of the algorithm, then $\mathcal{K}_n$ is such that

$$\mathrm{Th}(p) \cap \mathrm{Cn}(\mathcal{L}) \supseteq \mathrm{Cn}(\mathcal{K}_n) \supseteq \mathrm{Cn}(\mathrm{Th}(p) \cap \mathcal{L}).$$

So what this algorithm does is not computing a relative base $\mathcal{K}_n$ of $\mathrm{Cn}(\mathrm{Th}(p) \cap \mathcal{L})$, but a complete superset of it. However, this set $\mathcal{K}_n$ is not too far away from being sound for $\mathrm{Cn}(\mathrm{Th}(p) \cap \mathcal{L})$, as $\mathrm{Cn}(\mathcal{K}_n) \subseteq \mathrm{Th}(p) \cap \mathrm{Cn}(\mathcal{L})$. On the other hand, the set $\mathcal{K}_n$ is as small as possible for being sound and complete for itself.

We shall not prove the following result due to space restrictions, but instead refer the interested reader to [4].

Algorithm 2 (General Attribute Exploration)

Input: A domain expert p on a finite set M and sets $\mathcal{K}, \mathcal{L} \subseteq \mathrm{Imp}(M)$ such that $\mathcal{K} \subseteq \mathrm{Th}(p)$ and $\mathcal{K} \subseteq \mathcal{L}$.

Procedure
i. Initialize $i := 0, P_i := \mathcal{K}(\emptyset), \mathcal{K}_i := \mathcal{K}, \mathcal{L}_i := \mathcal{L}, \mathbb{L}_i := (\emptyset, M, \emptyset)$.
ii. Let P_{i+1} be the smallest $\mathcal{K}_i$-closed set lectically larger or equal to P_i, which is not $\mathcal{L}_i$-closed. If no such set exists, terminate.
iii. If p confirms $P_{i+1} \to \mathcal{L}_i(P_{i+1})$, then
     $- \mathcal{K}_{i+1} := \mathcal{K}_i \cup \{ P_{i+1} \to \mathcal{L}_i(P_{i+1}) \}$,
     $- \mathcal{L}_{i+1} := \mathcal{L}_i$,
     $- \mathbb{L}_{i+1} := \mathbb{L}_i$.
iv. If p provides a counterexample C for $P_{i+1} \to \mathcal{L}_i(P_{i+1})$, then
     $- \mathcal{K}_{i+1} := \mathcal{K}_i$,
     $- \mathcal{L}_{i+1} := \{ (A \to B) \in \mathcal{L}_i | C \models (A \to B) \}$,
     $- \mathbb{L}_{i+1} := \mathbb{L}_i + C$.
v. Set $i := i + 1$ and go to ii.

Output: Return $\mathcal{K}_i$ and $\mathbb{L}_i$.

Theorem 2. *Let $p, \mathcal{K}, \mathcal{L}$ be valid input for Algorithm 2. Then the algorithm with this input terminates after finitely many steps. Let n be the last iteration of the algorithm. Then*

1. $\mathrm{Th}(p) \cap \mathcal{L} \supseteq \mathrm{Cn}(\mathcal{K}_n) \supseteq \mathrm{Cn}(\mathrm{Th}(p) \cap \mathcal{L})$,
2. for each $(A \to B) \in \mathcal{L}$, either $(A \to B) \in \mathrm{Cn}(\mathcal{K}_n)$ or $(A \to B) \notin \mathrm{Th}(\mathbb{L}_n)$,
3. $\mathcal{K}_n \setminus \mathcal{K} = \mathrm{Can}(\mathcal{K}_n, \mathcal{K})$.

Since we do not have any control about whether the implications asked by Algorithm 2 are in the set $\mathcal{L}$ of interesting implications, we cannot expect that instantiating this algorithm with $\mathcal{L} = \mathrm{Th}_c(\mathbb{K})$ will indeed yield an algorithm for exploration by confidence. We shall therefore discuss another, even further generalized version of attribute exploration, which will allow for more freedom in which implications are asked to the expert. This generalization arises from Algorithm 2 by observing that instead of asking implications of the form $P_{i+1} \to \mathcal{L}_i(P_{i+1})$, it would be sufficient for the correctness of the algorithm to just ask implications of the form $P_{i+1} \to Q_{i+1}$, where Q_{i+1} is such that $P_{i+1} \subsetneq Q_{i+1} \subseteq \mathcal{L}_i(P_{i+1})$, $Q_{i+1} \nsubseteq \mathcal{K}_i(P_{i+1})$.

Applying this idea to Algorithm 2 yields Algorithm 3. The latter algorithm retains all properties of the former, except for the fact that it does not necessarily compute a minimal base anymore.

Theorem 3. *Let* $p, \mathcal{K}, \mathcal{L}$ *be valid input for Algorithm 3. Then the algorithm applied to this input terminates after finitely many steps. Let* n *be the last iteration of the algorithm. Then*

1. $\mathrm{Th}(p) \cap \mathcal{L} \supseteq \mathrm{Cn}(\mathcal{K}_n) \supseteq \mathrm{Cn}(\mathrm{Th}(p) \cap \mathcal{L})$,
2. for each $(A \to B) \in \mathcal{L}$, *either* $(A \to B) \in \mathrm{Cn}(\mathcal{K}_n)$ *or* $(A \to B) \notin \mathrm{Th}(\mathbb{L}_n)$.

However, as we shall see in Sect. 5.2, we can use Algorithm 3 to devise an algorithm for exploration by confidence, by choosing the sets Q_{i+1} appropriately.

Algorithm 3 (General Attribute Exploration, Weaker Version)

Input: A domain expert p on a finite set M and sets $\mathcal{K}, \mathcal{L} \subseteq \mathrm{Imp}(M)$ such that $\mathcal{K} \subseteq \mathrm{Th}(p)$ and $\mathcal{K} \subseteq \mathcal{L}$.
Procedure
    i. Initialize $i := 0, P_i := \mathcal{K}(\emptyset), \mathcal{K}_i := \mathcal{K}, \mathcal{L}_i := \mathcal{L}, \mathbb{L}_i := (\emptyset, M, \emptyset)$.
    ii. Let P_{i+1} be the smallest $\mathcal{K}_i$-closed set lectically larger or equal to P_i, which is not $\mathcal{L}_i$-closed. If no such set exists, terminate.
    iii. Choose $Q_{i+1} \subseteq M$ such that $P_{i+1} \subsetneq Q_{i+1} \subseteq \mathcal{L}(P_{i+1}), Q_{i+1} \nsubseteq \mathcal{K}_i(P_{i+1})$.
    iv. If p confirms $P_{i+1} \to Q_{i+1}$, then
        – $\mathcal{K}_{i+1} := \mathcal{K}_i \cup \{ P_{i+1} \to Q_{i+1} \}$,
        – $\mathcal{L}_{i+1} := \mathcal{L}_i$,
        – $\mathbb{L}_{i+1} := \mathbb{L}_i$.
    v. If p provides a counterexample C for $P_{i+1} \to Q_{i+1}$, then
        – $\mathcal{K}_{i+1} := \mathcal{K}_i$,
        – $\mathcal{L}_{i+1} := \{ (A \to B) \in \mathcal{L}_i \mid C \vDash (A \to B) \}$,
        – $\mathbb{L}_{i+1} := \mathbb{L}_i + C$.
    vi. Set $i := i + 1$ and go to ii.
Output: Return $\mathcal{K}_i$ and $\mathbb{L}_i$.

5   Exploration by Confidence

Based on the generalizations we have discussed in the previous section, we shall now turn our attention to our original question, namely to devise an algorithm for exploration by confidence. Recall that for this we are given a finite set M, a formal context $\mathbb{K}$ with attribute set M, an expert p on M, some background knowledge $\mathcal{K} \subseteq \mathrm{Th}(p)$, and some number $c \in [0, 1]$. What an algorithm for exploration by confidence now should achieve is to compute a base of $\mathrm{Th}(p) \cap \mathrm{Th}_c(\mathbb{K})$ with background knowledge $\mathcal{K}$. Ideally, for this it should invoke the expert p as few times as possible.

We shall start this section by presenting a first algorithm that is not precisely an algorithm for exploration by confidence, but instead is an *approximative* algorithm in the sense as discussed in the previous section. This first algorithm will be obtained by instantiating the generalized attribute exploration algorithm from Sect. 4. We shall do this in Sect. 5.1. A proper algorithm for exploration by confidence will then be discussed in Sect. 5.2, where we shall instantiate the weaker generalization of attribute exploration from Sect. 4.

5.1   An Approximative Exploration by Confidence

Our first idea is as simple as straightforward: we use Algorithm 2 and instantiate it with our setting of exploration by confidence, i.e., we set $\mathcal{L} = \mathrm{Th}_c(\mathbb{K})$. The resulting algorithm is shown as Algorithm 4. The properties of Algorithm 2, as given in Theorem 2, immediately yield the following result.

Algorithm 4 (Approximative Exploration by Confidence)

Input A domain expert p on a finite set M, a formal context $\mathbb{K}$, $c \in [0, 1]$ and a set $\mathcal{K} \subseteq \mathrm{Th}_c(\mathbb{K})$ such that $\mathcal{K} \subseteq \mathrm{Th}(p)$.

Procedure
   i. Initialize $i := 0$, $P_i := \mathcal{K}(\emptyset)$, $\mathcal{K}_i := \mathcal{K}$, $\mathcal{L}_i := \mathrm{Th}_c(\mathbb{K})$, $\mathbb{L}_i := (\emptyset, M, \emptyset)$.
   ii. Let P_{i+1} be the smallest $\mathcal{K}_i$-closed set lectically larger or equal to P_i, which is not $\mathcal{L}_i$-closed. If no such set exists, terminate.
   iii. If p confirms $P_{i+1} \to \mathcal{L}_i(P_{i+1})$, then
      – $\mathcal{K}_{i+1} := \mathcal{K}_i \cup \{\, P_{i+1} \to \mathcal{L}_i(P_{i+1}) \,\}$,
      – $\mathcal{L}_{i+1} := \mathcal{L}_i$,
      – $\mathbb{L}_{i+1} := \mathbb{L}_i$.
   iv. If p provides a counterexample C for $P_{i+1} \to \mathcal{L}_i(P_{i+1})$, then
      – $\mathcal{K}_{i+1} := \mathcal{K}_i$,
      – $\mathcal{L}_{i+1} := \{\, (A \to B) \in \mathcal{L}_i \mid C \models (A \to B) \,\}$,
      – $\mathbb{L}_{i+1} := \mathbb{L}_i + C$.
   v. Set $i := i + 1$ and go to ii.

Output Return $\mathcal{K}_i$ and $\mathbb{L}_i$.

Corollary 1. *Let $\mathbb{K} = (G, M, I)$ be a finite and non-empty formal context, $c \in [0, 1]$, p be a domain expert on M and $\mathcal{K} \subseteq \mathrm{Th}_c(\mathbb{K}) \cap \mathrm{Th}(p)$. Then Algorithm 4 terminates with input p, c and $\mathcal{K}$. Let n be the last iteration of this run of the algorithm. Then*

  i. $\mathrm{Th}(p) \cap \mathrm{Cn}(\mathrm{Th}_c(\mathbb{K})) \supseteq \mathrm{Cn}(\mathcal{K}_n) \supseteq \mathrm{Cn}(\mathrm{Th}(p) \cap \mathrm{Th}_c(\mathbb{K}))$,
  ii. $\mathrm{Can}(\mathcal{K}_n, \mathcal{K}) = \mathcal{K}_n \setminus \mathcal{K}$.

Evidently, Algorithm 4 does not guarantee that the implications asked are actually elements of $\mathcal{L} = \mathrm{Th}_c(\mathbb{K})$, i.e., those implications do not need to have a confidence of at least c in $\mathbb{K}$. This may or may not be an issue, depending on the application one is currently dealing with.

What is also important for Algorithm 4 to be practical is to be able to compute closures under $\mathcal{L}_i = \mathrm{Th}_c(\mathbb{K}) \cap \mathrm{Th}(\mathbb{L}_i)$. However, it is by far obvious how to compute closures under these sets of implications. Of course, one does not want to compute these sets explicitly, and indeed it is true that

$$\mathcal{L}_i(A) = A''_{\mathbb{L}_i} \cap \mathrm{Th}_c(\mathbb{K})(A),$$

for each $A \subseteq M$, where $A''_{\mathbb{L}_i}$ denotes double derivation in $\mathbb{L}_i$. Thus, to make Algorithm 4 practicably applicable, one only needs a way to compute closures of sets A under $\mathrm{Th}_c(\mathbb{K})$.

While it is possible to compute these closures effectively without computing the set $\mathrm{Th}_c(\mathbb{K})$ explicitly [4], the computational overhead might be unwelcomed. One may be tempted to think that we can eliminate the problem of computing closures under $\mathrm{Th}_c(\mathbb{K})$ by using the following approach: instead of asking implications of the form

$$P_{i+1} \to \mathcal{L}_i(P_{i+1}), \tag{1}$$

where $\mathcal{L}_i(P_{i+1}) = \mathrm{Th}_c(\mathbb{K})(P_{i+1}) \cap (P_{i+1})''_{\mathbb{L}_i}$, in Algorithm 4 we could just as well ask implications of the form

$$P_{i+1} \to \{\, m \in M \mid \mathrm{conf}_{\mathbb{K}}(P_{i+1} \to \{\, m \,\}) \geqslant c \,\}. \tag{2}$$

This would have the evident advantage that the right-hand side of the implication is easy to compute. However, it turns out that with this modification the algorithm is not correct anymore, in the sense that the set of implications accepted by the expert is not complete for $\mathrm{Th}_c(\mathbb{K})$.

$\mathbb{K}$	a	b	c
1	×	×	
2	×	×	
3	×	×	
4	×	×	×
5	×	×	×
6	×		×
7	×		×
8			
9			
10			

Fig. 1. Context which shows that a simple approach to exploration by confidence does not work

Example 1 (Example 6.2.2 from [4]).
    Consider the formal context $\mathbb{K}$ as given in Fig. 1, let $\mathcal{K} = \{\, \{\, a \,\} \to \{\, b \,\} \,\}$, and choose $c = \frac{1}{2}$. Suppose that we apply exploration by confidence in the simplified version as described before, i.e., we ask implications of the form of Eq. (2) instead of those in Eq. (1). Then since all sets P_i are closed under $\mathcal{K}$, the implication $\{\, a \,\} \to \{\, c \,\}$ is never asked to the expert, because $\{\, a \,\}$ is not closed under $\mathcal{K}$. On the other hand,

$$\mathrm{conf}_{\mathbb{K}}(\{\, a \,\} \to \{\, c \,\}) = \frac{4}{7} > \frac{1}{2},$$

i.e. $(\{\, a \,\} \to \{\, c \,\}) \in \mathrm{Th}_c(\mathbb{K})$, and is thus an interesting implication. Furthermore, the implication $\{\, a \,\} \to \{\, c \,\}$ also does not follow from other implications asked to the expert, as the implications $\{b\} \rightarrow \{c\}$, $\{a, b\} \rightarrow \{c\}$, and $\emptyset \rightarrow \{c\}$ will also not be asked to the expert, because

$$\mathrm{conf}_\mathbb{K}(\{b\} \rightarrow \{c\}) = \frac{2}{5} < \frac{1}{2}$$

$$\mathrm{conf}_\mathbb{K}(\{a, b\} \rightarrow \{c\}) = \frac{2}{5} < \frac{1}{2}$$

$$\mathrm{conf}_\mathbb{K}(\emptyset \rightarrow \{c\}) = \frac{4}{10} < \frac{1}{2}$$

Thus, if we assume that the expert p confirms all proposed implications, and if we denote the set of confirmed implications by $\mathcal{K}_n$, then

$$\mathcal{K}_n(\{a\}) = \{a, b\}.$$

But $\mathrm{Th}_c(\mathbb{K}) \cap \mathrm{Th}(p) = \mathrm{Th}_c(\mathbb{K})$, and

$$\mathrm{Th}_c(\mathbb{K})(\{a\}) = \{a, b, c\}.$$

Thus, the set $\mathcal{K}_n$ is not complete for $\mathrm{Th}_c(\mathbb{K}) \cap \mathrm{Th}(p)$.

5.2   An Exact Exploration by Confidence

The previous example shows that our simple idea of avoiding the computational overhead of computing closures under $\mathrm{Th}_c(\mathbb{K})$ did not work. In this section we shall show how we can make this idea work nonetheless, by further suitably modifying the algorithm. For this we shall use the weaker generalization of Algorithm 3. As a pleasant side-effect, by this we will obtain a proper algorithm for exploration by confidence, i.e., the new algorithm will indeed compute a base of $\mathrm{Th}(p) \cap \mathrm{Th}_c(\mathbb{K})$. On the downside, since this algorithm is based on the weaker generalization of attribute exploration, we cannot expect it to compute a base of minimal cardinality.

The main idea for this adaption is as follows: the weaker generalization of Algorithm 3 instantiated for our setting of exploration by confidence does not require us to compute closures under $\mathrm{Th}_c(\mathbb{K})$. Instead, all we need to check is whether a given set of attributes is closed under $\mathrm{Th}_c(\mathbb{K})$. The main problem with the latter is that in general we need to consider all subsets of $B \subseteq A$ and all elements $m \in M \setminus A$ checking whether they satisfy

$$\mathrm{conf}_\mathbb{K}(B \rightarrow \{m\}) \geqslant c.$$

This is because

$$A = \mathrm{Th}_c(\mathbb{K})(A) \iff (\forall B \subseteq A \forall m \in M: \mathrm{conf}_\mathbb{K}(B \rightarrow \{m\}) \geqslant c \implies m \in A).$$

On the other hand, if A would have the property that for each $m \in M$ and every $B \subsetneq A$ with $\mathrm{conf}_\mathbb{K}(B \rightarrow \{m\}) \geqslant c$ it is true that $m \in A$, then checking whether A is closed under $\mathrm{Th}_c(\mathbb{K})$ would be easy, as in this case

$$A = \mathrm{Th}_c(\mathbb{K})(A) \iff (\forall m \in M: \mathrm{conf}_\mathbb{K}(A \rightarrow \{m\}) \geqslant c \implies m \in A).$$

Let's make this more precise. In what follows, we shall write the context subposition of two contexts $\mathbb{K}_1 = (G_1, M, I_1), \mathbb{K}_2 = (G_2, M, I_2)$ as $\mathbb{K}_1 \div \mathbb{K}_2$, i.e.,

$$\mathbb{K}_1 \div \mathbb{K}_2 := (G_1 \cup G_2, M, I_1 \cup I_2).$$

Here we assume that G_1 and G_2 are disjoint. In the following proposition, we have $\mathbb{K}_1 = \mathbb{K}$ and $\mathbb{K}_2 = \mathbb{L}_i$, and we can think of the former as the initial formal context of our exploration process, while the latter contains all counterexamples collected up to iteration i. Then $\mathbb{K} \div \mathbb{L}_i$ is the currently known context of iteration i.

Proposition 1 (Proposition 6.2.5 from [4]). *Let $\mathbb{K} = (G, M, I)$ be a finite formal context, and let $c \in [0, 1]$. Let $\mathbb{L}_i = (G_i, M, I)$ be another finite formal context such that G_i and G are disjoint, and define $\mathcal{L}_i = \mathrm{Th}_c(\mathbb{K}) \cap \mathrm{Th}(\mathbb{L}_i)$. Let $A \subseteq M$ be such that for every intent $X \subsetneq A$ of $\mathbb{K} \div \mathbb{L}_i$ it is true that*

$$\forall m \in X_{\mathbb{L}_i}'' : \mathrm{conf}_{\mathbb{K}}(X \to \{ m \}) \geq c \implies m \in \mathcal{K}_i(X). \tag{3}$$

In addition, let A be $\mathcal{K}_i$-closed. Then it is true that A is $\mathcal{L}_i$-closed if and only if

$$A = A_{\mathbb{K} \div \mathbb{L}_i}'' \quad and \quad \forall m \in A_{\mathbb{L}_i}'' \setminus A : \mathrm{conf}_{\mathbb{K}}(A \to \{ m \}) < c. \tag{4}$$

Algorithm 5 (Exploration by Confidence)

Input: A domain expert p on a finite set M, a formal context $\mathbb{K}$, $c \in [0, 1]$ and a set $\mathcal{K} \subseteq \mathrm{Th}_c(\mathbb{K})$ such that $\mathcal{K} \subseteq \mathrm{Th}(p)$.
Procedure
   i. Initialize $i := 0, P_i := \mathcal{K}(\emptyset), \mathcal{K}_i := \mathcal{K}, \mathcal{L}_i := \mathrm{Th}_c(\mathbb{K}), \mathbb{L}_i := (\emptyset, M, \emptyset)$.
   ii. Let $P_{i+1} := \min_{\preceq}(P_{i+1}^1, P_{i+1}^2)$, where
     – P_{i+1}^1 is the lectically smallest intent P of $\mathbb{K} \div \mathbb{L}_i$ such that $P_i \preceq P$, and there exists some $m \in P_{\mathbb{L}_i}'' \setminus \mathcal{K}_i(P)$ with $\mathrm{conf}_{\mathbb{K}}(P \to \{ m \}) \geq c$.
     – P_{i+1}^2 is the lectically smallest set P such that $P_i \preceq P$, that is closed under $\mathcal{K}_i$, but is not an intent of $\mathbb{K} \div \mathbb{L}_i$.
   iii. If $P_{i+1} = P_{i+1}^1$, then set $Q_{i+1} := P_{i+1} \cup \{ m \}$, otherwise set $Q_{i+1} := (P_{i+1})_{\mathbb{K} \div \mathbb{L}_i}''$.
   iv. If p confirms $P_{i+1} \to Q_{i+1}$, then
     – $\mathcal{K}_{i+1} := \mathcal{K}_i \cup \{ P_{i+1} \to Q_{i+1} \}$,
     – $\mathcal{L}_{i+1} := \mathcal{L}_i$,
     – $\mathbb{L}_{i+1} := \mathbb{L}_i$.
   v. If p provides a counterexample C for $P_{i+1} \to Q_{i+1}$, then
     – $\mathcal{K}_{i+1} := \mathcal{K}_i$,
     – $\mathcal{L}_{i+1} := \{ (A \to B) \in \mathcal{L}_i | C \vDash (A \to B) \}$,
     – $\mathbb{L}_{i+1} := \mathbb{L}_i + C$.
   vi. Set $i := i + 1$ and go to ii.
Output: Return $\mathcal{K}_i$ and $\mathbb{L}_i$.

Based on this result, we shall now adapt our exploration algorithm to ensure that all sets of which we need to check closedness under $\mathrm{Th}_c(\mathbb{K})$ satisfy Eq. (3). We can do this as follows: as usual, we consider subsets X of M in lectic order, and for each such set X that is closed under the set $\mathcal{K}_i$ of currently known implications but is not an intent of $\mathbb{K} \div \mathbb{L}_i$, we ask the expert the implication

$$X \to X''_{\mathbb{K} \div \mathbb{L}_i}.$$

Additionally, in accordance with Proposition 1, for each X that is an intent of $\mathbb{K} \div \mathbb{L}_i$ we ask the implication

$$X \to \{\, m \in M \mid \operatorname{conf}_{\mathbb{K}}(X \to \{\, m \,\}) \geqslant c \,\}.$$

The resulting algorithm is shown in Algorithm 5. It is not hard to see that this algorithm is indeed an instance of Algorithm 3. Therefore, from the general results of Theorem 3 about Algorithm 3, we immediately obtain the following result. Moreover, the algorithm only asks implications with confidence at least c, wherefore Algorithm 5 is a proper algorithm for exploration by confidence.

Corollary 2. *Let* $\mathbb{K} = (G, M, I)$ *be a finite formal context,* p *a domain expert on* M, $c \in [0, 1]$, *and* $\mathcal{K} \subseteq \operatorname{Th}(p) \cap \operatorname{Th}_c(\mathbb{K})$. *Then Algorithm 5 applied to this input terminates after finitely many steps. Let* n *be the last iteration of the algorithm. Then* $\mathbb{K}_n$ *is a confident base of* $\operatorname{Th}(p) \cap \operatorname{Th}_c(\mathbb{K})$, *i.e.,* $\mathcal{K}_n \subseteq \operatorname{Th}_c(\mathbb{K})$ *and*

$$\operatorname{Cn}(\mathcal{K}_n) = \operatorname{Cn}(\operatorname{Th}(p) \cap \operatorname{Th}_c(\mathbb{K})).$$

Moreover, for each $(A \to B) \in \mathcal{L}$, *either* $(A \to B) \in \operatorname{Cn}(\mathcal{K}_n)$ *or* $(A \to B) \notin \operatorname{Th}(\mathbb{L}_n)$.

6   Outlook and Further Results

In this paper we have addressed the issue of applying attribute exploration to faulty data. We did this by extending the classical attribute exploration algorithm to not only ask implications that are valid in the data, but to ask also those implications that enjoy a *high confidence* therein. The motivation behind this approach was to assume that data which is only slightly faulty will invalidate important implications only with few counterexamples, compared to the number of examples where this implication does apply. Of course, this approach is purely heuristic, and should be treated as such.

In our discussion about how to design an exploration algorithm that takes the confidence of implications into account, we first formalized the notion of an expert. After that, we discussed how classical attribute exploration can be seen as an exploration of *sets of interesting implications*. For this more abstract view, we discussed a straight-forward generalization of the classical algorithm, as well as a weaker generalization which allowed for more freedom in the choice of the implications asked to the experts. Based on these generalization, we developed an approximative as well as an exact algorithm for exploration by confidence.

This paper deliberately avoids giving proofs for the statements it presents. Those proofs can be found in [4]. There we also discuss a generalization of the present results to *general concept inclusions* (GCIs). GCIs are logical formulas which provide a generalization of implications to the realm of *description logics* [2]. It is not hard to generalize the notion of confidence to GCIs, and one can then build upon the results presented in this paper and devise an algorithm for exploring general concept inclusions with high confidence. The immediate advantage of this would be the increase in expressivity provided by the use of description logics.

To generalize exploration by confidence to general concept inclusions, one has to extend the algorithm to also be able to work with *growing sets of attributes*. More precisely, during the exploration, the attribute set M, which is supposed to be fixed in this paper, may grow in a consistent way. Exploration by confidence can be adapted to this setting as well, much like [6] adapts classical attribute exploration to this setting.

A main motivation for considering the case of faulty data is that in real applications data is never free of errors. With respect to this, one could argue that the results of this paper contribute to making attribute exploration more usable in practice. However, this argumentation would be much more convincing if we could provide real-world use cases of exploration by confidence. Finding and evaluating such use cases is a main task for future research.

Another interesting application not discussed so far is the following. As soon as the expert accepts an implication with confidence not equal to 1, all counterexamples of this implication are false. Our algorithm could be adapted to propose these faulty objects to the expert for correction, thereby increasing the quality of the data-set during the course of the exploration. This form of error correction could be more efficient than walking through the whole data-set and correcting all errors. This is because an error-correcting exploration by confidence would only propose errors for correction that are relevant for the exploration process.

Acknowledgments. This work has been partially supported by the DFG Research Training Group 1763 "QuantLA", and by the Cluster of Excellence "Center for Advancing Electronics Dresden" (cfAED). Additionally, the author is grateful to the anonymous reviewers for the detailed and helpful comments.

References

1. Agrawal, R., Imielinski, T., Swami, A.N.: Mining association rules between sets of items in large databases. In: Proceedings of the ACM SIGMOD International Conference on Management of Data, pp. 207–216 (1993)
2. Baader, F., et al. (eds.): The Description Logic Handbook: Theory, Implementation, and Applications. Cambridge University Press, New York (2003)
3. Borchmann, D.: A general form of attribute exploration. LTCS-Report 13–02. Chair of Automata Theory, Institute of Theoretical Computer Science, Technische Universität Dresden (2013)
4. Borchmann, D.: Learning terminological knowledge with high confidence from erroneous data. Ph.D. thesis, Technische Universität Dresden (2014)
5. Burmeister, P., Holzer, R.: Treating incomplete knowledge in formal concept analysis. In: Ganter, B., Stumme, G., Wille, R. (eds.) Formal Concept Analysis. LNCS (LNAI), vol. 3626, pp. 114–126. Springer, Heidelberg (2005)

6. Distel, F.: Learning description logic knowledge bases from data using methods from formal concept analysis. Ph.D. thesis, Technische Universität Dresden (2011)
7. Ganter, B.: Attribute exploration with background knowledge. Theor. Comput. Sci. **217**(2), 215–233 (1999)
8. Ganter, B.: Two basic algorithms in concept analysis. In: Kwuida, L., Sertkaya, B. (eds.) ICFCA 2010. LNCS, vol. 5986, pp. 312–340. Springer, Heidelberg (2010)
9. Ganter, B., Wille, R.: Formal Concept Analysis: Mathematical Foundations. Springer, Heidelberg (1999)
10. Luxenburger, M.: Implikationen, Abhängigkeiten und Galois-Abbildungen. Ph.D. thesis, TH Darmstadt (1993)
11. Stumme, G.: Attribute exploration with background implications and exceptions. In: Bock, H.-H., Polasek, W. (eds.) Data Analysis and Information Systems: Studies in Classification, Data Analysis, and Knowledge Organization, pp. 457–469. Springer, Heidelberg (1996)

Automatic Validation of Terminology by Means of Formal Concept Analysis

Luis Felipe Melo Mora[✉] and Yannick Toussaint

Inria Nancy-Grand Est, BP 239 - 54506 Villers-lès-Nancy, France
{luis-felipe.melo-mora,yannick.toussaint}@inria.fr

Abstract. Term extraction tools extract candidate terms and anno-
tate their occurrences in the texts. However, not all these occurrences
are terminological and, at present, this is still a very challenging issue
to distinguish when a candidate term is really used with a terminological meaning. The validation of term annotations is presented as a
bi-classification model that classifies each term occurrence as a terminological or non-terminological occurrence. A context-based hypothesis
approach is applied to a training corpus: we assume that the words in
the sentence which contains the studied occurrence can be used to build
positive and negative hypotheses that are further used to classify undetermined examples. The method is applied and evaluated on a french
corpus in the linguistic domain and we also mention some improvements
suggested by a quantitative and qualitative evaluation.

1 Introduction

Terms in texts are important entities for any kind of document analysis:
information retrieval, knowledge extraction or ontology building, etc. They are
usually considered as linguistics entities that could be associated with meanings
or concepts, their mirror at the ontological level [4]. However, there is no formal
definition of what a term is, nor is there any reliable syntactic description that
could help term identification. Most of terms are noun phrases composed by
a word or several words. Moreover, terms depend on the domain of speciality,
and within a domain, terms are context-sensitive: a given string (a word or a
set of words) may be a term in a given context with some meaning, a term in
another context but with another meaning, or it could also be a non-term in a
third context. Term extraction tools [17,18] extract candidate terms, i.e. groups
of words that could be considered as terms. A candidate term fulfils linguistic
(mainly syntactic schema) and/or statistic (based on occurrences) criteria. Once
a candidate term is extracted by the tool, all its occurrences are annotated in
the corpus. However, some of its occurrences correspond to a terminological use
and some other correspond to a non-terminological use, *i.e.* these occurrences
should be considered as words from general language. Thus, candidate terms and
each of their occurrences should be manually validated by experts which makes
it difficult for large-scale applications.

© Springer International Publishing Switzerland 2015
J. Baixeries et al. (Eds.): ICFCA 2015, LNAI 9113, pp. 236–251, 2015.
DOI: 10.1007/978-3-319-19545-2_15

The paper presents how hypotheses built with formal concept analysis help to validate or invalidate candidate terms in texts of a specific domain. Some training data sets have been built on purpose but, as validation is time consuming, corpus are rather small and domain dependent. Thus, such a symbolic approach, based on itemset mining and classification, suits well the problem. In the longer term, linguists expect from this approach a better understanding of what term triggers are and how to find them.

The following examples with the candidate term subject remind us how ambiguous human language is: the same string may refer to different concepts. This is why term validation is so important for document indexing, automatic summarization, construction of ontologies and even for facilitating multilingual communication. The only help for meaning disambiguation is the context of the occurrences, *i.e.* the words that occurs with the term in the same sentence.

- (S_1) I **subject** him to a terrifying ordeal. $< Verb, non - term, general\ language >$.
- (S_2) This type of wound is highly **subject** to infection. $< Adjective, non - term, generallanguage >$.
- (S_3) What is the **subject** in a sentence? $< Noun, term, linguistics >$.
- (S_4) Maths is not my best **subject**. $< Noun, term, pedagogy >$.
- (S_5) *A* moving picture of a train is more dramatic than a still picture of the same **subject**.$< Noun, non - term, generallanguage >$.
- (S_6) The relation between the **subject** and predicate is identified by the use of: All, No, Some, ... $< Noun, term, logic >$.
- (S_7) The **subject of law** is a person (physical or juridical) who in law has the capacity to realize rights and juridical duties. $< Noun, complex\ term, law >$.

In the above examples, S_1, S_2, S_5 are contexts where the candidate subject is not a term, while S_3, S_4, S_6, S_7 are contexts where the candidate subject is a term in linguistics, in pedagogy, in logic or in law domains, respectively.

For each term candidate in a given domain, the goal is to validate or invalidate each of its occurrences. Each candidate term is studied separately and we propose a supervised learning method trained on a manually annotated corpus. For the learning phase, each occurrence of the candidate term is described by its textual context, *i.e.* the bag of the words of the sentence, and the occurrence is also tagged as "positive example" (belonging to the "T_+ class") if it is a terminological occurrence or as "negative example" ("T_- class") if it is not. Thus, from textual context our method extracts hypotheses, a notion that is formally introduced in the next section. Hypotheses are itemsets of words corresponding to the positive occurrences of a candidate term and, similarly, itemsets corresponding to negative occurrences of a candidate term. Then, during the test phase, a new occurrence of this candidate term in a new sentence is classified either as terminological occurrence or non-terminological occurrence according to the hypotheses that match the sentence.

The learning problem can be formulated in the paradigm of Formal Concept Analysis (FCA) [8], a formal method where ordered sets are classified in a lattice.

FCA builds a bi-classification model from positive and negative examples. In the binary matrix associated to each candidate term, the objects are occurrences of the candidate term. Attributes are words coming from the different contexts and a positive/negative flag is introduced in accordance with to the manual annotation. Hypotheses [12,13] are generalised descriptions of positive or negative examples. These itemsets are non-redundant descriptions of either the positive class or the negative class. There is a high demand from linguists for such human-readable sets that could be considered as triggers and distinguish terminological occurrences from non-terminological ones. Moreover, hypotheses are applied to new (unannotated) occurrences of a candidate term to discover its terminological or non-terminological nature in new texts.

The paper is organized as follows. Section 2 provides a brief overview of the problem of validating term occurrences. Section 3 introduces Formal Concept Analysis and its application to learning problems. Then, Sect. 4 describes how positive and negative hypotheses can be applied to textual contexts of term occurrences in order to validate or invalidate them as a terms. In Sect. 5, we describe the dataset *i.e.* the corpus, the experiments and their results. Then, Sect. 6 concludes the paper.

2 Terminology Extraction

Eugen Wüster [19] emphasized on the role of terms, their link with concepts, and the importance of normalization of terms to avoid ambiguity, to ease indexing, thesaurus building or translation. He was *the* author who defined the general theory of terminology and worked within a standardization perspective. At that time, terminology was initially a prerogative of translators, with a rather normative approach.

However, in the 90's, the renewal of corpus linguistics with some new robust tools such as part of speech taggers or syntactic parsers showed that terms are not restricted to set phrases in a previously defined list but they are full linguistic entities whose form may vary in the texts (plural forms, modifiers, etc.). New software applications in information retrieval, summarizing, or ontology construction stimulate this new conception of terminology. Thus, there has been several initiatives for developing term extractors. Among them some are term locators: they locate in texts terms belonging to a controlled vocabulary [1,9]. Some others are working *ab-nihilo*, looking for candidate terms [2,7,17,18].

Thereby, term extraction consists in a set of computational techniques that allow to identify the linguistic realizations of domain-specific concepts known as terms. Frequently it is seen as an intermediate phase of Natural Language Processing, that bridges the gap with the knowledge level and enables different kind of reasoning. Few term extractors use only statistics on word occurrences and co-occurrences to propose term candidates. Most of them are now combining linguistic rules and statistical filters [17]. However, despite this configuration, there is still a lot of noise both in candidate term identification and in the distinction between their terminological and non-terminological occurrences.

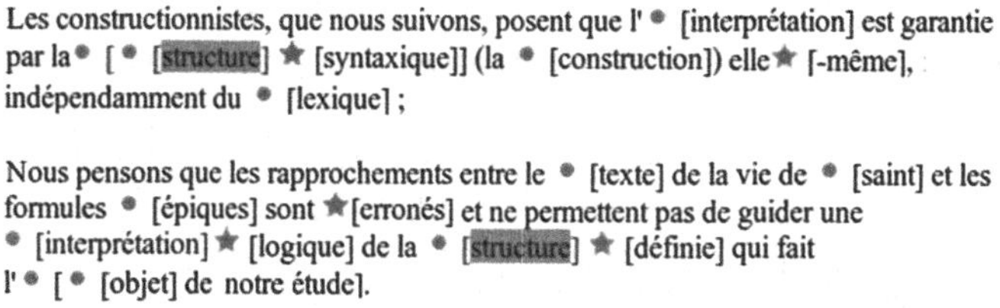

Les constructionnistes, que nous suivons, posent que l' ● [interprétation] est garantie
par la ● [● [structure] ★ [syntaxique]] (la ● [construction]) elle ★ [-même],
indépendamment du ● [lexique] ;

Nous pensons que les rapprochements entre le ● [texte] de la vie de ● [saint] et les
formules ● [épiques] sont ★ [erronés] et ne permettent pas de guider une
● [interprétation] ★ [logique] de la ● [structure] ★ [définie] qui fait
l' ● [● [objet] de notre étude].

Fig. 1. Chunks of texts where candidate terms (simple and multi-words) are located
with TTC Term Suite and represented by square brackets [ ]. The green dots indicate
validated candidates (terms), whereas the red stars define candidate terms refused by
the experts (Color figure online).

For instance, in Fig. 1, extracted candidate terms are represented between
square brackets. Some candidate terms include some others (nested brackets).
Thus, **structure syntaxique** (*syntactic structure*) is proposed as a candidate
term while **structure définie** (*defined structure*) is not. It should be noted
that both elements have the same grammatical structure and in both sentences
structure is also proposed as a candidate term. Occurrences which are marked
by a green dot have been manually validated as terminological occurrences while
non-terminological occurrences are marked by red stars.

In the next section, we introduce Formal Concept Analysis and its use for
bi-classification of term candidate occurrences.

3   Formal Concept Analysis (FCA)-Based Method

The main notions of Formal Concept Analysis (FCA) theory are introduced in
this section. Afterwards, concept-based hypotheses (called also JSM-hypotheses
from the John Stuart Mill method) are presented as a method for building a
bi-classification model from positive and negative examples. Similarly, Jumping
Emerging Patterns (JEPs) is an alternative formalism to identify the set of
discriminating attributes which only occur in one class and are absent in the
other.

3.1   Bases on FCA

FCA is a data analysis theory which builds conceptual structures defined by
means of the attributes shared by objects. Formally, this theory is based on the
triple $K = (G, M, I)$ called *formal context*, where G is a set of objects, M is a
set of attributes and I is the binary relation $I \subseteq G \times M$ between objects and
attributes. Therefore, $(g, m) \in I$ means that g has the attribute m. For instance,
some occurrences of the introductory examples with the candidate term **subject**
are encoded in the formal context given by Table 1.

Two derivation operators are then defined:

$$A' := \{m \in M \mid \forall g \in A : gIm\} \, for \, A \subseteq G,$$
$$B' := \{g \in G \mid \forall m \in B : gIm\} \, for \, B \subseteq M.$$

Table 1. An example of a formal context where each row represents an occurrence of the candidate term `subject` with the words appearing in its textual context.

	I	a	this	subject	him	to	terrifying	ordeal	type	of	wound	what	is	the	in	sentence	highly	infection
subject(S_1)	x			x	x	x	x	x										
subject(S_2)				x	x	x				x	x	x		x			x	x
subject(S_3)		x		x									x	x	x	x	x	

A *formal concept* is a pair *(A, B)*, satisfying $A \subseteq G$, $B \subseteq M$, $A' = B$ and $B' = A$. *A* is called the *extent* and *B* the *intent* of the (formal) concept." is a closure operator which means that for any X, Y, $X'''' = X''$ (idempotent), $X'' \subseteq X$ (extensive), and $X \subseteq Y \longrightarrow X'' \subseteq Y''$ (monotone). Thus, the intent of a concept is the maximum set of attributes shared by all the objects of its extent. Moreover, an itemset $X \subseteq M$ is a *generator* of a formal concept *(A, B)*, if $X \subseteq B$ and $X' = A$. Likewise, a *minimal generator* for a concept is defined as a minimal subset of its intent which can similarly characterize the concept in question.

Formal concepts are organized into a complete *concept lattice* denoted by $\mathcal{L}$ following a partial ordering, called subsumption, ($\sqsubseteq$), defined as follows: $(A_1, B_1) \sqsubseteq (A_2, B_2) \Leftrightarrow A_1 \subseteq A_2$ (*or* $B_2 \subseteq B_1$).

3.2   Classification by FCA

A learning model from a concept lattice has been extensively studied through the notion of *concept-based hypothesis* [12,13]. This model is based on positive and negative examples of a target attribute. The idea laying beyond this model is to discover the attribute combinations which are shared by positive examples, but not by negative examples.

Let us consider the target attribute $w \notin M$, which may have one of the three values: positive, negative and undetermined. Thereby, the input data for learning is composed by sets of positive and negative examples. Positive examples are objects that are known to have the target attribute and negative examples are objects that are known not to have this attribute. The learning results are rules supposed to classify a third set of objects called the undetermined examples.

With regard to FCA theory, this classification method can be described by three sub-contexts: a positive context $K_+ = (G_+, M, I_+)$, a negative context $K_- = (G_-, M, I_-)$ and an undetermined context $K_\tau = (G_\tau, M, I_\tau)$. M is a set of attributes, w is the target attribute and $w \notin M$, G_+ is the set of positive examples whereas G_- is the set of negative examples. Alternatively, G_τ denotes the set of new examples to be classified. The *learning context* is denoted by $K_\pm = (G_+ \cup G_-, M \cup w, I_+ \cup I_- \cup G_+ \times \{w\})$. In addition, $K_c = (G_+ \cup G_- \cup G_\tau, M \cup w, I_+ \cup I_- \cup I_\tau \cup G_+ \times \{w\})$ is called the *classification context*.

For generalizing the G_+ subset and defining the cause of target attribute, we are interested in finding the sets of attributes that are shared by only positive examples. In the best case, the membership to G_+ supposes a particular attribute combination. However, in most cases it is necessary to find several attribute combinations called *positive hypotheses* to characterize only G_+ examples. Ideally, we would like to find enough positive hypotheses to cover all G_+ examples.

A positive hypothesis H_+ for w is defined as a non empty intent of K_+ which is not contained in the intent g' of any negative example $g \in G_-$. A *negative hypothesis* H_-, is defined accordingly.

Thereby, hypotheses can be used to classify an undetermined example $x \in G_r$. If the intent of x contains at least one positive hypothesis and no negative hypothesis, then, x is classified as a positive example. If the intent of x contains at least one negative hypothesis and no positive hypothesis, then it is a negative example. Otherwise, x remains unclassified.

In addition, we can restrict the number of useful hypotheses with regard to subsumption in the lattice. Formally, a positive hypothesis H_+ is a *minimal positive hypothesis* if there is no positive hypothesis H such that $H \subset H_+$. *Minimal negative hypothesis* is defined similarly. Hypotheses which are not minimal should not be considered for classification because they do not improve discrimination between positive and negative examples.

In a not-so-far context of itemset mining, the notion of Jumping Emergent Pattern (JEP) is very similar to concept-based hypothesis [5]. A JEP is an itemset that occurs only in objects of one class and not in objects of the other class. Clearly, a hypothesis is a JEP. On the other hand, a JEP is a generator of some hypotheses for this class. We can also define a *minimal JEP* as an itemset that does not contain any other JEP. Consequently, searching the minimal JEPs is equivalent to finding the minimal generators of the concept-based hypotheses for a class. Other important class of patterns that represent a contrast between the classes are exposed in [16]. For instance, an *emerging pattern(EP)* is an itemset whose frequency changes significantly from one data set (G_+ for example) to another (G_- respectively). Similarly, the *constrained emerging patterns (CEPs)* are defined as the minimal set of items which occur at most α times in one data set and at least β in the other. Unlike concept-based hypotheses or JEPs, EPs and CEPs are potentially more resistant to noise because they are less restrictive patterns [16].

3.3   Relevant Hypotheses

A concept-based hypothesis generalizes a class of positive or a class of negative examples. Each hypothesis is a closed itemset, i.e. the intent of a concept. Nevertheless, because of the noise in the data, these hypotheses are not all relevant. *Stability* [14] is a measure that qualifies the tendency of a concept (and its intent) to persist when some objects are randomly removed from its extent. Thereby, stability measures how much a concept depends on each particular object of its extent. As a consequence, a stable concept will be independent of data noise.

Thus, it may happen that some minimal hypotheses have a low stability value. In that case, the intent of subsumed concepts $H_1, ..., H_n \supset H$, which are hypotheses but non-minimal, may have a higher stability value. These hypotheses are more restrictive when applied to undetermined object classification and the precision of the overall system could be improved. However, such a strategy may reduce the coverage of the positive (resp. negative) examples by the set of hypotheses, with a possible deterioration of the system recall.

Other measures have been proposed in [11] to recognise relevant concepts in noisy data. Among these measures are the support, concept probability or separation index which can be useful in different kinds of contexts. However this comparison concludes that stability is the most effective and the less independent of the type of the context. Stability is the only measure we kept to evaluate hypotheses relevance in our study.

Accordingly, we adopted a FCA classification model to identify the patterns that represent the largest shared textual contexts from the occurrences of a term on a specific domain. In the following section, we present practical aspects and some other considerations for our method.

4   Term Validation as a Bi-classification Problem

In order to minimize the human intervention and to improve the terminology validation scalability, Formal Concept Analysis (FCA) can be used for learning the hypotheses from positive and negative examples. As shown in [15], the textual context is the key for validating terminological occurrences. So, we assume that the textual context around each candidate occurrence gives us relevant information on its class. For a given domain and a given candidate term, we thus focus on the differentiation between a terminological use (T_+) and a non-terminological use (T_-). If the candidate term is multi-words, words are joined together. To build the context of a candidate term occurrence, its (textual) context, i.e. the sentence, is represented as a set of words $S_i = \{W_1, W_2, W_3, ...W_n\}$. Table 2 illustrates how occurrences of the candidate term subject in the linguistic domain is encoded as a formal context.

Table 2. Part of the subject formal context where each occurrence is defined by its textual context. The target attributes T_+ and T_- show the terminological nature of the occurrence in the linguistic domain.

	⊢	a	this	subject	him	to	terrifying	ordeal	type	of	wound	what	is	the	in	sentence	highly	infection	T_-	T_+
subject(S_1)	x			x	x	x	x	x											x	
subject(S_2)			x	x		x				x	x	x	x				x	x	x		
subject(S_3)		x		x									x	x	x	x	x			x

The lattice is built according to the formal context and then, positive and negative hypotheses are extracted. The preliminary results show that noise in data significantly reduces the quality of the results and increases drastically the size of the lattice. The next section is dedicated to noise reduction in the original data.

4.1 Reducing Noise in the Learning Process

In order to reduce the noise in data, we assume that some words in the textual context are more relevant than others. Such words should show an intrinsic semantic. Likewise, function words which semantics depends on the words they govern and the words they are governed by loose their semantics when placed within an unordered bag of words. Therefore, these function words are removed. Similarly, as we have a rather small number of examples for each term candidate, words are lemmatized to tackle the different forms of a word and reduce dispersion. A lemma is the canonical form shared by a set of words expressing the same meaning. For example, walk is the lemma of walking, walks and walked.

After several experimentations, the most relevant configuration to reduce the formal context for a candidate term is the following:

- The set of objects G: Each occurrence of the studied candidate;
- The set of attributes M: Lemmas of content words(nouns, verbs, adjectives and adverbs) for each textual context (*i.e.* the sentence) where the candidate term occurs;
- The binary relation I: It sets which lemma co-occurs with which candidate term occurrence.
- The target attributes (T_+ and T_-): Corresponding to the manual annotation in the corpus.

An example of such formal context is shown in the Table 3.

Table 3. An excerpt of the formal context with lemmas of content words for the candidate term **subject** and its class (T_+ or T_-).

	subject(cT)	terrify	ordeal	type	wound	be	sentence	highly	infection	T_-	T_+
subject(S_1)	x	x	x							x	
subject(S_2)	x			x	x	x		x	x	x	
subject(S_3)	x					x	x				x

5 Experiments

The experiments and evaluations of our method aim at demonstrating the quality and interest of extracted hypotheses as well as helping linguists in defining new features, *i.e.* new annotations, to improve term validation. In any corpus, there exist candidate terms whose occurrences are almost always terminological, some other candidate terms whose occurrences are mostly non-terminological and some other with a rather balanced distribution between the two classes as shown by the column *category* of Table 4. To ease reading, tables presented in this section are translated from French.

Table 4. Selected candidates for evaluation and values observed in the whole corpus.

Candidate	Frequency	Positive examples	Terminological degree	Category
Adjective	216	207	95.83 %	Highly terminological
Lexical relation	55	52	94.54 %	Highly terminological
Collocation	109	90	82.56 %	Highly terminological
Sentence	311	238	76.52 %	Enough terminological
Speaker	233	178	76.39 %	Enough terminological
Corpus	688	510	74.12 %	Enough terminological
Language	926	549	59.28 %	Ambiguous
Statement	289	164	56.74 %	Ambiguous
Context	302	147	48.67 %	Ambiguous
Text	568	266	46.83 %	Ambiguous
Speech	534	248	46.44 %	Ambiguous
Form	462	122	26.40 %	Slightly terminological
Relation	676	171	25.29 %	Slightly terminological
Expression	197	48	24.36 %	Slightly terminological
Semantic	413	80	19.37 %	Very slightly terminological
Lexical	477	84	17.61 %	Very slightly terminological
Model	250	13	5.20 %	Very slightly terminological

5.1   Dataset

The training corpus is composed of 60 free ScienceText documents in french from the linguistics domain. This corpus has been automatically enriched with different annotations: tokenization, sentence splitting and part-of-speech tagging (PoS) performed by the TreeTagger. For normalization issues, an XML-based format has been defined and applied to the documents. Subsequently, TTC Term Suite (Terminology Extraction, Translation Tools and Comparable Corpora project) extracted 5,038 different candidate terms and 69,007 occurrences of them. Finally, each occurrence of candidate terms was manually validated thanks to a dedicated annotation interface[1].

Two annotators evaluated each occurrence of a candidate term considering different linguistics aspects: syntagmatics considerations, membership to a scientific lexicon, membership to a linguistic lexicon and terminological nature. For each of these aspects, experts assign a class (positive or negative) to each occurrence as shown in [10]. To perform cross-validation, this corpus has been split into several parts for training and then, for classification of undetermined examples (testing).

---

[1] Smarties: The annotation interface by stickers (https://apps.atilf.fr/smarties/last visit 01.04.15).

In order to achieve a reliable evaluation of experiments, we selected a list of candidate terms which occur frequently and that belong to different categories as show in Table 4: adjectif (*adjective*), relation lexical (*lexical relation*), collocation, phrase (*sentence*), locuteur (*speaker*), corpus, langue (*language*), énnoncé (*statement*), contexte (*context*), texte (*text*), discours (*speech*), forme (*form*), relation, expression, sémantique (*semantic*), lexical, modèle (*model*). We also introduce a measure of the *terminological degree* of each candidate term (named *ambiguity rate* in [3]). This measure gives the ratio between the number of positive examples (terminological occurrences) of the candidate term with regards to all of its occurrences.

5.2 Implementation

For each experiment, a formal context is generated candidate per candidate. Attributes are the lemmas of content words (verbs, adverbs, nouns and adjectives) that co-occur with the candidate term in the same sentence.

Afterwards, concept-based hypotheses are extracted by means of Formal Concept Analysis to build a set of positive hypotheses (for terminological occurrences) and a set of negative hypotheses (for non-terminological occurrences).

To extract hypotheses, we developed a pipeline within the GATE Natural Language Engineering platform GATE [6]. This pipeline uses several plugins that deal with the specific XML-based format to represent a formal context and extract hypotheses. During the evaluation phase, these hypotheses are matched with sentences in the testing dataset in order to classify undetermined examples.

Table 5. Classification summary of candidate occurrences.

Word	Frequency	Positive Examples	Terminolog. Degree	Words used only in T_+	Hypotheses Generated in T_+	Proportion of Positive Hypotheses	Shared Words	Negative Examples	Words used only in T_-	Hypotheses Generated in T_-
Adjective	216	207	95.83%	966	301	97,41%	64	9	59	8
Corpus	688	510	74.12%	1035	1347	81,93%	713	178	535	297
Text	568	266	46.83%	735	913	52,32%	772	302	792	832
Relation	676	171	25.29%	159	183	11,48%	629	505	1427	1410
Semantic	413	80	19.37%	272	108	8,88%	560	333	1258	1107

Table 5 presents a summary of the results obtained on the whole corpus for some candidate terms selected among the different categories. We observed that certain words are shared by textual contexts of both positive and negative classes (Shared Words). The more ambiguous or frequent the candidate is, the bigger is the shared set. We also remark that the proportion of positive hypotheses with regards to the global number of hypotheses (positive and negative) is quite similar to the ratio of positive examples with regards to the whole set of examples (*i.e.* the Terminological Degree).

5.3   Results

This section presents our experimental results. Evaluation aims at measuring how good are hypotheses for classification of undetermined examples. We used a k-fold cross-validation over our annotated ScienceText corpus (partitioned in 8 folds with a length per fold of 7 texts). Thus, for each experiment, annotations of candidate terms in 7 texts were removed and texts were used for testing; the rest was used for training.

Table 6. Average of kept hypotheses and unnamed examples in k-fold cross-validation ($k = 8$).

Candidate	Cat.	Freq.	Ex2Cla	Hypotheses				Unclassified	
				Generated (+)	Projected (+)	Generated (-)	Projected (-)	Pos. Examp.	Neg. Examp.
Adjective	highly term.	216	27	263	46	7	0	1.375	0.125
Lexical relation	highly term.	55	4.71	58	3	1	0	3.14	0
Collocation	highly term.	109	13.62	132	10	16	0	3.5	0.875
Sentence	enough term.	311	38.87	394	94	77	14	4	1.875
Speaker	enough term.	233	29.12	397	96	64	10	2.875	0.25
Corpus	enough term.	688	86	1126	268	249	61	25.5	5.25
Language	ambig.	926	115,75	1201	418	617	231	21.5	13.125
Statement	ambig.	289	36.12	309	51	146	18	5.5	2.125
Context	ambig.	302	37.75	261	76	326	74	4.5	4.5
Text	ambig.	568	71	757	194	706	145	7.625	5.75
Speech	ambig.	534	66.75	396	80	535	83	5.75	6.125
Form	slightly term.	462	57.75	134	33	955	333	1.125	4.875
Relation	slightly term.	676	142,25	160	16	1172	244	1.5	9.5
Expression	slightly term.	197	24.62	44	5	275	67	1.25	3.0
Semantic	very slightly term.	413	51.62	92	20	915	288	0.5	5.5
Lexical	very slightly term.	477	59.62	77	12	950	312	0.375	2.625
Model	very slightly term.	250	31.25	8	0	384	80	0	1

Table 6 shows average values over the different runs. The Ex2Cla value is the number of undetermined examples to classify. Generated hypotheses is the number of hypotheses extracted from a training set. Accordingly, projected hypotheses is the number of hypotheses that matched undetermined examples. Positive (resp. neg.) unclassified examples are undetermined examples (know as being positive (resp. neg.) in the corpus) that do not contain any positive or negative hypothesis and thus, they have not been classified.

As could be expected, the amount of positive hypotheses is greater than the negative hypotheses if the candidate tends to have a terminological nature. Conversely, the number of negative hypotheses is greater than the positive hypotheses if the candidate tends to be not terminological. However, the ambiguous candidates contain a similar amount of positives and negatives hypotheses. The cause of this behaviour is related to the number of positive and negative occurrences (frequency) of each candidate by category.

The number of hypotheses (projected) used to classify examples is greater than the number of undetermined examples but the proportion between these two values varies a lot. Candidates at the top or at the bottom of the table have good results with a low number of unclassified examples. However, **corpus**, which is frequent, enough terminological, and with a very high number of positive hypotheses has a high number of unclassified positive examples. Thus, a high number of training examples does not always seem to garantee a better result. Candidate terms which are ambiguous are, of course, the most difficult to classify. Here again, one candidate term, **language**, seems apart: it is very frequent, generated lot of (+/-) hypotheses, but the number of unclassified examples (positive or negative) is high.

Table 7 gives the average of some performance measures (precision, recall and F-measure) over the 8 runs. In general, if a class (positive or negative) has a high number of training occurences, then this class gets a better precision and recall. On the opposite, the coverage of the training examples by hypotheses does not seem to impact precision and recal.

5.4 Qualitative Analysis

The second goal of this study is to help linguists to better understand what are the mechanisms that take part to the decision on the terminological status of an occurrence. The ideal process would be when validating occurrences of candidate terms is independent of the term candidate or, even better, when it is independent of the domain. To reach such a goal, we should identify new features that should be added to the initial annotation set. We still are far from reaching the goal but the qualitative analysis already helps us in interacting with linguists.

We carried out a qualitative analysis of patterns. We give here the way patterns are analysed looking abitrarily at positive and negative patterns for the candidate term **argument**. **argument** has 92 occurrences in the corpus and 66.30 % of them are positives (classify between the "enough terminological" and the "ambiguous" category), our method generates 48 positives and 40 negatives hypotheses. Tables 8 and 9 show positive hypotheses (resp. negative) ranked following support and stability.

The most stable positive hypotheses include, in addition to the candidate term itself, a "high" terminological term in linguistics **sdrt** (which stands for Segmented Discourse Representation Theory) and the meaningless verb **be**. There is no doubt that **sdrt**, a linguistic theory which study relation between arguments in a discourse, is a very good trigger for positive occurences.

Table 7. K-fold cross-validation over the collected ScienceText corpus ($k = 8$).

Candidate	Cat.	Freq.	Positive Examples covered by H_+	Precision in T_+	Recall in T_+	F1 in T_+	Negative Examples covered by H_-	Precision in T_-	Recall in T_-	F1 in T_-
Adjective	highly term.	216	99.57%	0.8229	0.8452	0.8339	100%	0.0	0.0	0.0
Lexical relation	highly term.	55	100%	0.2689	0.1820	0.2170	100%	0.0	0.0	0.0
Collocation	highly term.	109	96.11%	0.4864	0.4194	0.4504	95,39%	0.0	0.0	0.0
Sentence	enough term.	311	97.05%	0.8811	0.8828	0.8819	90.41%	0.1968	0.6	0.2963
Speaker	enough term.	233	98.52%	0.8436	0.7729	0.8067	95.22%	0.4375	0.3779	0.4055
Corpus	enough term.	688	78.11%	0.7087	0.3955	0.5076	84.05%	0.6626	0.4227	0.5161
Language	ambig.	926	82.60%	0.7879	0.590	0.6747	84.15%	0.7739	0.5256	0.6260
Statement	ambig.	289	87.19%	0.6572	0.5683	0.6095	85.30%	0.5177	0.3134	0.3904
Context	ambig.	302	98.21%	0.7649	0.5936	0.6684	98.30%	0.6897	0.5655	0.6214
Text	ambig.	568	97.03%	0.5675	0.4355	0.4928	96.77%	0.5353	0.4819	0.5071
Speech	ambig.	534	88.70%	0.7523	0.5108	0.6084	91.43%	0.4846	0.5220	0.5026
Form	slightly term.	462	94.97%	0.75	0.1124	0.1955	98.19%	0.8828	0.7252	0.7962
Relation	slightly term.	676	74.92%	0.25	0.0535	0.0881	91.70%	0.8337	0.8735	0.8531
Expression	slightly term.	197	94.53%	0.3125	0.0631	0.1049	97.06%	0.7490	0.8437	0.7935
Semantic	very slightly term.	413	90.93%	0.4583	0.1107	0.1783	97.86%	0.7735	0.8761	0.8216
Lexical	very slightly term.	477	89.58%	0.0	0.0	0.0	97.10%	0.8791	0.9408	0.9089
Model	very slightly term.	250	100%	0.0	0.0	0.0	100%	0.8780	0.9660	0.9199

Table 8. Set of the most representative positive hypotheses for the `argument` candidate term.

Support	Stability	Hypotheses in T_+	Hypotheses in T_+ -english-
7	0.7968	[sdrt, être, argument]	[sdrt, be, argument]
9	0.7792	[argument, plus]	[argument, more]
6	0.73437	[être, argument, aussi]	[be, argument, also]
6	0.7187	[argument, verbal]	[argument, verbal]
...	...	...	...
5	0.6562	[être, argument, indique]	[be, argument, denote]
4	0.5	[argument, syntaxique]	[argument, syntactic]
...	...	...	...
6	0.3281	[être, argument, rst]	[be, argument, rst]
8	0.25	[argument, nucleus]	[argument, nucleus]

Table 9. Set of the most representative negative hypotheses for the *argument* candidate term.

Support	Stability	Hypotheses in T_-	Hypotheses in T_- -*english*-
3	0.5	[argument, prendre]	[argument, assume]
1	0.5	[trancher, pas, ne, argument, permettre, décisif, position, avoir]	[settle, not, argument, allow, decisive, position, have]
...	...	...	...
4	0.375	[argument, hypothèse]	[argument, hypothesis]
4	0.3125	[dire, argument]	[say, argument]
...	...	...	...
2	0.25	[trouver, même, argument]	[find, same, argument]

Afterwards, some other high terminological terms `syntactic` or `verbal` also contribute to a positive validation. Others hypotheses with a lower support, but not less important, are related to Rhetorical Structure Theory (*rst*) representing the distinction between `nucleus` and satellite `arguments`.

However, we should notice that it is quite easy to find counter-examples. Considering the hypothesis [`sdrt, be, argument`] and the sentence *"An argument in favor of SDRT is also that ... "* (which is not in the initial corpus), the `argument` candidate term will be wrongly classified as positive. Similarly, the third positive hypothesis in the table, [`be, argument, also`], could fit the negative example *"An argument in favor of SDRT is also that ..."* and, of course, it could also fit the positive example like *"This relationship, which raises issues concerning the linear order of its arguments is studied in (Redeker and Egg, 2006) and also in (Hunter et al., 2006). ...".* The two last examples show that some additional information could probably produce better hypotheses: preserving order in the sentence (working with sequences instead of bag of words), using syntactic role (subject, object ...), syntactic dependencies between the studied occurrence and some other words, or keeping information about the type of determiner it is linked with, like definite (ex: *the*) or indefinite (*a*)...

By contrast, the sets of negative hypotheses showed in Table 9 showed another usage of the `argument` candidate. Mainly, `argument` refers to authors trying to convince the reader about an idea, an hypothesis or a theory by using an evidence. Consequently, the large diversity of situations leads to hypotheses which include meaningless words like `dire` (*to say*), `prendre` (*to assume*), `trouver` (*to find*).

6   Conclusion and Perspectives

In this paper, we describe a method for validating occurrences of candidate terms using Context-based Hypotheses. It starts with a corpus on a specific domain, where each occurrence of candidate terms has been manually annotated as terminological or non-terminological occurrence. We built a formal context for hypotheses extraction. Each positive hypothesis represents a textual context where the candidate is used as a term. Similarly, a negative hypothesis describes the textual context where the candidate is used as a non-terminological entity.

Some plugins have been developed to run under the GATE the open source solution for text processing. In that way, some higher-level linguistic annotations could be used to improve the process. Among them we could mention syntactic trees, dependencies or the use of linguistic resources such as a trans-disciplinary lexicon. As mentioned in Sect. 5.4, we have several options to improve annotations and better discriminate positive and negative occurrences defining hypotheses which are not only based on words (lexical level) but also on more elaborated linguistic features.

We would like to thank the ANR Agency for supporting this work which is part of the Termith project (ANR-12-CORD-0029-05) and we would like to thank all the linguists involved in the project for the work they did on the corpus.

References

1. Aronson, A., Lang, F.M.: An overview of metamap: historical perspective and recent advances. JAMIA **17**(3), 229–236 (2010)
2. Aubin, S., Hamon, T.: Improving term extraction with terminological resources. In: Salakoski, T., Ginter, F., Pyysalo, S., Pahikkala, T. (eds.) FinTAL 2006. LNCS (LNAI), vol. 4139, pp. 380–387. Springer, Heidelberg (2006)
3. Boumedyen, M., Camacho, J., Jacquey, E., Kister, L.: Annotation sémantique et validation terminologique en texte intégral en shs. In: Actes de la 21e Conférence sur le Traitement Automatique des Langues Naturelles (TALN'2014). Marseille, France (2014)
4. Bourigault, D., Jacquemin, C., L'Homme, M.: Searching for and identifyng conceptual relationships via a corpus-based approach to a terminological knowledge base (CTKB): method and results. In: Condamines, A., Rebeyrolle, J. (eds.) Recent Advances in Computational Terminology, Chap. 6. Natural Language Processing, J. Benjamins Publishing Company (2001)
5. Buzmakov, A., Kuznetsov, S., Napoli, A.: A new approach to classification by means of jumping emerging patterns. In: FCA4AI: International Workshop "What can FCA do for Artificial Intelligence?" - ECAI 2012 (2012)
6. Cunningham, H., Maynard, D., Bontcheva, K., Tablan, V., Aswani, N., Roberts, I., Gorrell, G., Funk, A., Roberts, A., Damljanovic, D., Heitz, T., Greenwood, M., Saggion, H., Petrak, J., Li, Y., Peters, W.: Text Processing with GATE (Version 6) (2011). http://tinyurl.com/gatebook
7. David, S., Plante, P.: Termino version 1.0. Rapport du Centre dAnalyse de Textes par Ordinateur. Université du Québec à Montréal (1990)
8. Ganter, B., Wille, R.: Formal Concept Analysis: Mathematical Foundations, 1st edn. Springer, Secaucus (1997)
9. Jacquemin, C.: Fastr: a unification-based front-end to automatic indexing. In: Funck-Brentano, J.L., Seitz, F. (eds.) RIAO, pp. 34–48. CID (1994)
10. Kister, L., Jacquey, E.: Relations syntaxiques entre lexiques terminologique et transdisciplinaire: analyse en texte intégral. In: Actes du Congrès Mondial de Linguistique Franaise, Lyon, France, pp. 909–919 (2012)

11. Klimushkin, M., Obiedkov, S., Roth, C.: Approaches to the selection of relevant concepts in the case of noisy data. In: Kwuida, L., Sertkaya, B. (eds.) ICFCA 2010. LNCS, vol. 5986, pp. 255–266. Springer, Heidelberg (2010)

12. Kuznetsov, S.: Machine learning on the basis of formal concept analysis. Autom. Remote Control **62**(10), 1543–1564 (2001)

13. Kuznetsov, S.: Complexity of learning in concept lattices from positive and negative examples. Discrete Appl. Math. **142**(13), 111–125 (2004)

14. Kuznetsov, S.: On stability of a formal concept. Ann. Math. Artif. Intell. **49**(1–4), 101–115 (2007)

15. Maynard, D., Ananiadou, S.: Term extraction using a similarity-based approach. In: Recent Advances in Computational Terminology. John Benjamins, pp. 261–278 (1999)

16. Ramamohanarao, K., Bailey, J.: Discovery of emerging patterns and their use in classification. In: Gedeon, T., Fung, L. (eds.) AI 2003: Advances in Artificial Intelligence. Lecture Notes in Computer Science, vol. 2903, pp. 1–11. Springer, Berlin Heidelberg (2003)

17. Rocheteau, J., Daille, B.: TTC termsuite: a uima application for multilingual terminology extraction from comparable corpora. In: Proceedings of the 5th International Joint Conference on Natural Language Processing (IJCNLP), Chiang Mai, Thailand (2011)

18. Sclano, F., Velardi, P.: Termextractor: a web application to learn the shared terminology of emergent web communities. In: Proceedings of the 3rd International Conference on Interoperability for Enterprise Software and Applications (I-ESA 2007) (2007)

19. Wüster, E.: La théorie générale de la terminologie un domaine interdisciplinaire impliquant la linguistique, la logique, l'ontologie, l'informatique et les sciences des objets. In: Actes du Colloque International de Terminologie (Québec, Manoir du lac Delage, 5–8 octobre 1975) (1976)

Towards a Navigation Paradigm for Triadic Concepts

Sebastian Rudolph[1]([⊠]), Christian Săcărea[2], and Diana Troancă[2]

[1] Technische Universität Dresden, Dresden, Germany
`sebastian.rudolph@tu-dresden.de`
[2] Babeş-Bolyai University, Cluj-Napoca, Romania
`csacarea@math.ubbcluj.ro`
`dianat@cs.ubbcluj.ro`

Abstract. The simple formalization and the intuitive graphical representation are main reasons for the growing popularity of Formal Concept Analysis (FCA). FCA gives the user the possibility to explore the structure of data and understand correlations and implications in the data set. Recently, triadic FCA (3FCA) has become increasingly popular, but exploring triadic conceptual landscapes is not easy, especially because of the less immediate structure of the space of triadic concepts. Even more, available graphical representations of trilattices are barely intelligible and hard to obtain even for small data sets. Driven by practical requirements, we propose a new navigation paradigm for triadic conceptual landscapes based on a neighborhood notion arising from appropriately defined dyadic concept lattices. Understanding the corresponding reachability relation gives also new theoretical insights about the behavior of triadic concepts and the corresponding triadic data sets.

1 Introduction

With the advent of the information society and the rise of data science, understanding big collections of information and knowledge and representing them in intuitive ways is more important than ever. Formal concept analysis is well-known for its capabilities addressing knowledge processing and knowledge representation as well as offering reasoning support for understanding the structure of large collections of data.

For dyadic FCA – the original version of FCA based on a binary incidence relation – this has proven to be the case through the graphical representation of the concept lattice that offers a intuitive visualization and hence understanding of the data. From this graphical representation, one can read any relation between objects and attributes, but also implications holding in the data. For cases where the concept lattice gets too big to be represented in a readable way, "local" navigational paradigms have been proposed, where only one concept and its direct neighbor concepts are visualized and the user can explore the concept lattice by successively moving to neighboring concepts [2,7].

© Springer International Publishing Switzerland 2015
J. Baixeries et al. (Eds.): ICFCA 2015, LNAI 9113, pp. 252–267, 2015.
DOI: 10.1007/978-3-319-19545-2_16

Dyadic FCA was extended in [11] by Rudolf Wille and Fritz Lehmann to the triadic case, featuring a ternary instead of a binary incidence relation. The use of FCA increased over the last years, still there was little focus on applications of triadic FCA (3FCA), mainly because of its higher complexity and unavailability of a graphical representation for trilattices which quickly become impossible to draw even for small data sets. Although there are a lot of data collections that map perfectly to a triadic representation, for instance collaborative tagging scenarios or folksonomies [10], there is no good support for helping humans understand the structure of the triconcepts in a tricontext. Despite the fact that 3FCA is just an extension of FCA, the graphical representation for the dyadic case does not have an intuitive extension to the triadic case [1,6,8]. Wille and Lehmann proposed a way to graphically represent a triadic context by using a triadic diagram in [11], inspired by the concept lattice from the dyadic case. However, the geometric representation obtained does not give much insight into the structure of the tricontext and cannot be easily read and understood. Furthermore, even a small set of triadic data can generate a large amount of triconcepts.

For the reasons mentioned above, we intend to present in this article a method to locally display a smaller part of the space of triconcepts, instead of displaying all of them at once. Our goal is to find an intuitive navigation strategy that allows for moving from one such local view to other, adjacent ones. Furthermore, we will formally analyze the properties of this strategy and ultimately suggest algorithms for producing the structures necessary for browsing the space of triconcepts using developed and theoretically well-understood methods.

Exploiting the fact that triconcepts are built three-dimensional, the navigation strategy we propose makes use of the elegance and the expressive power of dyadic concept lattices. Navigation starts local, with a triconcept. Herefrom, we fix what we call a *perspective*, i.e., one of the three dimensions (extent, intent or modus) and then collect all so-called *directly reachable triconcepts*. For each perspective, the triconcepts directly reachable via this perspective can be arranged in a dyadic concept lattice, hence navigating among them benefits from all advantages concept lattices are offering. After selecting a directly reachable triconcept, one may change the perspective and move towards another set of reachable triconcepts, exploring again another concept lattice. Despite of its apparent growth of computational complexity, this approach allows to cope with large sets of triconcepts. Moreover, the local navigation strategy discussed in this paper gives rise to a list of theoretical questions: reachability of all triconcepts, the existence and the number of reachability clusters, their structure and a method to navigate from one to another. Understanding these clusters proves to be not trivial and gives interesting insights about the inherent conceptual structure of triadic data.

2 Preliminaries

This section introduces the basic notions of triadic formal concept analysis. For further information about the dyadic case or more specific results about 3FCA we refer the interested reader to the standard literature [3,4,11,12].

Definition 1. *A* triadic context *(also:* tricontext*) is a quadruple* $(K_1, K_2,$ $K_3, Y)$, *where* K_1, K_2 *and* K_3 *are sets and* $Y \subseteq K_1 \times K_2 \times K_3$ *is a ternary relation between them. The elements of* K_1, K_2, K_3 *are called (formal) objects, attributes and conditions, respectively. An element* $(g, m, b) \in Y$ *is read object* g *has attribute* m *under condition* b.

The following definition shows how dyadic contexts can be obtained from a triadic one in a natural way.

Definition 2 (Derived Contexts). *Every triadic context* (K_1, K_2, K_3, Y) *gives rise to the following dyadic contexts:*

$$\mathbb{K}^{(1)} := (K_1, K_2 \times K_3, Y^{(1)}) \text{ with } gY^{(1)}(m, b) :\Leftrightarrow (g, m, b) \in Y,$$
$$\mathbb{K}^{(2)} := (K_2, K_1 \times K_3, Y^{(2)}) \text{ with } mY^{(2)}(g, b) :\Leftrightarrow (g, m, b) \in Y, \text{ and}$$
$$\mathbb{K}^{(3)} := (K_3, K_1 \times K_2, Y^{(3)}) \text{ with } bY^{(3)}(g, m) :\Leftrightarrow (g, m, b) \in Y.$$

For $\{i, j, k\} = \{1, 2, 3\}$ *and* $A_k \subseteq K_k$, *we define* $\mathbb{K}_{A_k}^{(ij)} := (K_i, K_j, Y_{A_k}^{(ij)})$, *where* $(a_i, a_j) \in Y_{A_k}^{(ij)}$ *if and only if* $(a_i, a_j, a_k) \in Y$ *for all* $a_k \in A_k$.

Intuitively, the contexts $\mathbb{K}^{(i)}$ represent "flattened" versions of the triadic context, obtained by putting the "slices" of (K_1, K_2, K_3, Y) side by side. Moreover, $\mathbb{K}_{A_k}^{(ij)}$ corresponds to the intersection of all those slices that correspond to elements of A_k.

In triadic FCA, there are two extensions for the dyadic derivation operators.

Definition 3 ((i)-Derivation Operators). *For* $\{i, j, k\} = \{1, 2, 3\}$ *with* $j < k$ *and for* $X \subseteq K_i$ *and* $Z \subseteq K_j \times K_k$ *the* (i)*-derivation operators are defined by:*

$$X \mapsto X^{(i)} := \{(a_j, a_k) \in K_j \times K_k \mid (a_i, a_j, a_k) \in Y \text{ for all } a_i \in X\}.$$
$$Z \mapsto Z^{(i)} := \{a_i \in K_i \mid (a_i, a_j, a_k) \in Y \text{ for all } (a_j, a_k) \in Z\}.$$

Obviously, these derivation operators correspond to the derivation operators of the dyadic contexts $\mathbb{K}^{(i)}, i \in \{1, 2, 3\}$.

Definition 4 ((i, j, X_k)-Derivation Operators). *For* $\{i, j, k\} = \{1, 2, 3\}$ *and* $X_i \subseteq K_i, X_j \subseteq K_j, X_k \subseteq K_k$, *the* (i, j, X_k)*-derivation operators are defined by*

$$X_i \mapsto X_i^{(i,j,X_k)} := \{a_j \in K_j \mid (a_i, a_j, a_k) \in Y \text{ for all } (a_i, a_k) \in X_i \times X_k\}$$
$$X_j \mapsto X_j^{(i,j,X_k)} := \{a_i \in K_i \mid (a_i, a_j, a_k) \in Y \text{ for all } (a_j, a_k) \in X_j \times X_k\}.$$

The (i, j, X_k)-derivation operators correspond to those of the dyadic contexts $(K_i, K_j, Y_{X_k}^{(ij)})$.

Similar to the notion of formal concepts in dyadic FCA, triadic concepts can be defined [11]. A triadic concept is a maximal box of incidences (Proposition 1) and can be generated using derivation operators (Proposition 2).

Definition 5. *A triadic concept (short:* triconcept*) of* $\mathbb{K} := (K_1, K_2, K_3, Y)$ *is a triple* (A_1, A_2, A_3) *with* $A_i \subseteq K_i$ *for* $i \in \{1, 2, 3\}$ *and* $A_i = (A_j \times A_k)^{(i)}$ *for every* $\{i, j, k\} = \{1, 2, 3\}$ *with* $j < k$. *The sets* $A_1, A_2,$ *and* A_3 *are called* extent, intent, *and* modus *of the triadic concept, respectively. We let* $\mathfrak{T}(\mathbb{K})$ *denote the set of all triadic concepts of* $\mathbb{K}$.

Proposition 1. *The triconcepts of a triadic context* (K_1, K_2, K_3, Y) *are exactly the maximal triples* $(A_1, A_2, A_3) \in \mathfrak{P}(K_1) \times \mathfrak{P}(K_2) \times \mathfrak{P}(K_3)$ *with* $A_1 \times A_2 \times A_3 \subseteq Y$, *with respect to the component-wise set inclusion.*

Proposition 2. *For* $X_i \subseteq K_i$ *and* $X_k \subseteq K_k$ *with* $\{i, j, k\} = \{1, 2, 3\}$, *let* $A_j := X_i^{(i,j,X_k)}$, $A_i := A_j^{(i,j,X_k)}$ *and* $A_k := (A_i \times A_j)^{(k)}$ *(if* $i < j$*) or* $A_k := (A_j \times A_i)^{(k)}$ *(if* $j < i$*). Then* (A_1, A_2, A_3) *is the triadic concept* $\mathfrak{b}_{ik}(X_i, X_k)$ *with the property that it has the smallest k-th component among all triadic concepts* (B_1, B_2, B_3) *with the largest j-th component satisfying* $X_i \subseteq B_i$ *and* $X_k \subseteq B_k$. *In particular,* $\mathfrak{b}_{ik}(A_i, A_k) = (A_1, A_2, A_3)$ *for each triadic concept* (A_1, A_2, A_3) *of* $\mathbb{K}$.

3 Motivating Example

In this section we present a small example, aiming to explain how the local navigation paradigm works in a set of triconcepts. The related theoretical aspects will be introduced in the following sections. For this, we consider the hostel tricontext from [5], whose trilattice is represented in Fig. 1. The objects of the triadic data set are hostels, the attributes services provided by the hostels, while the conditions are web portals where the hostels can be rated. The graphical representation as a 3-net displays all triconcepts and the equivalence classes to which they belong in a triadic diagram. The extent, intent and modus of a triconcept can be read by using the order diagrams displayed on the side of the

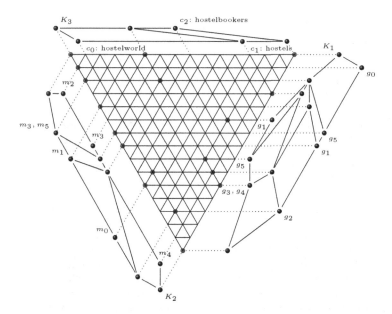

Fig. 1. Trilattice of the tricontext "Hostels".

trilattice. Global navigation in a 3-net becomes difficult for a (slightly) larger set of triconcepts and this is the case in many 3FCA applications. What graphical representation should be employed in the cases where a representation as a 3-net is not possible [12]? The complexity of the trillatice structure and that of the order diagrams of the extents, intents and modi set makes a global navigation approach quite difficult.

To cope with the complexity of larger data sets, we propose a local navigation paradigm which starts from a triconcept (A_1, A_2, A_3) and the selection of one of its components (extent, intent or modus), which we then call *perspective*. We build the projected context $\mathbb{K}_{A_k}^{(ij)}$ along perspective k and compute its concept lattice. It can be proved that every dyadic concept of this projected context corresponds to exactly one triconcept in the original trilattice. These triconcepts are called *directly reachable* and navigation among them is performed in the underlying dyadic concept lattice.

To start local navigation, choose $T := (\{g_3, g_4, g_5\}, \{m_0, m_1, m_2, m_3, m_5\}, \{c_1, c_2\})$ and consider perspective 3 (i.e., modus). By projecting along $\{c_1, c_2\}$, we obtain the concept lattice displayed in Fig. 2. Triconcept T corresponds to the leftmost dyadic concept in Fig. 2. Moreover, all dyadic concepts correspond to some triconcepts, having either the same modus or a larger one. The navigation can be continued herefrom by choosing one of the directly reachable triconcepts from T and a perspective, i.e., one of the concepts of $\mathbb{K}_{\{c_1, c_2\}}^{(12)}$ and then navigating

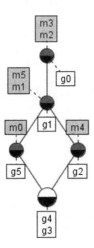

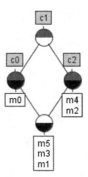

Fig. 2. Directly reachable triconcepts from T using perspective 3. The extent and intent of the triconcepts can be read from the concept lattice, while the modus is computed using the corresponding derivation operator $(\cdot)^3$ in the tricontext.

Fig. 3. Reachable triconcepts from T using perspective 3 and then 1. Only intent and modus are displayed in the concept lattice, the extent is computed using the corresponding derivation operator $(\cdot)^1$ in the tricontext.

within the new concept lattice. For example, the rightmost concept of this lattice corresponds to the triconcept $(\{g_2, g_3, g_4\}, \{m_2, m_3, m_4\}, \{c_1, c_2\})$. By choosing perspective 1 (i.e., extent), the triconcepts reachable herefrom are represented in Fig. 3.

This example shows how triconcepts can be clustered according to their reachability and how we can navigate from one triconcept to another. We might ask whether all concepts might be reachable by this approach or not, what are the maximal strongly connected components of the reachability relation, i.e., the reachability clusters, what are the properties of the set of reachability clusters and how can we set up a local navigation paradigm herefrom. By changing perspectives, all concepts in this example prove to be reachable (though not directly reachable). We will prove later on that this will not always be the case.

Motivated by this short example, we introduce in the following sections the theoretical aspects and considerations of the proposed navigation paradigm.

4   Reachable Triconcepts

This section aims to define the exploration paradigm exemplified in Sect. 3 and to discuss some theoretical issues. The following propositions are direct consequences of Proposition 2. For every triconcept, by projecting along one of the dimensions, we obtain a formal dyadic context, where the projection of the triconcept is a dyadic concept of the corresponding concept lattice (Proposition 3). Moreover, every dyadic concept herefrom generates a triconcept (Proposition 4).

Proposition 3. *Let* $(A, B, C) \in \mathfrak{T}(\mathbb{K})$ *be a triadic concept. Then* $(A, B) \in \mathfrak{B}(\mathbb{K}_C^{(12)})$.

Proposition 4. *Let* $(A, B, C) \in \mathfrak{T}(\mathbb{K})$ *be a triadic concept. Let* $(A_1, A_2) \in \mathfrak{B}(\mathbb{K}_C^{(12)})$. *Then* $(A_1, A_2, (A_1 \times A_2)^{(3)}) \in \mathfrak{T}(\mathbb{K})$.

By the above propositions, we conclude that given a triconcept (A, B, C), fixing either its extent, or its intent or modus, gives rise to a (dyadic) concept lattice, every concept of which can be deterministically turned into a triconcept by computing the missing component using an appropriate triadic derivation operator (for instance $(\cdot)^{(3)}$). Based on this, we are now able to define a reachability relation between triconcepts.

Definition 6. *For* (A_1, A_2, A_3) *and* (B_1, B_2, B_3) *triadic concepts, we say that* (B_1, B_2, B_3) *is* directly reachable *from* (A_1, A_2, A_3) *using perspective* (1) *and we write* $(A_1, A_2, A_3) \prec_1 (B_1, B_2, B_3)$ *if and only if* $(B_2, B_3) \in \mathfrak{B}(\mathbb{K}_{A_1}^{(23)})$. *Analogously, we can define direct reachability using perspectives* (2) *and* (3).

We say that (B_1, B_2, B_3) *is* directly reachable *from* (A_1, A_2, A_3) *if it is directly reachable using at least one of the three perspectives, that is, formally* $(A_1, A_2, A_3) \prec (B_1, B_2, B_3) :\Leftrightarrow [(A_1, A_2, A_3) \prec_1 (B_1, B_2, B_3)] \vee [(A_1, A_2, A_3) \prec_2 (B_1, B_2, B_3)] \vee [(A_1, A_2, A_3) \prec_3 (B_1, B_2, B_3)]$.

By Proposition 3, two triconcepts having the same extent, or the same intent, or modus are always mutually directly reachable. Hence, in a trilattice diagram, all triconcepts aligned on the same line (i.e., being equivalent with respect to one of the three preorders) are mutually directly reachable:

Proposition 5. *Let* $(A_1, A_2, A_3), (B_1, B_2, B_3)$ *be two triconcepts. If* $A_i = B_i$ *for an* $i \in \{1, 2, 3\}$ *then* $(A_1, A_2, A_3) \prec_i (B_1, B_2, B_3)$ *and* $(B_1, B_2, B_3) \prec_i (A_1, A_2, A_3)$.

Definition 7. *We define the* reachability *relation between two triconcepts as being the transitive closure of the direct reachability relation* $\prec$*. We denote this relation by* $\lhd$.

Definition 8. *The equivalence class of a triconcept* (A_1, A_2, A_3) *with respect to the preorder* $\lhd$ *on* $\mathfrak{T}(\mathbb{K})$ *will be called a* reachability cluster *and denoted by* $[(A_1, A_2, A_3)]$.

Intuitively, the reachability cluster of (A_1, A_2, A_3) contains all triconcepts which are mutually reachable from (A_1, A_2, A_3) When considering $\prec$ as directed edge relation of a graph, reachability clusters correspond to the strongly connected components of that graph.

The following results are providing a better understanding of the reachability clusters and their structure. We prove that there exist triconcepts which are always reachable (Proposition 6). Moreover, the induced order on the set of reachability clusters always has a greatest element.

Proposition 6. *The trivial triconcepts* $\theta_1 := (K_1, K_2, (K_1 \times K_2)^{(3)})$, $\theta_2 := (K_1, K_3, (K_1 \times K_3)^{(2)})$ *and* $\theta_3 := (K_2, K_3, (K_2 \times K_3)^{(1)})$ *are always reachable. Moreover, they are always directly reachable.*

Proof. Let us assume, without restricting generality that $(K_1 \times K_2)^{(3)} = (K_1 \times K_3)^{(2)} = (K_2 \times K_3)^{(1)} = \emptyset$. Let $(A, B, C) \in \mathfrak{T}(\mathbb{K})$. Using perspective (3), we have that $(A, B) \in \mathfrak{B}(\mathbb{K}_C^{(12)})$. The greatest and the lowest elements of $\mathbb{K}_C^{(12)}$ are $(K_1, \emptyset)$ and $(\emptyset, K_2)$, respectively. Hence $(A, B, C) \lhd \theta_2$ and $(A, B, C) \lhd \theta_3$. By choosing another perspective, θ_1 is directly reached from (A, B, C).

In particular, if $(A, B, C) = \theta_1$, then the trivial triconcepts θ_2 and θ_3 are reachable by perspective (1). □

Corollary 1. *The ordered set* $(\mathfrak{T}(\mathbb{K})/ \sim, \leq)$ *has always a greatest element, the reachability cluster of the trivial concepts. We denote this cluster by* ∇.

Proposition 7. *If* (A, B, C) *is a triconcept with either* $A = K_1$*, or* $B = K_2$*, or* $C = K_3$*, then* $(A, B, C) \in \nabla$.

Proof. Every trivial concept is reachable from (A, B, C). Let us assume that $A = K_1$. Take now $\theta_1 := (K_1, K_2, \emptyset)$ and choose perspective (1). We obtain the context $\mathbb{K}_{K_1}^{(23)} := (K_2, K_3, Y_{K_1}^{(23)})$. We want to prove that $(B, C) \in \mathfrak{B}(\mathbb{K}_{K_1}^{(23)})$.

We know that $B = (K_1 \times C)^{(2)} = \{m \in K_2 \mid \forall g \in K_1, \forall b \in C. (g, m, b) \in Y\}$. Also, by definition, $C^{(2,3,K_1)} = \{m \in K_2 \mid \forall g \in K_1, \forall b \in C. (g, m, b) \in Y\}$, hence $B = C^{(2,3,K_1)}$. Analogously, $C = B^{(2,3,K_1)}$. □

Remark 1

(1) If (A_1, B_1, C_1) and (A_2, B_2, C_2) are triconcepts with $A_1 = K_1$ or $B_1 = K_2$ or $C_1 = K_3$ and $(A_1, B_1, C_1) \lhd (A_2, B_2, C_2)$, then $(A_2, B_2, C_2) \in \nabla$.

(2) If $(A_1, B_1, C_1) \in \nabla$ and $(A_1, B_1, C_1) \lhd (A_2, B_2, C_2)$ then $(A_2, B_2, C_2) \in \nabla$. The converse does not hold true, i.e., more than one reachability cluster is possible. Take for example $K_1 := \{g_1, g_2\}, K_2 := \{m_1, m_2\}$, and $K_3 := \{b_1, b_2\}$ with $Y := \{(g_1, m_1, b_1)\}$. In this context there are exactly two reachability clusters, $\nabla = \{\theta_1, \theta_2, \theta_3\}$ and $\{(g_1, m_1, b_1)\}$.

Example 1. In general, triconcepts might be structured in more than one cluster, as the following examples show. A more profound discussion about the depth and width of the ordered set of reachability clusters will be given in Sect. 5.

(1) A tricontext with more than two clusters:

b1	m1	m2	m3
g1	×		
g2			
g3			

b2	m1	m2	m3
g1	×		
g2		×	
g3			

b3	m1	m2	m3
g1	×		
g2		×	
g3			×

The triconcepts are partitioned in clusters the following way:
$C_1 = \{(\{g_3\}, \{m_3\}, \{b_3\})\}$, $C_2 = \{(\{g_2\}, \{m_2\}, \{b_2, b_3\})\}$, $C_3 = \{(\{g_1\}, \{m_1\}, \{b_1, b_2, b_3\}), (\{g_1, g_2, g_3\}, \{m_1, m_2, m_3\}, \emptyset), (\{g_1, g_2, g_3\}, \emptyset, \{b_1, b_2, b_3\}), (\emptyset, \{m_1, m_2, m_3\}, \{b_1, b_2, b_3\})\}$, and $C_1 \le C_2 \le C_3$. Thereby, the triconcepts $(\{g_3\}, \{m_3\}, \{b_3\})$ and $(\{g_2\}, \{m_2\}, \{b_2, b_3\})$ have disjoint extents and intents, but $(\{g_3\}, \{m_3\}, \{b_3\}) \prec_3 (\{g_2\}, \{m_2\}, \{b_2, b_3\})$.

(2) A tricontext with exactly two clusters

b1	m1	m2
g1	×	
g2		×

b2	m1	m2
g1		
g2		×

b3	m1	m2
g1		
g2	×	

The triconcepts are partitioned in clusters the following way:
$C_1 = \{(\{g_1\}, \{m_1\}, \{b_1\}), (\{g_2\}, \{m_2\}, \{b_1, b_2\}), (\{g_2\}, \{m_1\}, \{b_3\})\}$, $C_2 = \{(\{g_1, g_2\}, \{m_1, m_2\}, \emptyset), (\{g_1, g_2\}, \emptyset, \{b_1, b_2, b_3\}), (\emptyset, \{m_1, m_2\}, \{b_1, b_2, b_3\})\}$, and $C_1 \le C_2$.

(3) A tricontext with a single cluster

b1	m1	m2	m3
g1	×		
g2			
g3			

b2	m1	m2	m3
g1		×	
g2			
g3		×	

b3	m1	m2	m3
g1	×	×	
g2			
g3			

The triconcepts are the following:
$C = \{(\{g_1\}, \{m_1\}, \{b_1, b_3\}), (\{g_1\}, \{m_2\}, \{b_2, b_3\}), (\{g_1\}, \{m_1, m_2\}, \{b_3\}), (\{g_1, g_3\}, \{m_2\}, \{b_2\}), (\{g_1, g_2, g_3\}, \{m_1, m_2, m_3\}, \emptyset), (\{g_1, g_2, g_3\}, \emptyset, \{b_1, b_2, b_3\}), (\emptyset, \{m_1, m_2, m_3\}, \{b_1, b_2, b_3\})\}$.

5   Reachability in Composed Tricontexts

There is a way of composing several tricontexts such that the reachability clusters of the composed tricontext coincide with the union of the reachability clusters of the constituents, except for the greatest cluster. We will exploit this correspondency later.

Definition 9. *Given tricontexts* $\mathbb{K}_1 := (K_1^1, K_2^1, K_3^1, Y^1), \ldots, \mathbb{K}_n := (K_1^n, K_2^n, K_3^n, Y^n)$, *with* K_j^i *and* K_k^i *being disjoint for all* $j \neq k$ *and all* $i \in \{1, 2, 3\}$, *their* composition $\mathbb{K}_1 \uplus \ldots \uplus \mathbb{K}_n$ *is the tricontext* $\mathbb{K} := (K_1, K_2, K_3, Y)$ *with* $K_i := \bigcup_{k=1}^n K_i^k$ *and* $Y := \bigcup_{k=1}^n Y^k$.

Proposition 8. *Let* $(K_1, K_2, K_3, Y) = \mathbb{K}_1 \uplus \ldots \uplus \mathbb{K}_n$ *with* $n \geq 2$ *and all* K_j^i *being non-empty. Then* (A_1, A_2, A_3) *is a triconcept of* (K_1, K_2, K_3, Y) *iff*

- A_1, A_2, A_3 *are all non-empty and* (A_1, A_2, A_3) *is a triconcept of some* $\mathbb{K}_j$ *or*
- (A_1, A_2, A_3) *is one of* $(\emptyset, K_2, K_3)$ *or* $(K_1, \emptyset, K_3)$ *or* $(K_1, K_2, \emptyset)$.

Proof. "If": First, consider a triconcept (A_1, A_2, A_3) of some $\mathbb{K}_j$ with A_1, A_2, and A_3 nonempty. Now suppose (A_1, A_2, A_3) were not a triconcept of $\mathbb{K}$, i.e., at least one of A_1, A_2, A_3 can be enlarged. W.l.o.g., assume some $a \in K_1 \setminus A_1$ with $(A_1 \cup \{a\}) \times A_2 \times A_3 \subseteq Y$. Now, for $a_2 \in A_2$ and $a_3 \in A_3$, we have $(a, a_2, a_3) \in Y$, implying $a \in K_j$ and thus $(A_1 \cup \{a\}) \times A_2 \times A_3 \in Y_j$, contradicting that (A_1, A_2, A_3) is a triconcept of $\mathbb{K}_j$.

Second, $(A_1, A_2, A_3) = (\emptyset, K_2, K_3)$ is maximal unless for some a holds $\{a\} \times K_2 \times K_3 \subseteq Y$. Yet this contradicts the construction of Y. The cases of $(K_1, \emptyset, K_3)$ and $(K_1, K_2, \emptyset)$ follow by symmetry.

"Only if": for any triconcept (A_1, A_2, A_3) of (K_1, K_2, K_3, Y) with nonempty A_1, A_2, A_3, we find an $(a_1, a_2, a_3) \in A_1 \times A_2 \times A_3$. By construction, for every such (a_1, a_2, a_3) must exist some j with $a_1 \in K_1^j$ and $a_2 \in K_2^j$ and $a_3 \in K_3^j$. Consequently, $A_1 \subseteq K_1^j$ and $A_2 \subseteq K_2^j$ and $A_3 \subseteq K_3^j$. Moreover, maximality of (A_1, A_2, A_3) in (K_1, K_2, K_3, Y) implies maximality in $\mathbb{K}_j$.

Finally if one of the components of (A_1, A_2, A_3) is empty, the other two must be maximal by definition. $\qquad\square$

Proposition 9. *Let* $\mathbb{K} = (K_1, K_2, K_3, Y) = \mathbb{K}_1 \uplus \ldots \uplus \mathbb{K}_n$ *with* $n \geq 2$ *and all* K_j^i *being non-empty. Then* (B_1, B_2, B_3) *is directly reachable from* (A_1, A_2, A_3) *in* $\mathbb{K}$ *iff*

- *they are triconcepts of the same* $\mathbb{K}_j$ *and* (B_1, B_2, B_3) *is directly reachable from* (A_1, A_2, A_3) *in* $\mathbb{K}_j$ *or*
- *one of* B_1, B_2, B_3 *is empty.*

Proof. "If": First assume (A_1, A_2, A_3) is directly reachable from (B_1, B_2, B_3) and both are are triconcepts of the same $\mathbb{K}_j$. W.l.o.g. let (1) be the corresponding perspective. Then $A_1 \subseteq B_1$. Moreover, none of A_1, A_2, A_3 is empty (otherwise (A_1, A_2, A_3) cannot be a triconcept of $\mathbb{K}_j$ due to Proposition 8). We find that $(B_2, B_3) \in \mathfrak{B}(\mathbb{K}_{j A_1}^{(23)})$. This implies $(B_2, B_3) \in \mathfrak{B}(\mathbb{K}_{A_1}^{(23)})$, thus (B_1, B_2, B_3) is directly reachable from (A_1, A_2, A_3) in $\mathbb{K}$.

Next, assume that one of B_1, B_2, B_3 is empty. W.l.o.g. assume $B_1 = \emptyset$. By Proposition 8, this entails $B_2 = K_2$ and $B_3 = K_3$. Then $(\emptyset, K_3) \in \mathfrak{B}(\mathbb{K}_{A_2}^{(13)})$ whenever $A_2 \neq \emptyset$ and $(\emptyset, K_2) \in \mathfrak{B}(\mathbb{K}_{A_3}^{(12)})$ whenever $A_3 \neq \emptyset$ (it is not possible that $A_2 = \emptyset = A_3$), therefore $(A_1, A_2, A_3) \prec (B_1, B_2, B_3)$ holds in $\mathbb{K}$.

"Only if": Assume $(A_1, A_2, A_3) \prec_i (B_1, B_2, B_3)$ in $\mathbb{K}$ and all of B_1, B_2, B_3 are nonempty. W.l.o.g. assume $i = 1$, i.e., $(B_2, B_3) \in \mathfrak{B}(\mathbb{K}_{A_1}^{(23)})$. Proposition 8 implies that (B_1, B_2, B_3) must be a triconcept of some $\mathbb{K}_j$. Then, due to $\emptyset \neq A_1 \subseteq B_1 \subseteq K_1^j$ we find that (A_1, A_2, A_3) cannot be a trivial triconcept, thus it is a triconcept of $\mathbb{K}_j$. Then $(B_2, B_3) \in \mathfrak{B}(\mathbb{K}_{A_1}^{(23)})$ implies $(B_2, B_3) \in \mathfrak{B}(\mathbb{K}_{j A_1}^{(23)})$ thus $(A_1, A_2, A_3) \prec_1 (B_1, B_2, B_3)$ holds in $\mathbb{K}_j$. □

Corollary 2. Let $\mathbb{K} = (K_1, K_2, K_3, Y) = \mathbb{K}_1 \uplus \ldots \uplus \mathbb{K}_n$ with $n \geq 2$ and all K_j^i being non-empty. Then (B_1, B_2, B_3) is reachable from (A_1, A_2, A_3) in $\mathbb{K}$ iff

- they are triconcepts of the same $\mathbb{K}_j$ and (B_1, B_2, B_3) is reachable from (A_1, A_2, A_3) in $\mathbb{K}_j$ or
- one of B_1, B_2, B_3 is empty.

Proof. This is a straightforward consequence of the previous proposition and the fact that all trivial triconcepts (those having one empty component) are together in the maximal cluster. □

Using the above results, we ask if there is any correlation between the cardinality of the three sets of a tricontext and the number of the reachability clusters we obtain. The first observation was that we can find qubic tricontexts (i.e., $|K_1| = |K_2| = |K_3| = n$), where the number of clusters equals $n + 1$.

Proposition 10. Let $\mathbb{K} = (K_1, K_2, K_3, Y)$ be a tricontext of size $n \times n \times n$ with $K_1 = \{k_i^1 \mid 1 \leq i \leq n\}$, $K_2 = \{k_i^2 \mid 1 \leq i \leq n\}$, $K_3 = \{k_i^3 \mid 1 \leq i \leq n\}$. Let the relation Y be the spatial main diagonal of the tricontext, meaning that a triple $(k_i^1, k_j^2, k_l^3) \in Y$ iff $i = j = k$. Then there are $n+1$ clusters, n minimal clusters and the maximal cluster.

Proof. Considering Proposition 9, the conclusion is immediate, since $\mathbb{K} = (\{k_1^1\}, \{k_1^2\}, \{k_1^3\}, \{(k_1^1, k_1^2, k_1^3)\}) \uplus \ldots \uplus (\{k_n^1\}, \{k_n^2\}, \{k_n^3\}, \{(k_n^1, k_n^2, k_n^3)\})$ □

Based on this example, we assumed that the number of clusters is bounded by the minimal dimension of the tricontext plus one. This assumption proved to be false due to the following example.

Example 2. Consider the following $4 \times 6 \times 6$ tricontext $\mathbb{K}_{466}$.

α	1	2	3	4	5	6
a	×					
b		×				
c			×			
d						
e						
f						

β	1	2	3	4	5	6
a	×					
b						
c						
d				×		
e					×	
f						

γ	1	2	3	4	5	6
a						
b		×				
c						
d				×		
e						
f						×

δ	1	2	3	4	5	6
a						
b						
c			×			
d						
e					×	
f						×

Besides the maximal cluster, we have six minimal ones which are all singletons consisting of the following triconcepts, respectively:

C_1 := $(\{a_1\},\{b_1\},\{c_1,c_2\})$, C_2 := $(\{a_2\},\{b_2\},\{c_1,c_3\})$, C_3 := $(\{a_3\},\{b_3\},\{c_1,c_4\})$, C_4 := $(\{a_4\},\{b_4\},\{c_2,c_3\})$, C_5 := $(\{a_5\},\{b_5\},\{c_2,c_4\})$, C_6 := $(\{a_6\},\{b_6\},\{c_3,c_4\})$.

Another assumption, about the number of cluster assumed to be the maximal dimension of the tricontext plus one, could be disproven.

Example 3. Given the tricontext $\mathbb{K}_{466} = (G,M,B,Y)$ from Example 2, we define $\mathbb{K}_{646} := (B,G,M,\{(b,g,m)|(g,m,b) \in Y\})$ as well as $\mathbb{K}_{664} := (M,B,G, \{(m,b,g)|(g,m,b) \in Y\})$, in words, we obtain $\mathbb{K}_{646}$ and $\mathbb{K}_{664}$ by rotating $\mathbb{K}_{466}$ twice. We now let $\mathbb{K}_{16^3} := \mathbb{K}_{466} \uplus \mathbb{K}_{646} \uplus \mathbb{K}_{664}$ be the $16 \times 16 \times 16$ context built by composing the three. Combining Example 2 with Corollary 2, we obtain that $\mathbb{K}_{16^3}$ has 19 clusters, viz. the maximal one and $6 + 6 + 6 = 18$ minimal ones.

Remark 2. The issue of whether the number of clusters or minimal clusters is bounded and what could be an estimation of that bound remains an open question.

6   Properties of Reachability Clusters

This section is devoted to the study of several properties of reachability clusters. We prove that reachability clusters can be found among some concepts of the context of reachable triconcepts, more exactly as object concepts.

Proposition 11. *Let* $(A_1,B_1,C_1),(A_2,B_2,C_2) \in \mathfrak{T}(\mathbb{K})$ *with* $(A_1,B_1,C_1) \prec_3$ (A_2,B_2,C_2). *Then* $Y_{C_2}^{12} \subseteq Y_{C_1}^{12}$.

Proof. Let $(g,m) \in Y_{C_2}^{12}$. Then, for every $b \in C_2$, we have $(g,m,b) \in Y$. Since $C_1 \subseteq C_2$, we have that for every $b \in C_1, (g,m,b) \in Y$, hence $(g,m) \in Y_{C_1}^{12}$.   □

Let $\mathbb{K} := (K_1,K_2,K_3,Y)$ be a triadic context. Let $\mathbb{K}_\triangleleft := (\mathfrak{T}(\mathbb{K}),\mathfrak{T}(\mathbb{K}),\triangleleft)$ be the formal context of reachable triconcepts. The concepts of $\mathbb{K}_\triangleleft$ are exactly the pairs (A,B) having the property that every triconcept from B is reachable from any of A and (A,B) is maximal with this property. If we take a look at the concepts of the symmetric kernel of $\triangleleft$, i.e., $\triangleleft \cap \triangleleft^{-1}$, we get exactly the reachability clusters of triconcepts without the ordering between them.

Proposition 12. *Let* $(A,B) \in \mathbb{K}_\triangleleft$ *be a concept and denote by* $C := A \cap B$. *If* $C \neq \emptyset$, *then* C *is a set of mutually reachable concepts, i.e.,* $C \times C$ *is a rectangle of crosses in* $\mathbb{K}_\triangleleft$.

Proof. From the definition, we have that $\forall T_1,T_2 \in C, T_1 \triangleleft T_2$ and $T_2 \triangleleft T_1$. It follows that all the triconcepts from C are part of the same cluster.   □

Remark 3. If we denote with $\mathcal{C}$ the set of clusters of triconcepts from $\mathbb{K}$ and with $\mathcal{I} := \{A \cap B \mid (A,B) \in \mathfrak{B}(\mathbb{K}_\triangleleft), A \cap B \neq \emptyset\}$, i.e., the set of all concepts having non disjoint extent and intent, then the previous proposition states that $\mathcal{I} \subset \mathcal{C}$.

Example 4. One might expect that there exist a one-to-one correspondence between concepts in the context of reachable triconcepts and reachability clusters. This would mean that the structure of reachability clusters is a concept lattice. The following example shows that there exist concepts in $\mathbb{K}_\lhd$, having disjoint extent and intent.

α	1	2	3	4
a	×			
b				
c		×		
d		×		

β	1	2	3	4
a				
b				
c		×		
d		×		

γ	1	2	3	4
a		×		
b		×		
c				
d				

δ	1	2	3	4
a		×		
b		×		
c				
d			×	

We have $(\{a\}, \{1\}, \{\alpha\}) \lhd (\{a, b\}, \{3\}, \{\gamma, \delta\})$, $(\{a\}, \{1\}, \{\alpha\}) \lhd (\{c, d\}, \{2\}, \{\alpha, \beta\})$, $(\{d\}, \{4\}, \{\delta\}) \lhd (\{a, b\}, \{3\}, \{\gamma, \delta\})$, $(\{d\}, \{4\}, \{\delta\}) \lhd (\{c, d\}, \{2\}, \{\alpha, \beta\})$, and the context $\mathbb{T}$ is given by:

	T_1	T_2	T_3	T_4	T_5	T_6	T_7
$T_1 = (\{a\}, \{1\}, \{\alpha\})$	×		×	×	×	×	×
$T_2 = (\{d\}, \{4\}, \{\delta\})$		×	×	×	×	×	×
$T_3 = (\{a, b\}, \{3\}, \{\gamma, \delta\})$			×		×	×	×
$T_4 = (\{c, d\}, \{2\}, \{\alpha, \beta\})$				×	×	×	×
$T_5 = (\{a, b, c, d\}, \{1, 2, 3, 4\}, \emptyset)$					×	×	×
$T_6 = (\{a, b, c, d\}, \emptyset, \{\alpha, \beta, \gamma, \gamma\})$					×	×	×
$T_7 = (\emptyset, \{1, 2, 3, 4\}, \{\alpha, \beta, \gamma, \gamma\})$					×	×	×

The concept $(\{T_1, T_2\}, \{T_3, T_4, T_5, T_6, T_7\})$ has disjoint extent and intent. In the following, we are going to characterize the reachability clusters as object concepts in the context of reachable triconcepts.

Proposition 13. *Let C be a reachability cluster of triconcepts. Then there exists a concept in $(A, B) \in \mathbb{K}_\lhd$ with $C = A \cap B$.*

Proof. Consider $(C^{\lhd\lhd}, C^\lhd)$.   □

Proposition 14. *If $(A, B), (C, D) \in \mathfrak{B}(\mathbb{K}_\lhd)$ are two different concepts of the context $\mathbb{K}_\lhd$ of reachable triconcepts with $A \cap B \neq \emptyset$ and $C \cap D \neq \emptyset$, then $A \cap B \neq C \cap D$.*

Proof. Let $(A, B), (C, D) \in \mathfrak{B}(\mathbb{K}_\lhd)$ be two different concepts. We assume $A \cap B = C \cap D = M \neq \emptyset$. Since they are different concepts, we can conclude that they have different extents and intents, so $A \neq C$ and $B \neq D$. It follows that at least one of the extents and one of the intents is bigger than M.

If $A \neq M$, $B \neq M$, $C = M$ and $D = M$ (or the other way around) it contradicts the fact that $(C, D) \in \mathfrak{B}(\mathbb{K}_\lhd)$ because it is not maximal. We can conclude that at least the extent of one concept and the intent of the other concept are bigger than M. Hence, we can assume $A \neq M$ and $D \neq M$. Let $T_1 \in$

$A \setminus M, T_2 \in M, T_3 \in D \setminus M$. Since $T_2 \in M \subseteq B$ it follows $(T_1, T_2) \in I \Rightarrow T_1 \triangleleft T_2$. Since $T_2 \in M \subseteq C$ it follows $(T_2, T_3) \in I \Rightarrow T_2 \triangleleft T_3$. From the transitivity of the relation $\triangleleft$ we have $T_1 \triangleleft T_3$. Herefrom we conclude that for every $T \in A \setminus M$, we have $(T, T_3) \in I$, but since $M \subseteq C$ and $T_3 \in D$, we also have that for every $T \in M$, we have $(T, T_3) \in I$. It follows that T_3 should be in the intent of the concept (A, B), so $T_3 \in B \Rightarrow T_3 \in B \cap D = M$ and we reach a contradiction since we chose $T_3 \in D \setminus M$. Therefore, the two different concepts in $\mathfrak{B}(\mathbb{K}_\triangleleft)$ cannot have the same intersection of the extent and intent. $\qquad\square$

Remark 4. The previous proposition proves that, by intersecting the extent and intent of the concepts in the context $\mathbb{K}_\triangleleft$ we cannot obtain the same cluster twice.

Proposition 15. *Let* $(A, B), (C, D) \in \mathfrak{B}(\mathbb{K}_\triangleleft)$. *Let* $M := A \cap B \neq \emptyset$ *and* $N := C \cap D \neq \emptyset$. *Then* $M \cap N = \emptyset$.

Proof. By the previous proposition, we know that $M \neq N$. Assume $M \cap N \neq \emptyset$. Let $a \in A \setminus M$, $b \in B \setminus M$, $x \in M \setminus (M \cap N)$, $y \in M \cap N$, and $z \in N \setminus (M \cap N)$ be arbitrary elements. Then we have $a \triangleleft y$ and $y \triangleleft z$ wherefrom follows that $a \triangleleft z$. Similarly, we have that from $x \triangleleft y$ and $y \triangleleft z$ follows $x \triangleleft z$. We also have that $y \triangleleft z$, hence for all $g \in A, g \triangleleft z$, i.e., $z \in A^\triangleleft = B$.

On the other hand, $z \triangleleft y \triangleleft x \triangleleft b$, hence $z \triangleleft b$ for all $b \in B$, i.e., $z \in B^\triangleleft = A$. We have that $z \in A \cap B = M$ and $z \in N$, thus $z \in M \cap N$. Contradiction! $\qquad\square$

Proposition 16. *The sets defined in Remark 3,* $\mathcal{C}$ *and* $\mathcal{I}$, *are equal:* $\mathcal{C} = \mathcal{I}$.

Proof. The first part of the equivalence was proved in Proposition 12 which showed that $\mathcal{I} \subseteq \mathcal{C}$. For the converse inclusion, let $(A, B) \in \mathfrak{B}(\mathbb{K}_\triangleleft)$ be a concept and $M := A \cap B$. Assume M is not maximal and build $M' := M^{\triangleleft\triangleleft} \cap M^\triangleleft$. Then $M \subseteq M'$ which is a contradiction. $\qquad\square$

Proposition 17. *Let* $(A, B) \in \mathfrak{B}(\mathbb{K}_\triangleleft)$ *with* $A \cap B = \emptyset$. *Then* (A, B) *is a concept of the contraordinal scale* $(\mathfrak{T}(\mathbb{K}), \mathfrak{T}(\mathbb{K}), \not\triangleright)$.

Proof. Since $\triangleleft$ is a preorder, it makes sense to speak about the contraordinal scale induced by $\triangleleft$. The concepts of the contraordinal scale are exactly the pairs (A, B) with A oder ideal, B order filter, $A \cap B = \emptyset$, and $A \cup B = \mathfrak{T}(\mathbb{K})$. Let $(A, B) \in \mathfrak{B}(\mathbb{K}_\triangleleft)$ with $A \cap B = \emptyset$. Then for every $a \in A$ and every $b \in B$, we have $x \triangleleft y$ and $y \not\triangleleft x$, i.e., $x \not\triangleright y$.

Let now $x \in A$ and $z \triangleleft x$. By transitivity, we get that for every $b \in B$, $z \triangleleft b$ and $z \in A$. Hence A is an order ideal. Dually, B is an order filter. We only have to prove that $B = \complement A$. Let $y \in \complement A$. Then for every $a \in A$, $y \not\triangleleft a$, which is equivalent to $a \not\triangleright y$, i.e., $y \in B$. $\qquad\square$

Concluding all the results obtained above, the following holds true:

Proposition 18. *Let* $T \in \mathfrak{T}(\mathbb{K})$ *be a triconcept. Then the cluster* $[T]$ *of* T *is generated by the object concept* $\gamma(T)$ *by* $T^{\triangleleft\triangleleft} \cap T^\triangleleft = [T]$. *Herefrom follows that the reachability clusters are generated by the object concepts of* $(\mathfrak{T}(\mathbb{K}), \mathfrak{T}(\mathbb{K}), \triangleleft)$. *If* (A, B) *is a proper concept which is not an object concept, then* $A \cap B = \emptyset$.

Remark 5. The above proposition states that reachability clusters are exactly the object concepts of the reachability context. This result gives a possibility to display all reachability clusters, along with a navigation support in a concept lattice, by highlighting the object concepts and deleting all the others, except the greatest concept.

7   Exploration Strategy and Algorithmics

Considering the theoretical aspects introduced in the previous paragraphs, we use reachability clusters to propose a strategy for navigating inside and between them. The purpose of this approach is to obtain a tool that can be used for navigation and visualization of a triadic context. Basically, starting from a tri-concept, one can browse through all the others from the reachability cluster of that triconcept, navigate to another triconcept (not necessarily in the same cluster), moving back and forth among these triconcepts in order to explore as much as possible the triadic conceptual knowledge landscape.

In order to be able to navigate through the data with the proposed paradigm the following steps are necessary:

(1) compute the triconcepts,
(2) compute the reachability relation between the triconcepts,
(3) compute the clusters of the tricontext,
(4) compute the partial order relation between the clusters.

The first step can be implemented by using Trias [9]. For the second step, we use the following procedure.

Listing 1.1. Procedure directlyReachable(T1, T2)

```
If  T1.extent ⊆ T2.extent  then
      B_e = extentProjectionContext(T1.extent)
      If  (T2.intent)'_{B_e}  =T2.modus  and
      (T2.modus)'_{B_e}=T2.intent  then
            Return true

If  T1.intent ⊆ T2.intent  then
      B_i = intentProjectionContext(T1.intent)
      If  (T2.extent)'_{B_i}  =T2.modus  and
      (T2.modus)'_{B_i}=T2.extent  then
            Return true

If  T1.modus ⊆ T2.modus  then
      B_m = modusProjectionContext(T1.modus)
      If  (T2.extent)'_{B_m}  =T2.intent  and
      (T2.intent)'_{B_m}=T2.extent  then
            Return true

Return false
```

The procedure checks whether the triconcept T_2 is directly reachable from the triconcept T_1. Thereby, $extentProjectionContext(T1.extent)$ represents the dyadic context obtained by projecting the triadic context on the extent dimension. This means that all the tricontexts selected have the extent equal or greater than $T1.extent$. The derivation $(T2.intent)'_{B_e}$ is considered to be a dyadic derivation in the obtained projection context B_e.

In order to obtain the reachability clusters, the most efficient method is to obtain the graph of the triconcepts with the direct reachability relation and compute the strongly connected components. This can be done by using existing algorithms for computing strongly connected components in directed graphs which have linear complexity. So for step 3, we consider the directed graph (since the direct reachability is not a symmetric relation) having the triconcept set as nodes and the edges given by the direct reachability relation. Then we obtain the clusters by computing all strongly connected components of the graph.

As proven earlier in the theoretical aspects of the navigation paradigm, the clusters correspond to nodes in a lattice, but not all the nodes in the lattice correspond to a cluster. Therefrom, the set of clusters is a partially ordered. The fact that they are object concepts in a particular concept lattice helps us navigate form one cluster to another. Also, this assures that we can reach any triconcept from the tricontext.

8   Conclusions and Further Work

We have proposed an approach to navigating in the space of triconcepts of a tricontext. To this end we defined three relatedness notions on the triconcepts based on extent, intent or modus. For each of these three perspectives, the triconcepts related to a given tricontext correspond to the concepts of a dyadic formal context, whence we can leverage the successful visualization approach of dyadic FCA by displaying, given a triconcept, all similar triconcepts in a lattice diagram. From such a diagram, a triconcept can be picked by a user, which will then be the starting point for the next visualization and navigation step.

We have investigated the reachability relation stemming from this navigation paradigm. As it turned out, for some tricontexts, not every triconcept can be reached from every other triconcept, although this seems to be the case in most practical scenarios. This gave rise to the notion of reachability clusters obtained as maximal sets of mutually reachable triconcepts. These clusters are in turn ordered by unidirectional reachability and form a partial order which always has a greatest element. Navigation can start either in one of the minimal clusters or the user can define its own constraints about included/excluded objects, attributes and/or conditions. By computing all triconcepts satisfying a given set of constraints, the user can choose them as navigation starting points. Not much more is known about the order of reachability clusters, some initial conjectures about upper bounds on their size or existence of suprema had to be refuted by counterexamples, which nevertheless provided some interesting structural insights and may pave the way to further investigations. As of yet, the only (and trivial) upper bound for the number of reachability clusters is the number of triconcepts, which may be exponential in the size of the tricontext. We, however, still conjecture that there is a polynomial bound.

Besides these open theoretical questions, future work on the topic has to include an implementation of the described navigation paradigm and user studies in order to confirm our hypothesis that this way of displaying and browsing the space of triconcepts is indeed accessible and intuitive for human users.

References

1. Cerf, L., Besson, J., Robardet, C., Boulicaut, J.: Closed patterns meet n-ary relations. TKDD **3**(1), 3:1–3:6 (2009)
2. Ferré, S., Ridoux, O.: Introduction to logical information systems. Inf. Process. Manage. **40**(3), 383–419 (2004)
3. Ganter, B., Obiedkov, S.: Implications in triadic formal contexts. In: Wolff, K.E., Pfeiffer, H.D., Delugach, H.S. (eds.) ICCS 2004. LNCS (LNAI), vol. 3127, pp. 186–195. Springer, Heidelberg (2004)
4. Ganter, B., Wille, R.: Formal Concept Analysis: Mathematical Foundations. Springer, Heidelberg (1999)
5. Glodeanu, C.V.: Tri-ordinal factor analysis. In: Cellier, P., Distel, F., Ganter, B. (eds.) ICFCA 2013. LNCS, vol. 7880, pp. 125–140. Springer, Heidelberg (2013)
6. Gnatyshak, D., Ignatov, D.I., Kuznetsov, S.O.: From triadic FCA to triclustering: Experimental comparison of some triclustering algorithms. In: Ojeda-Aciego, M., Outrata, J. (eds.) Proceedings of the Tenth International Conference on Concept Lattices and Their Applications, La Rochelle, France, October 15–18, 2013. CEUR Workshop Proceedings, vol. 1062, pp. 249–260. CEUR-WS.org (2013)
7. Godin, R., Missaoui, R., April, A.: Experimental comparison of navigation in a galois lattice with conventional information retrieval methods. Int. J. Man-Mach. Stud. **38**(5), 747–767 (1993)
8. Ignatov, D.I., Kuznetsov, S.O., Poelmans, J., Zhukov, L.E.: Can triconcepts become triclusters? Int. J. Gen. Syst. **42**(6), 572–593 (2013)
9. Jäschke, R., Hotho, A., Schmitz, C., Ganter, B., Stumme, G.: TRIAS - an algorithm for mining iceberg tri-lattices. In: Proceedings of the 6th IEEE International Conference on Data Mining (ICDM 2006), pp. 907–911. IEEE Computer Society (2006)
10. Jäschke, R., Hotho, A., Schmitz, C., Ganter, B., Stumme, G.: Discovering shared conceptualizations in folksonomies. J. Web Semant. **6**(1), 38–53 (2008)
11. Lehmann, F., Wille, R.: A triadic approach to formal concept analysis. In: Ellis, G., Rich, W., Levinson, R., Sowa, J.F. (eds.) ICCS 1995. LNCS, vol. 954, pp. 32–43. Springer, Heidelberg (1995)
12. Wille, R.: The basic theorem of triadic concept analysis. Order **12**(2), 149–158 (1995)

Graphs and FCA

A Proposal for Extending Formal Concept Analysis to Knowledge Graphs

Sébastien Ferré[✉]

IRISA/Université de Rennes 1,
Campus de Beaulieu, 35042 Rennes Cedex, France
`ferre@irisa.fr`

Abstract. Knowledge graphs offer a versatile knowledge representation, and have been studied under different forms, such as conceptual graphs or Datalog databases. With the rise of the Semantic Web, more and more data are available as knowledge graphs. FCA has been successful for analyzing, mining, learning, and exploring tabular data, and our aim is to help transpose those results to graph-based data. Previous FCA approaches have already addressed relational data, hence graphs, but with various limits. We propose G-FCA as an extension of FCA where the formal context is a knowledge graph based on n-ary relationships. The main contributions is the introduction of "n-ary concepts", i.e. concepts whose extents are n-ary relations of objects. Their intents, "projected graph patterns", mix relationships of different arities, objects, and variables. In this paper, we lay first theoretical results, in particular the existence of a concept lattice for each concept arity, and the role of relational projections to connect those different lattices.

Keywords: Formal concept analysis · Knowledge graph · Semantic web · Graph pattern · Relation · Projection

1 Introduction

Since the dawn of artificial intelligence, graphs have been used to represent knowledge as a set of interlinked entities. Notable formalisms are semantic networks, conceptual graphs [5], description logics [2], and more recently the Semantic Web [11]. In the last ten years, the number and size of knowledge graphs has exploded with the development of the Semantic Web, and its W3C standards (e.g., RDF, SPARQL). Its open side, called Linked Open Data (LOD), is now made of more than 1000 datasets [17], which contain about 70 billions semantic links (called triples). This effort has been joined by Web giants such as the Google Knowledge Graph or Facebook Graph Search.

Formal Concept Analysis (FCA) [10] is concerned with the definition of concepts from factual data, and their organization into a generalization ordering, the concept lattice. It serves many purposes such as knowledge discovery, machine

This research is supported by ANR project IDFRAud (ANR-14-CE28-0012-02).

J. Baixeries et al. (Eds.): ICFCA 2015, LNAI 9113, pp. 271–286, 2015.
DOI: 10.1007/978-3-319-19545-2_17

learning, information retrieval, or software refactoring. It seems therefore important to investigate the application of FCA to knowledge graphs. Only a few works consider its direct application to graphs. A power context family [18] is a form of knowledge graph, but there is a distinct concept lattice for each relation arity. Relational Concept Analysis (RCA) [16] defines concepts that mix unary and binary relationships, but only in tree-shaped patterns. Graphs have also been used as object descriptions (e.g., for molecules) as an application of pattern structures [9,13], but we do not consider those as knowledge graphs because graph nodes (e.g., individual atoms) are not formal objects. In all thoses approaches, only unary concepts are defined, i.e., extents are sets of objects, not relations. Concept lattices of relational structures [12] is an approach based on category theory that shares with us the use of n-ary relations, and relations as extents. However, the representation of an extent is not self-contained because it shares variables with the intent. Similarly, in RCA, an intent contains relational attributes that refer to other concepts, and so on, possibly in a circular way.

In this paper, we propose an extension of FCA, called Graph-FCA (G-FCA for short), that is directly applicable to knowledge graphs. Graph entities play the role of FCA objects, and graph relationships play the role of FCA attributes. The consequence is that the incidence relation relates tuples of objects (with various arities) to attributes, rather than single objects to attributes. A key novelty of our proposal is that an extensional representation is not a set of objects, but a set of *tuples of* objects, i.e. a n-ary relation. The particular case of unary relations corresponds to sets of objects. An intensional representation is defined as a *projected graph pattern* (PGP), i.e. as a graph pattern plus a *projection tuple*. The projection tuple can have any arity, and the graph pattern can mix relations with various arities. Both extensional and intensional representation are self-contained. A G-FCA concept is a pair (extent, intent) where the extent is an object relation, and the intent is a PGP. This significantly extends previous FCA approaches because power context families do not mix arities in concept definitions, and RCA only defines unary concepts based on unary and binary relations. In fact, PGPs are analogous to Datalog (non-recursive) predicate definitions [4], and to SPARQL queries. It suggests that G-FCA could be the basis for discovering or learning n-ary predicate definitions, and for querying knowledge graphs. The former is akin to Inductive Logic Programming (ILP) [15], and for the latter, we have already worked out a solution [6]. This paper aims at providing a formal basis and starting point for those applications.

After some technical preliminaries (Sect. 2), we formalize knowledge graphs as *graph contexts* (Sect. 3). We then introduce *projected graph patterns (PGP)* and *object relations*, and define mappings from one to the other based on *PGP inclusion* and *PGP intersection* (Sect. 4). From those definitions, we organize PGPs into a bounded lattice, and object relations into a complete lattice, from which we prove the existence of a concept lattice for each concept arity (Sect. 5). We relate the different concept lattices through projections (Sect. 6). Finally, we conclude and discuss future work on G-FCA (Sect. 7).

2   Tuples, Substitutions, and Projections

A *tuple* is noted $\overline{x} = (x_1, \ldots, x_k)$, where $|\overline{x}| = k$ is its *arity*. To avoid confusion with other kinds of indices, $\overline{x}[i]$ can be used as an alternate notation for x_i. The set of all k-tuples over a domain E is noted E^k. The set of all tuples is defined by $E^* = \bigcup_{k \geq 0} E^k$. There is only one 0-tuple, denoted by $()$. We use $1..k$ to denote the set of integers from 1 to k.

We assume an infinite set of variables $\mathcal{V}$, and we use letters to denote them (e.g., x, y). A *substitution* $\sigma \in \Sigma_E$ is a mapping from variables to elements of E. A substitution σ can be applied as a function to any structure, and returns that same structure with any variable x in it replaced by $\sigma(x)$. For example, given the substitution $\sigma = \{x \mapsto 1, y \mapsto z\}$, we have $\sigma((x, y, z)) = (1, z, z)$. The empty substitution is denoted by id, and the composition of two substitutions is denoted by $\sigma_2 \circ \sigma_1$, where $(\sigma_2 \circ \sigma_1)(x) = \sigma_2(\sigma_1(x))$. Given two k-tuples $\overline{x}, \overline{y}$, the notation $\sigma_{\overline{x}}^{\overline{y}}$ denotes the substitution that maps x_i to y_i, for every $i \in 1..k$, and any other variable not in $\overline{x}$ to itself. It is only well-defined when $\overline{y}$ is compatible with $\overline{x}$, i.e. for all $i \neq j \in 1..k$, $x_i = x_j \Rightarrow y_i = y_j$, and for all $i \in 1..k$, $x_i \notin \mathcal{V} \Rightarrow y_i = x_i$.

A *projection* $\pi \in \Pi_k^l$ is a function from target indices $1..l$ to source indices $1..k$. The projection $\pi_1 = \{1 \mapsto 3, 2 \mapsto 1\}$ can be more concisely represented by the tuple $(3, 1)$. Projections are used to map a tuple to another tuple according to the following formula: $\pi(\overline{x})[i] = \overline{x}[\pi(i)]$, i.e. the i-th element of a projected tuple is the element at index $\pi(i)$. The identity projection is denoted by $id_k \in \Pi_k^k$, and the composition of two projections is denoted by $\pi_2 \circ \pi_1$, where $(\pi_2 \circ \pi_1)(\overline{x}) = \pi_2(\pi_1(\overline{x}))$ (i.e., $(\pi_2 \circ \pi_1)(i) = \pi_1(\pi_2(i))$). A *permutation* is a bijective projection π, and has an inverse projection π^{-1} that is the inverse permutation. Note that the combined applications of a substitution σ and a projection π commute, i.e. $\pi(\sigma(\overline{x})) = \sigma(\pi(\overline{x}))$ for every substitution σ, projection $\pi \in \Pi_k^l$, and tuple $\overline{x} \in E^k$.

3   Knowledge Graphs as G-FCA Contexts

The first step is to formalize a knowledge graph as a formal context, which we call a *graph context*. The only difference with the classical FCA definition lies in the use of object tuples (O^*) instead of objects (O) in the incidence relation.

Definition 1 (Graph Context). *A* graph context *is a triple* $K = (O, A, I)$, *where* O *is a set of* objects, A *is a set of* attributes, *and* $I \subseteq O^* \times A$ *is an* incidence relation *between object tuples and attributes. The maximum cardinality of incidences is denoted by* $|K|$.

Figure 1 shows the graphical representation of a small graph context about USA presidents. It uses a notation similar to conceptual graphs, using rectangles for entities, and ellipses for relations [5]. A graph context is a directed multi-hypergraph, where each node is labelled by an object $o \in O$ (e.g., "Obama", "Hawaii", "2009"), and where each directed hyperedge is an incidence $(\overline{o}, a) \in I$. A hyperedge connects a number of objects $\overline{o}$ in a fixed order, and is labelled by an attribute $a \in A$ (e.g., "has president", "is a country"). If $|\bar{o}| = 2$, it is equivalent to a classical edge linking two nodes: e.g., $((USA, Obama), president)$. If $|\bar{o}| = 1$, it is equivalent to a classical node labelling: e.g., $((USA), country)$. The advantage of this definition is therefore to treat uniformly classical node labels and edge labels, and to support hyperedges, i.e. n-ary relationships: e.g., $((Obama, PeaceNobelPrize, 2009), awardInYear)$. Hyperedges are directed such that each position in the tuple of objects $\bar{o}$ corresponds to a particular role in the relationship: e.g., $((USA, Hawaii), state)$ holds while $((Hawaii, USA), state)$ does not. Nothing forbids to use the same attribute with different arities, but this amounts to have different relationships with the same name[1].

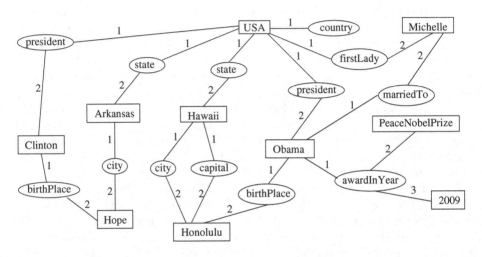

Fig. 1. Graphical representation of a graph context about USA presidents.

Graph contexts can easily be translated to/from other well-known relation-based representations. For example, a graph context is equivalent to the union of all elements of a power context family. The main limit when translating to other representations is for n-ary relationships with $n \geq 2$. That limit can be addressed by reifying hyperedges as nodes. In the Semantic Web, objects are RDF nodes (URIs, literals, and blank nodes), attributes are class and property URIs, and incidences are triples. In relational databases, objects are keys and values, attributes are table names, and incidences are table rows. In Inductive Logic Programming and Datalog, objects are Prolog atoms, attributes are predicates, and incidences are background knowledge facts. In RCA, objects are objects, attributes are either context attributes or relation names, and incidences are either $((o), a)$ when o has attribute a in some context, or $((o_1, o_2), r)$ when (o_1, o_2) are in relation r.

---

[1] Similarly to Prolog where predicates are identified by their name *and* arity.

4   Projected Graph Patterns and Object Relations

In this section, we introduce the intensional and extensional representations of G-FCA, as an extension of the sets of attributes and sets of objects of FCA. Section 4.1 defines *projected graph patterns* as an extension of the sets of attributes of FCA. Section 4.2 defines *object relations* as an extension of the sets of objects of FCA. Section 4.3 then defines a mapping from projected graph patterns to object relations, and Sect. 4.4 defines an inverse mapping from object relations to projected graph patterns. The two mappings are shown to form a Galois connection in Sect. 5.3, and are the basis of G-FCA concept lattices.

4.1   Projected Graph Patterns as Intensional Representations

A concept intent is an intensional representation that describes everything that a set of objects have in common. As a particular case, the intent of a single object is the description of that object. Depending on the FCA variant, an intensional representation can be a set of attributes, a logical formula [7], or a structure [9]. So, in order to identify G-FCA intensional representations, we may start by asking what is an adequate object description in G-FCA. An object (e.g., *USA*) should at least be described by its adjacent hyperedges (e.g., *having a president*) and adjacent objects (e.g., *Obama*). Then, if adjacent objects are interlinked (e.g., *USA's president* and *USA's first lady* are *married*), this should also appear in the description. Similarly, the descriptions of adjacent objects (e.g., *Obama being born in Honolulu*) should be included as they indirectly impact what the object is (e.g., *USA having a president born in Hawaii*).

All in all, this implies that the description of an object is the entire knowledge graph, or at least the connected component it belongs to if the knowledge graph is disconnected. Given that knowledge graphs, like the Web, are generally not disconnected, this seems to imply that all objects have the same description! In fact, it is like if all objects were represented by the same knowledge graph, the same world, but each from a different point of view. The different points of view can be interpreted as different phrasings of the same information: e.g., "Obama is the president of USA, and was born in Honolulu", and "Honolulu is the birth place of the president of USA, Obama". Therefore, the description of an object o in a graph context $K = (O, A, I)$ can be defined as the couple (o, I).

It is possible to generalize descriptions from objects to tuples of objects. For example, the description of the 2-tuple (*Honolulu, USA*) should contain the properties of each object, and the relationships (direct and indirect) that link them (e.g., Honolulu "is the capital of a state of" USA, Honolulu "is the birth place of a president of" USA). Similarly to single objects, the description of a tuple of objects $\bar{o}$ can be defined as $(\bar{o}, I)$.

Object descriptions need to be generalized to form concept intents shared by several objects, or several tuples of objects. For instance, we want to describe what *Obama* and *Clinton* have in common, or what (*Honolulu, Hawaii*) and (*Houston, Texas*) have in common. Like in ILP, generalization is obtained in two ways: (1) replacing objects by variables, and (2) removing hyperedges. This leads us to the following definitions of a *graph pattern* as a generalized incidence relation, and of a *projected graph pattern* (PGP) as a generalized description. In the following definitions, we use $\mathcal{X} = O \cup \mathcal{V}$ to denote the domain of nodes in graph patterns, which can be objects or variables. We assume a graph context $K = (O, A, I)$ that defines objects and attributes.

Definition 2 (Graph Pattern). *A graph pattern $P \subseteq \mathcal{X}^* \times A$ is a set of directed hyperedges with variables and/or objects as nodes, and attributes as labels. Substitutions are extended to patterns: $\sigma(P) = \{(\sigma(\overline{v}), a) \mid (\overline{v}, a) \in P\}$.*

Graph patterns have the same type as incidence relations, only allowing variables in addition to objects as nodes. For instance, the pattern $P_{ex} = \{((x, y), president), ((y, Honolulu), birthPlace)\}$ describes any situation where "some entity x has as a president another entity y, which has birth place *Honolulu*". Every graph pattern can be seen as a small graph context, abstracted over some objects by variables. We are primarily interested in connected patterns, but disconnected patterns are not excluded.

Definition 3 (Projected Graph Pattern). *A projected graph pattern (PGP) is a couple $Q = (\overline{x}, P)$ where P is a graph pattern, and $\overline{x} \in \mathcal{X}^*$, called projection tuple, is a tuple of variables and possibly objects. $|Q| = |\overline{x}|$ denotes the arity of the PGP. We note $\mathcal{Q}$ the set of PGPs, and $\mathcal{Q}_k$ the subset of PGPs having arity k. Projections are extended to PGPs: $\pi(Q) = (\pi(\overline{x}), P)$.*

A projection tuple can be seen as a tuple of objects abstracted with variables. For instance, the PGP $Q_{ex} = ((x), P_{ex})$ using the above pattern describes "any entity x having a president born in Honolulu". Pattern hyperedges can be seen as constraints on variables. A variable that occurs in the projection tuple but not in the pattern is unconstrained, and can take any object as value. A variable that occurs in the pattern but not in the projection tuple is existentially quantified with respect to projected variables. In Q_{ex}, there must exist an entity y that the president of x and is born in Honolulu, but which one and how many does not matter.

Objects and duplicates in the projection tuple $\overline{x}$ define equality constraints between the indices of $\overline{x}$.

Definition 4 (Equality Constraints). *Let $\overline{x}$ be a projection tuple of arity k. Its set of equality constraints is defined by*

$$Eq(\overline{x}) = \{(i, j) \mid i < j \in 1..k, x_i = x_j\} \cup \{(i, o) \mid i \in 1..k, o \in O, x_i = o\}.$$

A set of equality constraints generates an equivalence relation between projection tuple indices and objects, and is confused with it in the following.

For example, the set of equality constraints of the projection tuple (x, o, x, y) is $\{(1, 3), (2, o)\}$, and generates three equivalence classes: $\{1, 3\}$, $\{2, o\}$, and $\{4\}$.

Equality constraints come in addition to hyperedge constraints from the graph pattern P. By allowing objects in projection tuples, we allow object descriptions in the form $(\overline{o}, I)$ to act as fully instantiated PGPs. By allowing duplicates, we allow a single entity to play different roles, like when searching a common PGP between the pairs (*Canberra*, *Sydney*) and (*Paris*, *Paris*): e.g. $((x, y), \{((z), Country), ((z, x), capital), ((z, y), biggestCity)\})$.

As said above, PGPs are an extension of the sets of attributes of FCA. For instance, the set of attributes $\{a, c, d\}$ corresponds to the PGP $((x), \{((x), a), ((x), c), ((x), d)\})$, where x is a variable. Indeed, in FCA, $\{a, c, d\}$ covers all objects that have at least attributes a, c, d. In G-FCA, this can be expressed as a node labelled by a, c, d, hence the introduction of the variable x. In general, a set of attributes Y corresponds to the PGP $((x), \{((x), a) \mid a \in Y\})$.

PGPs are comparable to non-recursive predicate definitions, and to SPARQL ASK/SELECT conjunctive queries. For example, the definition of the 'uncle' predicate

$$uncle(x, y) :\Leftrightarrow \exists z.parent(x, z) \wedge brother(z, y)$$

is equivalent to the PGP $((x, y), \{((x, z), parent), ((z, y), brother)\})$. Similarly, the SPARQL SELECT query

```
SELECT ?x ?y WHERE
{ ?x a :Film . ?x :genre :ScienceFiction . ?x :director ?y }
```

is equivalent to the PGP $((x, y), \{((x), Film), ((x, ScienceFiction), genre),$ $((x, y), director)\})$. The SPARQL ASK queries (yes/no questions) correspond to PGPs whose arity is zero $(\overline{x} = ())$.

4.2   Object Relations as Extensional Representations

In our introduction of PGPs as intensional representations, we made a shift from single objects to tuples of objects. That implies that extensional representations are sets of tuples of objects. With the constraint that all member tuples have the same arity, we obtain that extensional representations are *object relations*.

Definition 5 (Object Relation). *An* object relation *is a set* $R \subseteq O^k$, *for some arity* $|R| = k$, *of object tuples. We note* $\mathcal{R}$ *the set of object relations, and* $\mathcal{R}_k$ *the subset of relations having arity* k. *Projections are extended to relations:* $\pi(R) = \{\pi(\overline{o}) \mid \overline{o} \in R\}$.

$\mathcal{R}_0$ has only two relations: $\{()\}$ and $\{\}$. It can be seen as the Boolean type, with the two relations meaning "true" and "false" respectively. $\mathcal{R}_1$ has one relation for each set of objects $X \subseteq O$. It therefore corresponds to FCA extensional representations. For instance, the set of objects $\{o_1, o_3, o_4\}$ corresponds to the object relation $\{(o_1), (o_3), (o_4)\}$, simply embedding each object into a 1-tuple. In general, a set of objects X corresponds to the object relation $\{(o) \mid o \in X\}$. Object relations are comparable to the interpretations of a predicate in classical logic, and to SPARQL query results.

4.3   From Patterns to Relations

We here define a mapping from PGPs to object relations, i.e. from intensional representations to extensional representations. It defines for each PGP its *extension*, i.e. its set of instances in a given graph context. An instance is a tuple of objects whose description $(\overline{o}, I)$ "contains" the PGP modulo a substitution.

Definition 6 (PGP Inclusion). *Let $Q_1 = (\overline{x}_1, P_1)$, $Q_2 = (\overline{x}_2, P_2)$ be two PGPs with same arity: $|Q_1| = |Q_2|$. Q_1 is included in Q_2, or equivalently Q_2 contains Q_1, which is denoted by $Q_1 \subseteq_q Q_2$ iff there exists an substitution σ s.t. $\sigma(\overline{x}_1) = \overline{x}_2$ and $\sigma(P_1) \subseteq P_2$.*

$$Q_1 \subseteq_q Q_2 :\Leftrightarrow \exists \sigma \in \Sigma_{\mathcal{X}} : \sigma(\overline{x}_1) = \overline{x}_2 \wedge \sigma(P_1) \subseteq P_2$$

That definition is careful to account for variable renamings by introducing a substitution from Q_1-variables to Q_2-nodes. Indeed, like in logical formulas, variable names are irrelevant to the meaning of PGPs.

Definition 7 (Extension). *Let $K = (O, A, I)$ be a graph context. The extension of a k-PGP $Q = (\overline{x}, P)$, denoted by $ext(Q)$, is defined by*

$$ext(Q) := \{ \overline{o} \in O^k \mid Q \subseteq_q (\overline{o}, I) \}$$

The above definitions say that for every occurence of the pattern P in the graph context ($\sigma(P) \subseteq I$), there is an instance of Q ($\sigma(\overline{x})$). Conversely, if $\overline{o}$ is an instance of Q, then $Q = (\overline{x}, P)$ must be a generalization of its description $(\overline{o}, I)$, i.e. replacing some objects by variables and relaxing some constraints.

For example, in the graph context of Fig. 1, the PGP $((x, y), \{((USA, x), president), ((x, z), birthPlace), ((USA, y), state), ((y, z), city)\})$ has the following extension: $\{(Obama, Hawaii), (Clinton, Arkansas)\}$. That PGP retrieves the list of USA presidents along with the state of their birth place.

The above definition is compatible with the interpretation of a predicate definition in classical logic: our incidence relation I corresponds to a model, and our substitution σ corresponds to a variable assignment. It is also compatible with SPARQL query results: our incidence relation I corresponds to a RDF graph, and our substitution corresponds to a solution mapping. Finally, it is consistent with classical FCA in the case where only 1-tuples are used, i.e., when $Q = ((x), \{((x), a) \mid a \in Y\})$ for some set of attributes $Y \subseteq A$. Indeed, by casting 1-tuples to their element, and picking $\sigma = \{x \mapsto o\}$, we obtain the classical FCA definition: $ext(Y) = \{o \in O \mid \forall a \in Y : (o, a) \in I\}$.

Note that substitutions used in PGP inclusion are homomorphisms, and not isomorphisms, because two different variables can be substituted by a single node. This departs from previous work in graph mining and FCA [13,19] which are based on isomorphisms, but this follows classical logic and SPARQL querying as explained above.

4.4   From Relations to Patterns

We here define a mapping from object relations to PGPs, i.e. from extensional representations to intensional representations. It defines for each object relation its *intension* as the "PGP intersection" of the description of all tuples $\bar{o} \in R$.

Definition 8 (PGP Intersection). *Let* $\{Q_i = (\bar{x}_i, P_i)\}_{i \in 1..n}$ *be a finite and non-empty collection of n PGPs of same arity k. Let ν be a fixed bijection from* $\mathcal{X}^n$ *to $\mathcal{X}$ s.t. $\nu(x, \ldots, x) = x$ for all $x \in \mathcal{X}$. The PGP intersection $\cap_q \{Q_i\}_{i \in 1..n}$ is the PGP $Q = (\bar{x}, P)$ of arity k, where:*

- $\bar{x} = (x_1, \ldots, x_k)$, *where for all* $j \in 1..k$, $x_j = \nu(\bar{x}_1[j], \ldots, \bar{x}_n[j])$,
- $P = \{((\nu(\bar{v}_1), \ldots, \nu(\bar{v}_k)), a) \mid k \in 1..|K|, \forall j \in 1..k : \bar{v}_j \in \mathcal{X}^n, a \in A,$
   $\forall i \in 1..n : ((\bar{v}_1[i], \ldots, \bar{v}_k[i]), a) \in P_i\}$.

Definition 9 (Intension). *Let $K = (O, A, I)$ be a graph context. The* intension *of a non-empty object relation $R \in \mathcal{R}_k$, denoted by $int(R)$, is defined by*

$$int(R) = \cap_q \{(\bar{o}, I)\}_{\bar{o} \in R}$$

The idea of PGP intersection is to define a node $\nu(\bar{v}_j)$ for every possible alignment of n-nodes $\{\bar{v}_j[i]\}_{i \in 1..n}$, one from each PGP Q_i of the collection. Then, if an hyperedge holds at every position $i \in 1..n$ of k alignment nodes $(\bar{v}_1, \ldots, \bar{v}_k)$, then it is a shared structure and it belongs to the pattern of the PGP intersection. The projection tuple is then derived from the alignment of the projection tuples of the collection. In practice, the connected components of the graph pattern P that do not contain any element of the projection tuple can be omitted (reduced intension) because they do not affect the extension of the PGP intersection. For the same reason, hyperedges that are in the incidence relation I can be omitted.

For example, in the graph context of Fig. 1, the object relation $R = \{(USA, Hope), (USA, Honolulu)\}$ has the following reduced intension: $((USA, x),$ $\{((USA, y), state), ((y, x), city), ((USA, z), president), ((z, x), birthPlace)\})$. The obtained variables are derived from the following alignments: $x = \nu(Hope, Honolulu)$, $y = \nu(Arkansas, Hawaii)$, $z = \nu(Clinton, Obama)$. Through alignments, PGP intersection does not only provide a common PGP, but also an explanation of how each instance relates to the others. PGP intersection also applies to PGPs with variables. For example, the intersection of $Q_1 = ((x_1), \{((x_1, y_1), a), ((x_1, z_1), c), ((y_1, z_1), b)\})$ and $Q_2 = ((x_2),$ $\{((x_2, y_2), a), ((x_2, y_2), c), ((y_2, y_2), b)\}$ is $Q = ((x), \{((x, y), a), ((x, z), c), ((y, z),$ $b)\})$, which is isomorphic to Q_1. Here, both y_1, z_1 are aligned with y_2, hence generating two variables $y = \nu(y_1, y_2)$ and $z = \nu(z_1, y_2)$. Note that Q is not subgraph isomorphic to Q_2. Indeed, under isomorphism, PGP intersection would be the problem of Maximum Common Subgraphs (MCS). A drawback of MCS is that there is generally not a unique solution, so that sets of graph patterns have to be used for intensional representations [13]. Moreover, the MCSs can be less specific. For example, the MCSs of Q_1 and Q_2 patterns are $\{((x, y), a)\}$ and $\{((x, z), c)\}$.

The above definitions of PGP intersection and intension are consistent with classical FCA, where only unary relations (sets) are used. In this case, every PGP needs only one variable. Therefore, given a set of objects $X = \{o_i\}_{i \in 1..n} \subseteq O$, the intension of X has one projected variable $x = \nu(o_1, \ldots, o_n)$, and a graph pattern like $P = \{((x), a) \mid a \in A, \forall i \in 1..n : ((o_i), a) \in I\}$. By casting 1-tuples to their element, we obtain the classical FCA definition: $int(X) = \{a \in A \mid \forall o \in X : (o, a) \in I\}$.

5   A Family of Graph Concept Lattices

In order to prove that (ext, int) forms a Galois connection, and hence the existence of a *graph concept lattice* for each concept arity, we first define partial orderings for each arity, over both PGPs and object relations. We also show that PGPs form a bounded lattice, and object relations a complete lattice.

5.1   Lattices of k-PGPs

The partial ordering over PGPs should correspond to a generalization ordering over them. Intuitively, a PGP Q_1 is more general than a PGP Q_2 if Q_1 is included in Q_2: $Q_1 \subseteq_q Q_2$ (see Definition 6). Indeed, assume $Q_2 = ((x, y), \{((x), country), ((x, y), president)\})$ representing the relationship between countries and their president. Then, $Q_1 = ((x, y), \{((x, y), president)\})$ representing the relationship between different kinds of organizations and their president is more general than Q_2 because it relaxes the constraint saying that the organization should be a country. Generalization by constraint relaxation is also found in ILP to define subsumption between learning hypotheses.

Recall that PGP inclusion is defined modulo a substitution, and that a substitution can map two different nodes in Q_1 to a single node in Q_2. The latter corresponds to adding an equality constraint between two entities, which is indeed a specialization. It enables to have

$$((x, y), \{((x, y), president), ((x', y), president)\}) \subseteq_q ((x, y), \{((x, y), president)\}),$$

by mapping both x, x' to x. Note that the first PGP does not state that y is the president of two organizations, but rather states twice that y is a president, which is equivalent to the second PGP. As the reverse inclusion trivially holds, the two PGPs are equivalent representations of the same thing. We note $Q_1 \equiv_q Q_2$ when $Q_1 \subseteq_q Q_2$ and $Q_2 \subseteq_q Q_1$. PGP inclusion is compatible with inclusion between sets of attributes in FCA. Indeed, as a single variable is involved in FCA, the substitution must be the identity function, and the definition of $Q_1 \subseteq_q Q_2$ boils down to $P_1 \subseteq P_2$.

We prove that $\subseteq_q$ is a preorder, and hence that a partially ordered set is obtained for patterns by considering equivalence classes of PGPs modulo $\equiv_q$.

Lemma 1. *PGP inclusion $\subseteq_q$ is a preorder over PGPs.*

Proof. **Reflexivity.** Given a PGP Q, it suffices to take $\sigma = id$ to verify $\sigma(P) \subseteq P$ and $\sigma(\overline{x}) = \overline{x}$, and hence $Q \subseteq_q Q$.

Transitivity. Assume PGPs Q_1, Q_2, Q_3 s.t. $Q_1 \subseteq_q Q_2$ and $Q_2 \subseteq_q Q_3$. Hence, there exists two substitutions σ_1, σ_2 s.t. $\sigma_1(P_1) \subseteq P_2$, $\sigma_2(P_2) \subseteq P_3$, $\sigma_1(\overline{x}_1) = \overline{x}_2$, and $\sigma_2(\overline{x}_2) = \overline{x}_3$. Then, it suffices to take $\sigma = \sigma_2 \circ \sigma_1$ to verify $\sigma(P_1) = \sigma_2(\sigma_1(P_1)) \subseteq \sigma_2(P_2) \subseteq P_3$, and also $\sigma(\overline{x}_1) = \sigma_2(\sigma_1(\overline{x}_1)) = \sigma_2(\overline{x}_2) = \overline{x}_3$. Hence $Q_1 \subseteq_q Q_3$. ∎

Before showing that the pre-ordering over k-PGPs forms a bounded lattice modulo $\equiv_q$, for every arity k, we first need to define *PGP union* to act as a supremum. To this purpose, we need to introduce an additional *maximal PGP*, denoted by Ω_q, that is defined as containing all PGPs ($\forall Q : Q \subseteq_q \Omega_q$), and only included in itself ($\forall Q : \Omega_q \subseteq_q Q \Rightarrow Q = \Omega_q$). It can therefore be used to extend the definition of PGP intersection to empty collections: $\cap_q \emptyset := \Omega_q$.

Definition 10 (PGP Union). *Let $\{Q_i = (\overline{x}_i, P_i)\}_{i \in 1..n}$ be a finite collection of n PGPs of same arity k, using disjoint sets of variables. The PGP union $\cup_q \{Q_i\}_{i \in 1..n}$ is either the PGP $(\overline{x}, P)$ of arity k verifying*

– $Eq(\overline{x}) = \bigcup_{i \in 1..n} Eq(\overline{x}_i)$
– $P = \bigcup_{i \in 1..n} \{\sigma_{\overline{x}_i}^{\overline{x}}(P_i)\}$

when $Eq(\overline{x})$ has no two different objects in a same equivalence class; or else Ω_q.

The assumption that PGPs do not share any variable is there to avoid variable capture. It entails no loss of generality because variable can be renamed freely. PGP union corresponds to add both equality and edge constraints of all PGPs Q_i, and is logically equivalent to a conjunction. When a projection tuple that satisfies all equality constraints can be formed, it is used as a projection tuple of the PGP union, and also to merge variables ($\sigma_{\overline{x}_i}^{\overline{x}}$) from the different projection tuples $\overline{x}_i$ in the collection of PGPs. Given a set of equality constraints, e.g. $\{(1, 3), (2, o)\}$, a projection tuple is formed by using a single node for all indices of an equivalence class (e.g., 1 and 3), and by choosing as a node the object in the equivalence class if there is one (e.g., o for 2), or a fresh variable otherwise (e.g., x for 1 and 3). The case where several objects belong to a same equivalence class corresponds to a contradiction between the differents PGPs, and the maximal PGP Ω_q is used to denote such a contradiction. The extension of Ω_q is always the empty relation because it is only included in itself, and not in any object tuple description. The PGP union of an empty collection corresponds to an empty set of constraints, and defines the *minimal PGP* $\emptyset_q = (\overline{x}, \emptyset)$, where $\overline{x}$ is a tuple of k distinct variables. Finally, PGP union is compatible with the union of sets of attributes in FCA.

We first prove two lemmas stating that $\cup_q$ and $\cap_q$ are respectively the supremum and infimum of k-PGPs, before stating the main theorem about bounded lattices of k-PGPs.

Lemma 2. *Let Q_1, Q_2 be two PGPs. Their PGP union $Q_1 \cup_q Q_2$ is their supremum relative to query inclusion $\subseteq_q$.*

Proof. Let $Q = Q_1 \cup_q Q_2$. To prove that Q is an upper bound, it suffices to prove that it contains both Q_1 and Q_2. If $Q = \Omega_q$, then it contains both Q_1 and Q_2 by definition. Otherwise $Q = (\overline{x}, P)$. To prove $Q_1 \subseteq_q Q$, it suffices to choose $\sigma_1 = \sigma_{\overline{x}_1}^{\overline{x}}$, which is well-defined because $Eq(\overline{x}_1) \subseteq Eq(\overline{x})$, and to prove $\sigma_1(\overline{x}_1) = \overline{x}$ and $\sigma_1(P_1) \subseteq P$. This is easily obtained from the definition of Q. The proof of $Q_2 \subseteq_q Q$ is identical with $\sigma_2 = \sigma_{\overline{x}_2}^{\overline{x}}$.

To prove that Q is the *least* upper bound (the supremum), we have to prove that every PGP Q' that contains both Q_1 (via σ_1) and Q_2 (via σ_2) also contains Q. If $Q' = \Omega_q$, then $Q \subseteq_q Q'$ by definition of Ω_q. Otherwise, $Q' = (\overline{x}', P')$. From hypotheses $\sigma_1(\overline{x}_1) = \overline{x}'$ and $\sigma_2(\overline{x}_2) = \overline{x}'$, we obtain that $Eq(\overline{x}_1) \subseteq Eq(\overline{x}')$ and $Eq(\overline{x}_2) \subseteq Eq(\overline{x}')$. Then, we have $Eq(\overline{x}) = Eq(\overline{x}_1) \cup Eq(\overline{x}_2) \subseteq Eq(\overline{x}')$, and hence that $\sigma = \sigma_{\overline{x}}^{\overline{x}'}$ is well-defined. We can then easily prove that $\sigma(\overline{x}) = \overline{x}'$ and $\sigma(P) \subseteq P'$, and hence that $Q \subseteq_q Q'$.   ∎

Lemma 3. *Let Q_1, Q_2 be two PGPs. Their PGP intersection $Q_1 \cap_q Q_2$ is their infimum relative to query inclusion $\subseteq_q$.*

Proof. To prove that $Q_1 \cap_q Q_2$ is a lower bound, it suffices to prove that Q is included in both Q_1 and Q_2. To prove $Q \subseteq_q Q_1$, it suffices to choose the substitution $\sigma_1(x) = (\nu^{-1}(x))[1]$, and to prove that $\sigma_1(\overline{x}) = \overline{x}_1$ and $\sigma_1(P) \subseteq P_1$. This is easily obtained from the definition of Q. The proof of $Q \subseteq_q Q_2$ is identical with $\sigma_2(x) = (\nu^{-1}(x))[2]$.

To prove that $Q_1 \cap_q Q_2$ is the *greatest* lower bound (the infimum), we have to prove that every PGP Q' that is included in both Q_1 (via σ_1) and Q_2 (via σ_2) is also included in Q. To that purpose, it suffices to choose $\sigma(x') = \nu(\sigma_1(x'), \sigma_2(x'))$, and to prove that $\sigma(\overline{x}') = \overline{x}$ and $\sigma(P') \subseteq P$. This can be obtained from the definition of Q, and from the hypotheses $\sigma_1(\overline{x}') = \overline{x}_1$, $\sigma_1(P') \subseteq P_1$, $\sigma_2(\overline{x}') = \overline{x}_2$, and $\sigma_2(P') \subseteq P_2$.   ∎

Theorem 1. *For every arity k, the algebraic structure $(\mathcal{Q}_k, \subseteq_q, \cap_q, \cup_q, \Omega_q, \emptyset_q)$ forms a bounded lattice, module $\equiv_q$.*

Proof. The proof follows immediately from above lemmas and definitions.   ∎

5.2   Complete Lattices of Object k-Relations

The partial ordering over object relations should be consistent with the partial ordering on PGPs if we are to obtain concept lattices. Therefore, it should correspond to a form of generalization at the extensional level. A PGP can be made more general by relaxing constraints, which entails a larger extension. As object relations are sets of object tuples, we simply use set inclusion to partially order them. Given that $\mathcal{R}_k$ is the powerset of O^k, the poset $(\mathcal{R}_k, \subseteq, \cap, \cup, O^k, \emptyset)$ is a complete lattice, with set intersection $\cap$ as infimum, set union $\cup$ as supremum, full relation O^k as top, and empty relation $\emptyset$ as bottom.

5.3   Lattices of Graph k-Concepts

In order to prove the existence of a concept lattice for each arity, it suffices to prove that the two mappings between extensional and intensional representations form a Galois connection.

Theorem 2 (Galois Connection). *Let $K = (O, A, I)$ be a graph context. For every arity k, the pair of mappings (ext, int) forms a Galois connection between $(\mathcal{R}_k, \subseteq)$ and $(\mathcal{Q}_k, \subseteq_q)$, i.e. for every object relation $R \in \mathcal{R}_k$ and PGP $Q \in \mathcal{Q}_k$,*

$$R \subseteq ext(Q) \Longleftrightarrow Q \subseteq_q int(R)$$

Proof. $R \subseteq ext(Q) \iff \forall \bar{o} \in R : \bar{o} \in ext(Q)$
$\iff \forall \bar{o} \in R : Q \subseteq_q (\bar{o}, I)$ (Definition 7)
$\iff Q \subseteq_q \cap_q \{(\bar{o}, I)\}_{\bar{o} \in R}$ (Lemma 3)
$\iff Q \subseteq_q int(R)$ (Definition 9)   ∎

Corollary 1. *From (ext, int) being a Galois connection and from $(\mathcal{R}_k, \subseteq, \cap, \cup, O^k, \emptyset)$ and $(\mathcal{Q}_k, \subseteq_q, \cap_q, \cup_q, \Omega_q, \emptyset_q)$ being lattices, we have the following propositions for every relations $R, R_1, R_2 \in \mathcal{R}_k$, and every PGP $Q, Q_1, Q_2 \in \mathcal{Q}_k$, for any arity k:*

(1a) $Q_1 \subseteq_q Q_2 \Rightarrow ext(Q_1) \supseteq ext(Q_2)$ (1b) $R_1 \subseteq R_2 \Rightarrow int(R_1) \supseteq_q int(R_2)$
(2a) $Q \subseteq_q int(ext(Q))$ (2b) $R \subseteq ext(int(R))$
(3a) $int(R) \equiv_q int(ext(int(R)))$ (3b) $ext(Q) = ext(int(ext(Q))$
(4a) $int(R_1 \cup R_2) \equiv_q int(R_1) \cap_q int(R_2)$ (4b) $ext(Q_1 \cup_q Q_2) = ext(Q_1) \cap ext(Q_2)$
(5a) $int(\emptyset) \equiv_q \Omega_q$ (5b) $ext(\emptyset_q) = O^k$

From the Galois connection, *graph concepts* can be defined and organized into concept lattices, like in classical FCA, with one concept lattice for each arity k.

Definition 11 (Graph Concept). *Let $K = (O, A, I)$ be a graph context. A graph concept of K is a pair (R, Q), made of an object relation (the extent) and a PGP (the intent), such that $R = ext(Q)$ and $Q \equiv_q int(R)$. The arity of a graph concept is the arity of its extent and intent, which have to be equal.*

Theorem 3 (Graph Concept Lattices). *The set of graph k-concepts $\mathcal{C}_k$, partially ordered by $\leq$, which is defined by $(R_1, Q_1) \leq (R_2, Q_2) : \iff R_1 \subseteq R_2 \iff Q_2 \subseteq_q Q_1$, forms a bounded lattice $(\mathcal{C}_k, \leq, \wedge, \vee, \top, \bot)$ where:*

1. $(R_1, Q_1) \wedge (R_2, Q_2) = (R_1 \cap R_2, int(ext(Q_1 \cup_q Q_2)))$,
2. $(R_1, Q_1) \vee (R_2, Q_2) = (ext(int(R_1 \cup R_2)), Q_1 \cap_q Q_2)$,
3. $\top = (O^k, int(ext(\emptyset_q)))$,
4. $\bot = (\emptyset, \Omega_q)$.

In the example context of Fig. 1, the most interesting graph concept has as an extent the set of triples (president, city, state): {(*Obama, Honolulu, Hawaii*), (*Clinton, Hope, Arkansas*)}. Its intent is the PGP ((p,c,s), {(((*USA*,p),president), ((p,c), birthPlace), ((*USA*,s),state), ((s,c),city)}). Other graph concepts are either projections of it (see Sect. 6), concepts with singleton extents (having one tuple), the top concepts (having all tuples), and the bottom concepts (having no tuple).

6   Projections Between Concept Lattices

In the previous section, we have shown the existence of a family of graph concept lattices, one for each arity $k \geq 0$. This is analogous to previous work with power context families [18], with the important difference that each k-concept has as an intent a PGP that may combine relationships of different arities. As FCA lattices are generally used as search spaces for knowledge discovery, it is useful to relate concepts from different lattices, i.e. having different arities. To that purpose, we use *projections*, a fundamental operation in relational algebra (see Sect. 2 for definitions and notations). A projection enables to permute, duplicate, and remove *columns* in relations and PGPs. Our projections differ from those of pattern structures, which are used to simplify graph patterns [9].

We first demonstrate that the set of all concept extents is closed by projection because the projection of the extension of a PGP is the extension of the projection of the PGP.

Lemma 4. *For all $Q \in \mathcal{Q}_k$, and $\pi \in \Pi_k^l$, we have: $\pi(ext(Q)) = ext(\pi(Q))$.*

Proof. $\pi(ext(Q)) = \pi(\{\overline{o} \mid \exists \sigma : \sigma(\overline{x}) = \overline{o} \wedge \sigma(P) \subseteq I\})$
$= \{\pi(\overline{o}) \mid \exists \sigma : \sigma(\overline{x}) = \overline{o} \wedge \sigma(P) \subseteq I\} = \{\overline{o}' \mid \exists \sigma : \pi(\sigma(\overline{x})) = \overline{o}' \wedge \sigma(P) \subseteq I\}$
$= \{\overline{o}' \mid \exists \sigma : \sigma(\pi(\overline{x})) = \overline{o}' \wedge \sigma(P) \subseteq I\} = ext((\pi(\overline{x}), P)) = ext(\pi(Q))$ ■

Theorem 4. *Let $\pi \in \Pi_k^l$ be a projection. For every k-concept (R, Q), $(\pi(R), int(ext(\pi(Q))))$ is a l-concept. The latter is called the π-projection of concept (R, Q), denoted by $\pi(R, Q)$.*

Proof. From Lemma 4, we have $\pi(R) = \pi(ext(Q)) = ext(\pi(Q))$, so that $\pi(R)$ is a l-concept extent. The corresponding l-concept intent is $int(\pi(R)) = int(\pi(ext(Q))) = int(ext(\pi(Q)))$.   ■

For example, let $P = \{((x), country), ((x, y), president)\}$ be a graph pattern relating a country to its president. The PGP $Q = ((x, y), P)$ returning pairs (country,president) can be projected to the PGP $((y), P)$ returning all presidents of a country, or to the PGP $((x), P)$ returning all countries having a president. In the particular case where $k = l$, the two concepts belong to the same lattice. For example, the PGP $((y, x), P)$ is a permutation of Q. Therefore, there may be up to $k!$ permutations of a single k-concept in the same concept lattice. As those permutations are equivalent from the point of view of knowledge discovery, a concept lattice could in principle be made smaller by retaining only one of the permutations. Note that a concept can sometimes be equal to some of its permutations. For example, The query $((x, y), \{((x, z), parent), ((y, z), parent)\})$, which defines the sibling relationship, has the same extension as its permutation (y, x). This equality comes from a symmetry in the PGP.

The existence of a projection between two concepts defines a pre-ordering $\leq_\pi$ on the set of all concepts $\mathcal{C} = \bigcup_{k \geq 0} \mathcal{C}_k$. Indeed, it satisfies transitivity (by composing projections), reflexivity (by using the identity projection), but not antisymmetry (consider a permutation and its inverse). Two concepts are then equivalent $(=_\pi)$ if they are a permutation one of the other. The example concept from Sect. 5.3, which contains triples (president, city, state), has 6 distinct permutations, and 2×3 projections of arity 2, and 3 projections of arity 1. The fact that all those projections have the same number of instances as the example concept reveals functional dependencies from any column to the others. For instance, the president determines the city and state. The functional dependency from state to president would be violated if two presidents in the concept extent were associated to the same state. This example suggests that the partial ordering $\leq_\pi$ can support the discovery of functional dependencies, and may generalize previous work on multi-valued contexts [1].

7   Conclusion and Future Work

We have proposed an extension of FCA, G-FCA, where objects are replaced by tuples of objects. In G-FCA, the context is a *knowledge graph*, concept intents are *projected graph patterns (PGP)*, and concept extents are *object relations*. A set-like algebra of PGPs is defined with inclusion, intersection, and union. PGP inclusion is related to graph matching, and hence to query answering. PGP intersection is related to finding common subgraphs under homomorphism, and hence to data mining and machine learning. The constructive definitions of PGP operations already allow for a direct implementation, but more efficient algorithm have to be devised for practical use. Another objective is to clarify the relationship between G-FCA and previous FCA works, notably power content families and concept graphs [18], Relational Concept Analysis [16], $\mathcal{EL}$-implication bases [3], and concept lattices of relational structures [12].

The results presented in this paper have yet a limited utility, and it remains to show how FCA applications can be transposed to G-FCA. The most common application is to compute and visualize concept lattices. The main difficulty is the huge number of graph patterns, even closed ones [19]. Restrictions can be applied to PGPs (e.g., fully connected patterns, bounded arity, graph isomorphism), but those will probably not be enough in practice due to the combinatorial explosion of graph patterns. Another common application is the discovery of implication rules. In G-FCA, this would correspond to unsupervised ILP, but limited to exact rules. Given how costly ILP is in the supervised setting, the computation of all implication rules could be a challenge. Alternately, those implications could be computed on the need, specifically for each (tuple of) object(s) to be classified [8,14]. Yet another application is to use the concept lattice as a search space for information retrieval. In fact, we have already formalized and implemented such an application [6], and it was the inspiration for the current work.

References

1. Allard, P., Ferré, S., Ridoux, O.: Discovering functional dependencies and association rules by navigating in a lattice of OLAP views. In: Kryszkiewicz, M., Obiedkov, S. (eds.) Concept Lattices and Their Applications, pp. 199–210. CEUR-WS, Sevilla (2010)

2. Baader, F., Calvanese, D., McGuinness, D.L., Nardi, D., Patel-Schneider, P.F. (eds.): The Description Logic Handbook: Theory, Implementation, and Applications. Cambridge University Press, Cambridge (2003)
3. Baader, F., Distel, F.: A finite basis for the set of EL-implications holding in a finite model. In: Medina, R., Obiedkov, S. (eds.) ICFCA 2008. LNCS (LNAI), vol. 4933, pp. 46–61. Springer, Heidelberg (2008)
4. Ceri, S., Gottlob, G., Tanca, L.: What you always wanted to know about datalog (and never dared to ask). IEEE Trans. Knowl. Data Eng. 1(1), 146–166 (1989)
5. Chein, M., Mugnier, M.L.: Graph-Based Knowledge Representation: Computational Foundations of Conceptual Graphs. AIKP. Springer, London (2008)
6. Ferré, S.: Conceptual navigation in RDF graphs with SPARQL-like queries. In: Kwuida, L., Sertkaya, B. (eds.) ICFCA 2010. LNCS, vol. 5986, pp. 193–208. Springer, Heidelberg (2010)
7. Ferré, S., Ridoux, O.: A logical generalizationof formal concept analysis. In: Ganter, B., Mineau, G.W. (eds.) ICCS 2000. LNCS, vol. 1867, pp. 371–384. Springer, Heidelberg (2000)
8. Ferré, S., Ridoux, O.: The use of associative concepts in the incremental building of a logical context. In: Priss, U., Corbett, D.R., Angelova, G. (eds.) ICCS 2002. LNCS (LNAI), vol. 2393, p. 299. Springer, Heidelberg (2002)
9. Ganter, B., Kuznetsov, S.O.: Pattern structures and their projections. In: Delugach, H.S., Stumme, G. (eds.) ICCS 2001. LNCS (LNAI), vol. 2120, p. 129. Springer, Heidelberg (2001)
10. Ganter, B., Wille, R.: Formal Concept Analysis – Mathematical Foundations. Springer, Heidelberg (1999)
11. Hitzler, P., Krötzsch, M., Rudolph, S.: Foundations of Semantic Web Technologies. Chapman & Hall/CRC, London (2009)
12. Kötters, J.: Concept lattices of a relational structure. In: Pfeiffer, H.D., Ignatov, D.I., Poelmans, J., Gadiraju, N. (eds.) ICCS 2013. LNCS, vol. 7735, pp. 301–310. Springer, Heidelberg (2013)
13. Kuznetsov, S.O., Samokhin, M.V.: Learning closed sets of labeled graphs for chemical applications. In: Kramer, S., Pfahringer, B. (eds.) ILP 2005. LNCS (LNAI), vol. 3625, pp. 190–208. Springer, Heidelberg (2005)
14. Kuznetsov, S.O.: Fitting pattern structures to knowledge discovery in big data. In: Cellier, P., Distel, F., Ganter, B. (eds.) ICFCA 2013. LNCS, vol. 7880, pp. 254–266. Springer, Heidelberg (2013)
15. Muggleton, S., Raedt, L.D.: Inductive logic programming: theory and methods. J. Logic Program. 19(20), 629–679 (1994)
16. Rouane-Hacene, M., Huchard, M., Napoli, A., Valtchev, P.: Relational concept analysis: mining concept lattices from multi-relational data. Annals Math. Artif. Intell. 67(1), 81–108 (2013)
17. Schmachtenberg, M., Bizer, C., Paulheim, H.: Adoption of the linked data best practices in different topical domains. In: Mika, P., Tudorache, T., Bernstein, A., Welty, C., Knoblock, C., Vrandečić, D., Groth, P., Noy, N., Janowicz, K., Goble, C. (eds.) ISWC 2014, Part I. LNCS, vol. 8796, pp. 245–260. Springer, Heidelberg (2014)
18. Wille, R.: Conceptual graphs and formal concept analysis. In: Lukose, D., Delugach, H., Keeler, M., Searle, L., Sowa, J. (eds.) Conceptual Structures: Fulfilling Peirce's Dream. LNCS (LNAI), vol. 1257, pp. 290–303. Springer, Springer (1997)
19. Yan, X., Han, J.: Closegraph: mining closed frequent graph patterns. In: ACM International Conference on Knowledge Discovery and Data Mining (SIGKDD). pp. 286–295. ACM (2003)

Simple Undirected Graphs as Formal Contexts

Giampiero Chiaselotti[2], Davide Ciucci[1](✉), and Tommaso Gentile[2]

[1] DISCo, University of Milano – Bicocca,
Viale Sarca 336 – U14, 20126 Milan, Italy
ciucci@disco.unimib.it

[2] Department of Mathematics and Informatics, University of Calabria,
Via Pietro Bucci, Cubo 30B, 87036 Arcavacata di Rende (CS), Italy
giampiero.chiaselotti@unical.it, gentile@mat.unical.it

Abstract. The adjacency matrix of a graph is interpreted as a formal context. Then, the counterpart of Formal Concept Analysis (FCA) tools are introduced in graph theory. Moreover, a formal context is seen as a Boolean information table, the structure at the basis of Rough Set Theory (RST). Hence, we also apply RST tools to graphs. The peculiarity of the graph case, put in evidence and studied in the paper, is that both FCA and RST are based on a (different) binary relation between objects.

1 Introduction

The aim of this work is to define a framework that enables us to apply Formal Concept Analysis (FCA) tools, and to some extent also Rough Set Theory (RST) tools, to graphs. In order to do so, we will view the adjacency matrix of a graph as a formal context (Boolean Information Table in case of RST). It is well known that RST and FCA are similar but complementary disciplines that can be integrated in several ways. A key difference between the two theories is the binary relation on which they are based, in the RST case it is a relation between objects and in the FCA case between objects and properties. However, in this particular framework the two theories are even closer, since objects coincide with attributes. The two relations remain different but they can be interpreted in the same setting, understanding their complementarity. We will consider not only the standard operators: formal concepts in FCA and lower/upper approximations in RST but a more general framework arising from the theory of oppositions [2].

The relationship between graphs and FCA is not new, however it has not yet been clearly outlined and developed. The paper [7] defines a bipartite graph from a formal context and proves that (X, Y) is a concept iff $X \cup Y$ is a maximal bi-clique of the corresponding graph. The same result is mentioned briefly in [3]. Here, we work in the other direction: starting from a general graph, we use the adjacency matrix to define a formal context. Then, we show that concepts coincide with bipartitions of the maximal bi-cliques (see Theorem 3.1). This result is also mentioned in [9], but with no formalization nor proof. We focus then on complete and complete bipartite graphs studying their concept lattice. In Sect. 3.3 some considerations on other Galois connections than the standard

© Springer International Publishing Switzerland 2015
J. Baixeries et al. (Eds.): ICFCA 2015, LNAI 9113, pp. 287–302, 2015.
DOI: 10.1007/978-3-319-19545-2_18

one are given. Section 4 is devoted to rough sets: we study the partition and the approximations that can be introduced on a given graph.

2   Preliminary Notions

The basic notions of Graph Theory, Formal Concept Analysis and Rough Set Theory are recalled.

2.1   Graphs

We denote by $G = (V(G), E(G))$ a finite simple (i.e. no loops and no multiple edges are allowed) undirected graph, with vertex set $V(G) = \{v_1, \ldots, v_n\}$ and edge set $E(G)$. If $v, v' \in V(G)$, we will write $v \sim v'$ if $\{v, v'\} \in E(G)$ and $v \nsim v'$ otherwise. We denote by $Adj(G)$ the adjacency matrix of G. We recall that $Adj(G)$ is a $n \times n$ matrix (a_{ij}) such that $a_{ij} := 1$ if $v_i \sim v_j$ and $a_{ij} := 0$ otherwise. If $v \in V(G)$, we set

$$N_G(v) := \{w \in V(G) : \{v, w\} \in E(G)\}.$$

$N_G(v)$ is usually called *neighborhood* of v in G. Graphs of particular interest for our discussion will be complete and bipartite ones.

Definition 2.1. *The* complete graph *on n vertices, denoted by K_n, is the graph with vertex set $\{v_1, \ldots, v_n\}$ and such that $\{v_i, v_j\}$ is an edge, for each pair of indexes $i \neq j$.*

Definition 2.2. *A graph $B = (V(B), E(B))$ is said* bipartite *if there exist two non-empty subsets B_1 and B_2 of $V(B)$ such that $B_1 \cap B_2 = \emptyset$, $B_1 \cup B_2 = V(B)$ and $E(B) \subseteq \{\{x, y\} : x \in B_1, y \in B_2\}$. In this case the pair (B_1, B_2) is called a* bipartition *of B and we write $B = (B_1 | B_2)$. It is said that $B = (B_1 | B_2)$ is a* complete bipartite graph *if $E(B) = \{\{x, y\} : x \in B_1, y \in B_2\}$. If p and q are two positive integers and $B_1 = \{x_1, \ldots, x_p\}$, $B_2 = \{y_1, \ldots, y_q\}$, we denote by $K_{p,q}$ the complete bipartite graph having bipartition (B_1, B_2).*

Definition 2.3. *A* biclique *B of G is a complete bipartite subgraph of G. We say that a biclique $B = (B_1 | B_2)$ of G is* maximal *if for any biclique $B' = (B_1' | B_2')$ of G such that $B_1 \subseteq B_1'$ and $B_2 \subseteq B_2'$ it results that $B_1 = B_1'$ and $B_2 = B_2'$.*

2.2   Formal Concept Analysis

We start by recalling the general definition of formal contexts and their basic properties (see [6]).

Definition 2.4. *A* Formal Context *is a triple $\mathbb{K} = (Z, M, \mathcal{R})$, where Z and M are sets and $\mathcal{R} \subseteq Z \times M$ is the binary relation involving them. The elements of*

Z and M are called objects and attributes (or properties) respectively. We write $g\mathcal{R}m$ instead of $(g,m) \in \mathcal{R}$. If $O \subseteq Z$ and $Q \subseteq M$, we set

$$O^\uparrow := \{m \in M : (\forall g \in O)\, g\mathcal{R}m\} \subseteq M$$

and

$$Q^\downarrow := \{g \in Z : (\forall m \in Q)\, g\mathcal{R}m\} \subseteq Z.$$

In this way the following two mappings are defined: $^\uparrow : \mathcal{P}(Z) \to \mathcal{P}(M)$, $O \mapsto O^\uparrow$ and $^\downarrow : \mathcal{P}(M) \to \mathcal{P}(Z)$, $Q \mapsto Q^\downarrow$. By suitable compositions of these two mappings we are able to construct the two new mappings $^* : \mathcal{P}(Z) \to \mathcal{P}(Z)$, $O \mapsto O^{\uparrow\downarrow}$ and $^\circ : \mathcal{P}(M) \to \mathcal{P}(M)$, $Q \mapsto Q^{\downarrow\uparrow}$, which are closure operators on, respectively, $\mathcal{P}(Z)$ and $\mathcal{P}(M)$ [6].

Definition 2.5. *A* concept *of the Formal Context* $\mathbb{K} = (Z, M, \mathcal{R})$ *is a pair* (O, Q), *where* $O \subseteq Z$, $Q \subseteq M$, $O^\uparrow = Q$ *and* $Q^\downarrow = O$. *If* (O,Q) *is a concept, O is called* extent *of* (O, Q) *and Q is called* intent *of* (O, Q). *We denote by* $\mathfrak{B}(\mathbb{K})$ *the set of all the concepts of the Formal Context* $\mathbb{K}$.

If (O_1, Q_1) and (O_2, Q_2) are two concepts in $\mathfrak{B}(\mathbb{K})$, it is usual to consider the relation $(O_1, Q_1) \sqsubseteq (O_2, Q_2)$ if and only if $O_1 \subseteq O_2$ (that is equivalent to $Q_1 \supseteq Q_2$). Then $\sqsubseteq$ is a partial order on $\mathfrak{B}(\mathbb{K})$ and $(\mathfrak{B}(\mathbb{K}), \sqsubseteq)$ is a complete lattice, called *concept lattice* (or also *Galois lattice*) of the Formal Context $\mathbb{K}$, whose meet and join operations on an arbitrary family of formal concepts $\{(O_\alpha, Q_\alpha) : \alpha \in A\}$ are the following:

$$\bigwedge_{\alpha \in A} (O_\alpha, Q_\alpha) = \left(\bigcap_{\alpha \in A} O_\alpha, \left(\bigcup_{\alpha \in A} Q_\alpha \right)^\circ \right)$$

$$\bigvee_{\alpha \in A} (O_\alpha, Q_\alpha) = \left(\left(\bigcup_{\alpha \in A} O_\alpha \right)^*, \bigcap_{\alpha \in A} Q_\alpha \right)$$

2.3 Rough Set Theory

In the context of RST a table representing a formal context is named Boolean information table (or Boolean information system) [10]. More formally, a *Boolean information table* is a structure $\mathcal{I} = \langle U, Att, Val, F \rangle$, where U (called *universe set*) is a non empty set of *objects*, Att (called *attribute set*) is a non empty set of *attributes*, $Val = \{0, 1\}$ is called the *value set* (in the general case it is not assumed to be Boolean) and $F : U \times Att \to Val$ (called *information map*) is an application from the direct product $U \times Att$ into the value set Val.

If $A \subseteq Att$, it is usual to consider the binary relation I_A on the universe set U defined as follows: if $u, u' \in U$ then

$$u I_A u' \iff F(a, u) = F(a, u'), \forall a \in A. \tag{1}$$

The binary relation I_A is an equivalence relation on U and it is called *A-indiscernibility relation*. If $u \in U$, we denote by $[u]_A$ the equivalence class of u with respect to I_A. We also set $\pi_A(\mathcal{I}) := \{[u]_A : u \in U\}$ and we call $\pi_A(\mathcal{I})$ the A-indiscernibility partition of the information system $\mathcal{I}$.

Definition 2.6. *Let* $\mathcal{I} = \langle U, Att, Val, F \rangle$ *be an information table,* $A \subseteq Att$ *and* $Y \subseteq U$. *The* A-*lower approximation of* Y *is the following subset of* U:

$$\mathbf{l}_A(Y) := \{x \in U : [x]_A \subseteq Y\} = \bigcup \{C \in \pi_A(\mathcal{I}) : C \subseteq Y\}.$$

The A-*upper approximation of* Y *is defined as:*

$$\mathbf{u}_A(Y) := \{x \in U : [x]_A \cap Y \neq \emptyset\} = \bigcup \{C \in \pi_A(\mathcal{I}) : C \cap Y \neq \emptyset\}.$$

The subset Y *is called* A-*exact if and only if* $\mathbf{l}_A(Y) = \mathbf{u}_A(Y)$ *and* A-*rough otherwise.*

The lower approximation represents the elements that *certainly*, with respect to our knowledge expressed by A, belongs to Y. On the other hand, the upper approximation is the set of objects *possibly* belonging to A.

We will denote by $\mathbb{CO}_A(\mathcal{I})$ the set of all the A-exact subsets. The following result is well known (where $\hat{s} = \{1, 2, \ldots s\}$).

Proposition 2.1. *(i) If* $\pi_A(\mathcal{I})$ *contains exactly* s *elements (i.e. equivalence classes), then* $\mathbb{CO}_A(\mathcal{I})$ *is a Boolean algebra isomorphic to* $\langle \mathcal{P}(\hat{s}), \subseteq, \cap, \cup, {}^c, \emptyset, \hat{s} \rangle$. *(ii) More specifically, a non-empty subset* Y *of the universe* U *is* A-*exact if and only if* Y *is a set theoretical union of blocks of the set-partition* $\pi_A(\mathcal{I})$.

2.4 The Cube of Oppositions

Starting from a binary relation $R \subseteq X \times Y$ and generalizing the Aristotelian square of oppositions, it is possible to define a cube of oppositions [5]. Given a subset $S \subseteq Y$, the eight vertices of the cube are defined by $R(S) = \{x \in X | \exists s \in S, xRs\}$ and all the interaction of three kinds of negation: the complement on X, on Y and the negation of the relation R. More in detail, let us assume that R and its negation $\overline{R}$ ($x\overline{R}y$ if and only if $\neg(xRy)$) are both not empty and serial, and define $xR = \{y \in Y | xRy\}$. Then, we can obtain from R four vertices, that form a classical square of oppositions (in what follows $\overline{S} := Y \setminus S$):

(I) $R(S) = \{x \in X | \exists s \in S, xRs\} = \{x \in X | S \cap xR \neq \emptyset\}$
(O) $\underline{R(\overline{S})} = \{x \in X | \exists s \in \overline{S}, xRs\}$
(E) $\overline{R(S)} = \{x \in X | \forall s \in S, \neg(xRs)\}$
(A) $\overline{R(\overline{S})} = \{x \in X | \forall s \in \overline{S}, \neg(xRs)\} = \{x \in X | xR \subseteq S\}$

We remark that E and A are the complement of I and O, respectively, and that A is a subset of I and E a subset of O. The other four corners are obtained using the complementary relation $\overline{R}$:

(o) $\overline{R}(S) = \{x \in X | \exists s \in S, \neg(xRs)\}$
(i) $\underline{\overline{R}(\overline{S})} = \{x \in X | \exists s \in \overline{S}, \neg(xRs)\} = \{x \in X | S \cup xR \neq Y\}$
(a) $\overline{\overline{R}(S)} = \{x \in X | \forall s \in S, xRs\} = \{x \in X | S \subseteq xR\}$
(e) $\overline{\overline{R}(\overline{S})} = \{x \in X | \forall s \in \overline{S}, xRs\}$

All these sets can have a nice interpretation both in FCA and RST [2]. In the case of FCA, R is the standard relation $\mathcal{R}$ defining a formal context, and in the case of RST, R is the indiscernibility relation I_A, hence it is defined on the same domain $X \times X$. As we will discuss, in our particular case, also for FCA we have $X = Y$, so both relations are defined on $X \times X$ even if they are not the same relation.

3   Simple Undirected Graphs Viewed as Formal Contexts

We begin now the study of the finite simple undirected graphs as particular types of formal contexts.

Definition 3.1. *Let $G = (V(G), E(G))$ be a finite simple undirected graph, with vertex set $V(G) = \{v_1, \ldots, v_n\}$ and edge set $E(G)$. We call* Formal Context of the graph G *the Formal Context $\mathbb{K}[G] := (V(G), V(G), \mathcal{R}_G)$, where $v\mathcal{R}_G v'$ if and only if $\{v, v'\} \in E(G)$ for all $v, v' \in V(G)$.*

Hence the object subset and the attribute subset of the Formal Context $\mathbb{K}[G]$ are both equal to the vertex set $V(G)$, whereas the binary relation which defines this formal context is exactly the incidence relation between vertices of the graph G. Let us also note that, since the graph G is undirected, the relation $\mathcal{R}_G$ is symmetric.

3.1   Concepts of a Graph

Given the above considerations, in the Formal Context $\mathbb{K}[G]$ induced by a simple undirected graph G, the maps $^{\uparrow} : \mathcal{P}(V(G)) \rightarrow \mathcal{P}(V(G))$ and $^{\downarrow} : \mathcal{P}(V(G)) \rightarrow \mathcal{P}(V(G))$ are coincident. Therefore in the sequel we denote with the same symbol $'$ the map $' : \mathcal{P}(V(G)) \rightarrow \mathcal{P}(V(G))$ such that $O \mapsto O' := O^{\uparrow} = O^{\downarrow}$, when O is any vertex subset of G. This implies obviously that also the two operators $^* : \mathcal{P}(V(G)) \rightarrow \mathcal{P}(V(G))$ and $^{\circ} : \mathcal{P}(V(G)) \rightarrow \mathcal{P}(V(G))$ coincide. Therefore in the sequel we set $O \mapsto O'' := O^* = O^{\circ}$, for all $O \subseteq V(G)$.

Let us now see how O' and O'' are defined in terms of neighborhood of vertices.

Proposition 3.1. *If $O \subseteq V(G)$ then*

$$O' = \bigcap_{v \in O} N_G(v) = \{w \in V(G) : O \subseteq N_G(w)\} \qquad (2)$$

Proof. We have that

$$O' := \{w \in V(G) : (\forall v \in O)v\mathcal{R}_G w\} = \{w \in V(G) : (\forall v \in O)w \in N_G(v)\},$$

that is $O' = \bigcap_{v \in O} N_G(v)$. For the other set equality, if $w \in V(G)$ is such that $O \subseteq N_G(w)$ and $v \in O$, then $w \in N_G(w)$, therefore $w \in \bigcap_{v \in O} N_G(v)$. On the other hand, if $w \in \bigcap_{v \in O} N_G(v)$ and v_0 is an arbitrary vertex in O then $w \in N_G(v_0)$. Hence $v_0 \in N_G(w)$, and this shows that $O \subseteq N_G(w)$. $\square$

Remark 3.1. The identity in (2) is valid also when the subset $O = \emptyset$. In fact, in this case, we always have $\emptyset^\uparrow = M$ and $\emptyset^\downarrow = Z$, that is $O' = O^\uparrow = O^\downarrow = V(G)$ in the formal context $\mathbb{K}[G]$. On the other hand, it is usual (in elementary set theory) to interpret the intersection $\bigcap_{v \in O} N_G(v)$ as coincident with the whole set $V(G)$ when O is the empty set.

Corollary 3.1. *If $O \subseteq V(G)$ then*

$$O' \subseteq V(G) \setminus O \tag{3}$$

Proof. If $w \in O'$ and $w \in O$, by (2) it follows that $w \in N_G(w)$, i.e. $\{w, w\} \in E(G)$, but this contradicts the hypothesis that G is a simple graph. This proves (3). □

Remark 3.2. If G is a finite simple undirected graph, a vertex subset $O \subseteq V(G)$ is the extent [intent] of some concept of the Formal Context $\mathbb{K}[G]$ if and only if $O'' = O$. In this case, both the pairs (O, O') and (O', O) are concepts of $\mathbb{K}[G]$. Moreover, since G has no loops, the cross table of the Formal Context $\mathbb{K}[G]$ (that is, the adjacency matrix of G) has zeroes in all its diagonal places, and this obviously implies that $V(G)' = \emptyset$. Hence both the pairs $(\emptyset, V(G))$ and $(V(G), \emptyset)$ are always concepts of the Formal Context $\mathbb{K}[G]$.

We re-interpret now the notion of concept in the case of the formal context $\mathbb{K}[G]$. Recalling the definition of biclique of a graph (see Definition 2.3), we have then the following characterization.

Theorem 3.1. *Let O and Q be two subsets of $V(G)$. Then, the pair (O, Q) is a concept in $\mathbb{K}[G]$ if and only if (O, Q) is a bipartition of some maximal biclique of G. On the other hand, if $B = (B_1 | B_2)$ is a maximal biclique of G, then the pair (B_1, B_2) is a concept in $\mathbb{K}[G]$. Hence the concepts in $\mathbb{K}[G]$ are exactly the bipartitions of the maximal bicliques of G.*

Proof. Let (O, Q) be a concept in $\mathbb{K}[G]$. By definition of concept we have in this case that:

$$O' := \{v \in V(G) : (\forall u \in O)\, u \sim v\} = Q$$

and

$$Q' := \{u \in V(G) : (\forall v \in Q)\, u \sim v\} = O.$$

Since G has no loops, the subsets O and Q are disjoint. Moreover, if $u \in O$ and $v \in Q$, then $u \sim v$. Thus $(O | Q)$ is a biclique of G.

Let $B = (B_1 | B_2)$ be a biclique of G such that $O \subseteq B_1$ and $Q \subseteq B_2$. By definition of bipartite graph, if $u \in O \subseteq B_1$ and $v \in B_2$, $u \sim v$, then $B_2 \subseteq O' = Q$ and thus $B_2 = Q$. Similarly if $v \in Q \subseteq B_2$ and $u \in B_1$, $u \sim v$, then $B_1 \subseteq Q' = O$ and thus $B_1 = O$. It follows that $(O | Q)$ is a maximal biclique of G.

Let now $(O | Q)$ be a maximal biclique of G. Then, by definition of biclique, $Q \subseteq O'$ and $O' \subseteq Q$. Moreover we have:

$$O'' := \{u \in V(G) : (\forall v \in O')\, u \sim v\}.$$

It follows that $O \subseteq O''$ and that $(O''|O')$ is a biclique of G. Then, by maximality of $(O|Q)$, we obtain that $Q = O'$ and $O = O'' = Q'$, so (O, Q) is a concept in $\mathbb{K}[G]$. □

One of the consequences of this result is the possibility to apply algorithms developed for formal concept generation [8] to improve results to compute maximal bicliques on graphs [1]. We leave this comparison to a future study.

When $G = K_n$ is a complete graph, the context coincide with the contra-nominal scale $(V(G), V(G), \neq)$ [6], hence we obtain that the map $'$ behaves as the set complement.

Proposition 3.2. *If $G = K_n$ and $O \subseteq V(G)$ we have that $O' = V(G) \setminus O$ and $O'' = O$.*

In the case of a complete bipartite graph $G = K_{p,q}$ we obtain:

Proposition 3.3. *If $G = K_{p,q} = (B_1|B_2)$ and O is a non-empty subset of $V(G)$ then*

$$O' = \begin{cases} B_1 & \text{if } O \subseteq B_2 \\ B_2 & \text{if } O \subseteq B_1 \\ \emptyset & \text{otherwise} \end{cases} \tag{4}$$

and

$$O'' = \begin{cases} B_1 & \text{if } O \subseteq B_1 \\ B_2 & \text{if } O \subseteq B_2 \\ V(G) & \text{otherwise} \end{cases} \tag{5}$$

Proof. Let $B_1 = \{x_1, \ldots, x_p\}$ and $B_2 = \{y_1, \ldots, y_q\}$. By definition of $K_{p,q}$ we have that $N_G(x_i) = B_2$ for $i = 1, \ldots, p$ and $N_G(y_j) = B_1$ for $j = 1, \ldots, q$. Therefore, if $O \subseteq B_2$, then $\bigcap_{v \in O} N_G(v) = B_1$, hence $O' = B_1$ by (2). Analogously if $O \subseteq B_1$. Finally, we assume that $x_i \in O$, for some $i = 1, \ldots, p$, and also $y_j \in O$, for some $j = 1, \ldots, q$. Then, by (2) it follows that $O' \subseteq N_G(x_i) \cap N_G(y_j) = B_2 \cap B_1 = \emptyset$ since $B_1|B_2$ is a set-partition of the vertex set of G. This proves (4). On the other hand, if $O \subseteq B_1$, by (4) we deduce that $O' = B_2$, therefore $O'' = (O')' = B_2' = B_1$ again by (4). Analogously, we obtain $O'' = B_2$ if $O \subseteq B_2$. Finally, if $O \cap B_1 \neq \emptyset$ and $O \cap B_2 \neq \emptyset$, by (4) we have that $O' = \emptyset$, hence $O'' = (\emptyset)' = V(G)$. This proves (5). □

3.2 The Concept Lattice of a Graph

We explicitly introduce now the notion of concept lattice for a finite simple undirected graph.

Definition 3.2. *We call* concept lattice *(or also* Galois lattice*) of the graph G the concept lattice of the Formal Context $\mathbb{K}[G]$ and we denote it simply by $(\mathfrak{B}(G), \sqsubseteq)$ instead of $(\mathfrak{B}(\mathbb{K}[G]), \sqsubseteq)$.*

At first let us recall some basic notions about posets. If $P = (X, \leq)$ is a partially ordered set (briefly *poset*), we can consider the usual *dual poset* of P, that is the poset $P^* = (X, \leq^*)$, where $\leq^*$ is the partial order on X defined by $x \leq^* y : \iff y \leq x$, for all $x, y \in X$. A poset $P = (X_1, \leq_1)$ is said *isomorphic* to another poset $P_2 = (X_2, \leq_2)$ if there exists a bijective map $\phi : X_1 \to X_2$ such that $x \leq_1 y \iff \phi(x) \leq_2 \phi(y)$, for all $x, y \in X_1$. A poset P is called *self-dual* if P is isomorphic to its dual poset P^*.

Then, the following basic result about concept lattices of a graph holds.

Proposition 3.4. *Let G be a finite simple undirected graph. Then the concept lattice $(\mathfrak{B}(G), \sqsubseteq)$ is self-dual.*

Proof. By Remark 3.2 we know that a pair $(O, Q) \in \mathcal{P}(V(G)) \times \mathcal{P}(V(G))$ is a concept if and only if also (Q, O) is a concept, that is, $(O, Q) \in \mathfrak{B}(G)$ if and only if $(Q, O) \in \mathfrak{B}(G)$. We define then the map $\phi : \mathfrak{B}(G) \to \mathfrak{B}(G)$ such that $\phi((O, Q)) := (Q, O)$. Obviously the map ϕ is surjective, therefore, since the set $\mathfrak{B}(G)$ is finite, it is also bijective. Finally, if (O_1, Q_1) and (O_2, Q_2) are any two concepts in $\mathfrak{B}(G)$, by definition of the partial order $\sqsubseteq$ and definition of dual order $\sqsubseteq^*$ we have that

$$(O_1, Q_1) \sqsubseteq (O_2, Q_2) \iff (Q_2, O_2) \sqsubseteq (Q_1, O_1) \iff \phi((O_1, Q_1)) \sqsubseteq^* \phi((O_2, Q_2))$$

Hence the map ϕ is an order-isomorphism between the concept lattice $(\mathfrak{B}(G), \sqsubseteq)$ and its dual lattice $(\mathfrak{B}(G), \sqsubseteq^*)$. $\square$

In the next result we determine the concept lattice when G is the complete graph K_n.

Proposition 3.5. *If $n \geq 1$ then $\mathfrak{B}(K_n) = \{(O, O^c) : O \subseteq V(K_n)\}$ and $(\mathfrak{B}(K_n), \sqsubseteq) \cong (\mathcal{P}(V(K_n)), \subseteq)$.*

Proof. It is a consequence of the equivalence of $\mathbb{K}(K_n)$ with the contranominal scale [6]. $\square$

For the complete bipartite graph we have the following result.

Proposition 3.6. *Let $K_{p,q} = (B_1 | B_2)$ and $V = V(K_{p,q})$. Then*

$$\mathfrak{B}(K_{p,q}) = \{(\emptyset, V), (B_1, B_2), (B_2, B_1), (V, \emptyset)\} \tag{6}$$

and the Hasse diagram of the concept lattice $(\mathfrak{B}(K_{p,q}), \sqsubseteq)$ is the following:

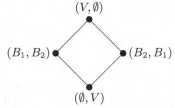

Hence $(\mathfrak{B}(K_n), \sqsubseteq) \cong (\mathcal{P}(\hat{2}), \subseteq)$.

Proof. By Remark 3.2, a concept of $\mathfrak{B}(K_{p,q})$ is a pair (O, O'), where $O'' = O$. Therefore, by (5) we deduce that the unique concepts of $\mathfrak{B}(K_{p,q})$ are $(\emptyset, V)$, $(B_1, B_2), (B_2, B_1), (V, \emptyset)$. This proves (6). Finally, by definition of the partial order $\sqsubseteq$ we immediately deduce that the Hasse diagram of $(\mathfrak{B}(K_{p,q}), \sqsubseteq)$ is that given above. $\qquad\square$

3.3   Other Operations in FCA

The operation $'$ is one of the four operations that can be introduced in FCA in analogy with possibility theory [4]. These four operations generate the sets A,I,a,i defined in Sect. 2.4 (the other four are just their complement). In the particular case of formal contexts induced by graphs, they read as:

- $R^{\Delta}(O) := \overline{\overline{R(O)}} = O'$;
- $R^{\nabla}(O) := \overline{R(\overline{O})} = \{v \in V | N_G(v) \cup O \neq V\}$ the set of vertices that are missing at least a link outside O;
- $R^{\Pi}(O) := R(O) = \{v \in V | N_G(v) \cap O \neq \emptyset\}$ the set of vertices connected with at least one vertex in O;
- $R^N(O) := \overline{R(\overline{O})} = \{v \in V | N_G(v) \subseteq O\}$ the set of vertices connected with no vertex outside O.

As discussed above, the Galois connection induced by R^{Δ} is of particular interest in the case of graphs. The interpretation of the Galois connections induced by the other operations in terms of graphs is not so easy. In [7], the Galois connection induced by R^{Π} is nicely interpreted in terms of maximal connected components. However, this result can be hardly translated to our framework (let us remark that the graph in [7] is obtained from a given formal context, we operate in the other direction). The problem lies in the fact that X and $Y = R^{\Pi}(X)$ are generally not disjoint hence they do not form a bipartition of $X \cup Y$ as it happens in [7]. More constraints needs to be considered on the starting graph in order to have some geometrical interpretation of this kind of operator. We deserve this issue to a further investigation.

Finally, let us notice that as an easy consequence of the definitions of R^{Π} and $N_G(v)$, R^{Π} can be expressed in terms of neighborhoods as

$$O' \subseteq R^{\Pi}(O) = \bigcup_{v \in O} N_G(v) \tag{7}$$

4   Simple Graphs as Boolean Information Tables

Analogously to the formal context case, the adjacency matrix of a graph G can be interpreted as a Boolean information table $\mathcal{I}[G]$, where the universe set and the attribute set are both V and the information map is defined as $F(v_i, v_j) := 1$ if $v_i \sim v_j$ and $F(v_i, v_j) := 0$ otherwise.

The equivalence relation I_A (where A is a set of verteces) is in relation with the notion of neighborhood as can be seen in the following theorem.

Theorem 4.1. *Let $A \subseteq V(G)$ and $v, v' \in V(G)$. The following conditions are equivalent:*

(i) $v I_A v'$.
(ii) For all $z \in A$ it results that $v \sim z$ if and only if $v' \sim z$.
(iii) $N_G(v) \cap A = N_G(v') \cap A$.

Proof. $(i) \implies (ii)$: Let $z \in A$ and $v \sim v'$, we show that $v' \sim z$. By (i) we have that $F(v, a) = F(v', a)$ for all $a \in A$, therefore $F(v, z) = F(v', z)$. Since $v \sim z$ it follows that $F(v, z) = 1$, and hence also $F(v', z) = 1$, that is $v' \sim z$. By symmetry of the relation I_A, if we assume that $v' \sim z$, we obtain $v \sim z$. This proves (ii)

$(ii) \implies (iii)$: By symmetry of the condition (ii), it is sufficient to prove that $N_G(v) \cap A \subseteq N_G(v') \cap A$. Let therefore $z \in N_G(v) \cap A$, then $v \sim z$ and $z \in A$. By (ii) we have then that $v' \sim z$, that is $z \in N_G(v')$. Hence $z \in N_G(v') \cap A$.

$(iii) \implies (i)$: Let $a \in A$. We show that $F(v, a) = F(v', a)$. Let us note that

$$F(v, a) = F(v', a) \iff (v \sim a \iff v' \sim a). \tag{8}$$

Then, if $v \sim a$, we have that $a \in N_G(v) \cap A =$(by (iii))$= N_G(v') \cap A$, hence $a \in N_G(v')$, that is $v' \sim a$. Analogously, by symmetry of (iii), if $v' \sim a$ then $v \sim a$. By (8) we deduce therefore that $F(v, a) = F(v', a)$. Since $a \in A$ is arbitrary, this proves (i). □

Corollary 4.1. *If $v \in V(G)$ and $A \subseteq V(G)$, then $[v]_A = \{v' : N_G(v) \cap A = N_G(v') \cap A\}$.*

That is two vertices are equivalent if they have the same neighborhood (relatively to A). The Theorem 4.1 also provides a sufficient condition for two vertices of the graph to have no common edges.

Corollary 4.2. *If $v I_A v'$ and $\{v, v'\} \cap A \neq \emptyset$, then $v \nsim v'$.*

Proof. It follows directly by Theorem 4.1 because there are no loops into G. □

4.1 The Partitions of a Graph

Now, we turn our attention to the partition generated by the relation I_A on complete and bipartite graphs. Let us start with an example.

Example 4.1. Let us consider now the complete graph K_4 and the corresponding information table in Fig. 1.

In this case we can easily compute all the set partitions $\pi_A(K_4)$, where $A \subseteq \{1, 2, 3, 4\}$. Once denoted a partition $\pi_A = X_1 | \cdots | X_n$ with X_i the equivalence classes induced by I_A, we have :

$\pi_\emptyset = 1234$, $\pi_{\{1\}} = 1|234$, $\pi_{\{2\}} = 2|134$, $\pi_{\{3\}} = 3|124$, $\pi_{\{4\}} = 4|123$, $\pi_{\{1,2\}} = 1|2|34$, $\pi_{\{1,3\}} = 1|3|24$, $\pi_{\{1,4\}} = 1|4|23$, $\pi_{\{2,3\}} = 14|2|3$, $\pi_{\{2,4\}} = 13|2|4$, $\pi_{\{3,4\}} = 12|3|4$, $\pi_{\{1,2,3\}} = \pi_{\{1,2,4\}} = \pi_{\{1,3,4\}} = \pi_{\{2,3,4\}} = \pi_{\{1,2,3,4\}} = 1|2|3|4$.

Fig. 1. The complete graph K_4.

As the previous example suggests, we can determine the general form of any partition $\pi_A(K_n)$, for all $n \geq 1$ and all $A \subseteq V(K_n)$.

Proposition 4.1. *Let $n \geq 1$ and let $A = \{w_1, \ldots, w_k\}$ be a subset of $V(K_n) = \{v_1, \ldots, v_n\}$. Then*

$$\pi_A(K_n) = w_1|w_2|\ldots|w_k|A^c, \tag{9}$$

where A^c is the complementary subset of A in $V(K_n)$.

Proof. Let $v, v' \in V(K_n)$, with $v \neq v'$. By Corollary 4.2, since $v \sim v'$, it holds that if vI_Av', then $v, v' \in A^c$. On the other hand, if $v, v' \in A^c$, then $\forall z \in A$, $F(z, v) = F(z, v') = 1$, namely vI_Av'. The proposition is proved. $\square$

Example 4.2. Let us consider now the complete graph $K_{3,4}$ in Fig. 2.

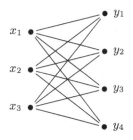

	x_1	x_2	x_3	y_1	y_2	y_3	y_4
x_1	0	0	0	1	1	1	1
x_2	0	0	0	1	1	1	1
x_3	0	0	0	1	1	1	1
y_1	1	1	1	0	0	0	0
y_2	1	1	1	0	0	0	0
y_3	1	1	1	0	0	0	0
y_4	1	1	1	0	0	0	0

Fig. 2. The graph $K_{3,4}$ and the corresponding information table.

It is easy to verify then that in this case we have only two possibilities: $\pi_\emptyset = x_1x_2x_3y_1y_2y_3y_4$ and $\pi_A = x_1x_2x_3|y_1y_2y_3y_4$ if $A \neq \emptyset$.

Also in this case we can generalize the previous example to any complete bipartite graph.

Proposition 4.2. *Let p and q be two positive integers. Let $K_{p,q} = (B_1|B_2)$, where $B_1 = \{x_1, \ldots, x_p\}$ and $B_2 = \{y_1, \ldots, y_q\}$. Then $\pi_A(K_{p,q}) = x_1 \ldots x_p|y_1 \ldots y_q$ for each subset $A \subseteq V(K_{p,q})$ such that $A \neq \emptyset$.*

Proof. Let $A \subseteq V(G)$ be a non-empty subset of $V(G)$ and let v, $v' \in V(G)$. If v, $v' \in B_1$ or v, $v' \in B_2$, then for each $z \in A$ we have $F(z,v) = F(z,v')$, so vI_Av'. If $v \in B_1$ and $v' \in B_2$, then for each $z \in A$ we have $F(z,v) \neq F(z,v')$, so $\neg(vI_Av')$. Thus $\pi_A(G) = B_1|B_2$.   □

4.2   Upper and Lower Approximations

In this section we provide some results and discussion on rough set approximations, at first in the general graph case and, then, in the case of complete and bipartite graphs.

Proposition 4.3. *Let $G = (V(G), E(G))$ be a simple undirected graph and let $\mathcal{I}[G]$ be the Boolean information system associated to G. Let A and Y be two subsets of $V(G)$. Then:*

(i) $\mathbf{l}_A(Y) = \{v \in V(G) : (u \in V(G) \wedge N_G(u) \cap A = N_G(v) \cap A) \Longrightarrow u \in Y\}$.
(ii) $\mathbf{u}_A(Y) = \{v \in V(G) : \exists u \in Y : N_G(u) \cap A = N_G(v) \cap A\}$.

Proof. It follows directly by (iii) of Theorem 4.1 and the definitions of the approximations.   □

The lower approximation of a set of vertices Y represents a subset of Y such that there are no elements outside Y with the same connections of any vertex in $\mathbf{l}_A(Y)$ (relatively to A). The upper approximation of Y is the set of vertices with the same connections (w.r.t. A) of at least one element in Y.

We study now the cases of complete $G = K_n$ and bipartite $G = K_{p,q}$ graphs.

Proposition 4.4. *Let $G = K_n$ be the complete graph on n vertices and let A and Y be two subsets of $V(G) = \{v_1, \ldots, v_n\}$. Then:*

(i) the A-lower approximation of Y is

$$\mathbf{l}_A(Y) = \begin{cases} Y \cup A^c & \text{if } A^c \subseteq Y \\ A \cap Y & \text{otherwise .} \end{cases}$$

(ii) The A-upper approximation of Y is

$$\mathbf{u}_A(Y) = \begin{cases} Y & \text{if } Y \subseteq A \\ Y \cup A^c & \text{otherwise} \end{cases}$$

(iii) Y is A-exact if and only if $Y \subseteq A$ or $A^c \subseteq Y$.

Proof. In this proof we denote $V(G)$ simply by V. If $v \in V$, by definition of K_n we have $N_G(v) = V \setminus \{v\}$, therefore $N_G(v) \cap A = A \setminus \{v\}$. By Corollary 4.1 we obtain then $[v]_A = \{v' \in V : A \setminus \{v\} = A \setminus \{v'\}\}$, hence

$$[v]_A = \begin{cases} \{v\} & \text{if } v \in A \\ A^c & \text{otherwise .} \end{cases} \tag{10}$$

By definition of A-lower approximation of Y and by (10) we have then

$$l_A(Y) = \{v \in V : (v \in A \Longrightarrow v \in Y) \vee (v \in A^c \Longrightarrow A^c \subseteq Y)\}. \quad (11)$$

It is immediate to note then that (11) is equivalent to (i). This proves (i). By definition of A-upper approximation of Y and by (10) we have then

$$\mathbf{u}_A(Y) = \{v \in V : (v \in A \Longrightarrow v \in Y) \vee (v \in A^c \Longrightarrow A^c \cap Y \neq \emptyset)\}. \quad (12)$$

It is immediate to note then that (12) is equivalent to (ii). This proves (ii). In order to prove (iii), if $Y \subseteq A$ then $A^c \not\subseteq Y$, therefore $\mathbf{u}_A(Y) = Y$ by (ii) and $l_A(Y) = A \cap Y = Y$ by (i), hence Y is A-exact. If $A^c \subseteq Y$ and $A^c \neq \emptyset$ then $Y \not\subseteq A$ therefore $l_A(Y) = (A \cap Y) \cup A^c$ by (i) and $\mathbf{u}_A(Y) = (A \cap Y) \cup A^c$ by (ii), hence Y is A-exact. If $A^c = \emptyset$ then $A = V(G)$, therefore $l_A(Y) = Y$ by (i) and $\mathbf{u}_A(Y) = Y$ by (ii), hence Y is A-exact. On the other hand, if $Y \not\subseteq A$ and $A^c \not\subseteq Y$, then $A^c \neq \emptyset$ and $\mathbf{u}_A(Y) = (A \cap Y) \cup A^c$ by (ii), $l_A(Y) = A \cap Y$ by (ii). Since $A^c \neq \emptyset$, we obtain then $l_A(Y) \neq \mathbf{u}_A(Y)$, hence Y is not A-exact. This proves (iii). $\qquad\square$

We now examine for $K_{p,q}$ the results similar to those described previously for K_n.

Proposition 4.5. *Let $K_{p,q} = (B_1|B_2)$, where $B_1 = \{x_1, \ldots, x_p\}$ and $B_2 = \{y_1, \ldots, y_q\}$. Let A and Y be two non-empty subsets of $V = V(K_{p,q})$ such that $Y \neq V$. Then:*

(i) the A-lower approximation of Y is

$$l_A(Y) = \begin{cases} B_1 & \text{if } B_1 \subseteq Y \\ B_2 & \text{if } B_2 \subseteq Y \\ \emptyset & \text{otherwise .} \end{cases}$$

(ii) The A-upper approximation of Y is

$$\mathbf{u}_A(Y) = \begin{cases} B_1 & \text{if } Y \subseteq B_1 \\ B_2 & \text{if } Y \subseteq B_2 \\ V & \text{otherwise .} \end{cases}$$

(iii) Y is A-exact if and only if $Y = B_1$ or $Y = B_2$.

Proof

(i) Let $B_1 \subseteq Y$. If $x \in B_1$, by Proposition 4.2 follows that $[x]_A = B_1 \subseteq Y$, therefore, by definition of $l_A(Y)$, we obtain $B_1 \subseteq l_A(Y)$. On the other hand, if it were $x \in B_2 \cap l_A(Y)$, for some vertex $x \in V$, then, again by Proposition 4.2 and by definition of $l_A(Y)$, we would have that $B_2 = [x]_A \subseteq Y$. Since $B_1|B_2$ is a set-partition of V, the last inclusion implies that $Y = V$, which is contrary to our hypothesis. Hence $B_1 \subseteq l_A(Y)$ and $B_2 \cap l_A(Y) = \emptyset$, and since $B_1|B_2$ is a set-partition of V we deduce that $l_A(Y) = B_1$ if $B_1 \subseteq Y$. A similar reasoning also shows that if $B_2 \subseteq Y$ then $l_A(Y) = B_2$. Finally, let $B_1 \not\subseteq Y$ and $B_2 \not\subseteq Y$. Since each vertex $x \in V$ is such that $x \in B_1$ or $x \in B_2$, by Proposition 4.2 we have respectively $[x]_A = B_1 \not\subseteq Y$ and $[x]_A = B_2 \not\subseteq Y$, that is $x \notin l_A(Y)$ for each vertex $x \in V$, hence $l_A(Y) = \emptyset$.

(ii) Let $Y \subseteq B_1$. If $x \in B_1$, by Proposition 4.2 follows that $[x]_A = B_1 \cap Y \neq \emptyset$ because Y is non-empty subset of B_1. Hence $x \in \mathbf{u}_A(Y)$. On the other hand, if $x \in \mathbf{u}_A(Y)$ by definition of $\mathbf{u}_A(Y)$ we have $[x]_a \cap Y \neq \emptyset$. Let $y \in [x]_A \cap Y$. Since $y \in Y \subseteq B_1$, by Proposition 4.2 we obtain $B_1 = [y]_A = [x]_A$, therefore, again by Proposition 4.2 we deduce that $x \in B_1$. Hence $\mathbf{u}_A(Y) = B_1$. The case $Y \subseteq B_2$ is exactly similar. Finally, let $Y \not\subseteq B_1$ and $Y \not\subseteq B_2$. Since $B_1|B_2$ is a set-partition of V, we deduce that $B_1 \cap Y \neq \emptyset$ and $B_2 \cap Y \neq \emptyset$. Now, if we take an arbitrary vertex $x \in V$, then $x \in B_1$ or $x \in B_2$. If $x \in B_1$, then, by Proposition 4.2 it follows that $[x]_A \cap Y = B_1 \cap Y \neq \emptyset$, therefore $x \in \mathbf{u}_A(Y)$. Analogously if $x \in B_2$. This shows that $V \subseteq \mathbf{u}_A(Y)$, that is $V = \mathbf{u}_A(Y)$.

(iii) It follows at once by Proposition 2.1 (ii) and by Proposition 4.2. □

4.3 Other Operations in RST

Let us consider the sets introduced in Subsect. 2.4. The vertex (A) corresponds to the lower approximation and the corner (I) to the upper one [2]. Then, (E) is the negation of the upper approximation, called the *exterior region* **e** and it represents the objects (vertices in our case) surely not belonging to the set under approximation. In the graph case, a vertex x belongs to $\mathbf{e}(O)$ if there is no vertex in O sharing all the connections with x. As a simple corollary of the results on the upper approximation we get the following.

Corollary 4.3. *Let* $G = (V(G), E(G))$ *be a simple undirected graph and let* $\mathcal{I}[G]$ *be the Boolean information system associated to* G. *Let* A *and* Y *be two subsets of* $V(G)$. *Then:*

(i) $\mathbf{e}_A(Y) = \{v \in V(G) : \nexists u \in V(G) : N_G(u) \cap A = N_G(v) \cap A)\}$.
(ii) If G *is complete, then*

$$\mathbf{e}_A(Y) = \begin{cases} Y^c & \text{if } Y \subseteq A \\ A \cap Y^c & \text{otherwise} \end{cases}$$

(iii) If the graph is bipartite, i.e., $G = K_{p,q} = (B_1|B_2)$, *then*

$$\mathbf{e}_A(Y) = \begin{cases} B_2 & \text{if } Y \subseteq B_1 \\ B_1 & \text{if } Y \subseteq B_2 \\ \emptyset & \text{otherwise .} \end{cases}$$

The corner (a) is named in RST a *sufficiency* operator and (i) is the dual sufficiency. In case of R being an equivalence relation, the sufficiency operator is trivial since it gives either the emptyset or the set O under approximation. Similarly, the dual sufficiency either results in the complement of O or in the universe. Both the operators become more interesting in a generalized setting, for instance when R is a similarity instead of an equivalence relation. However, this generalized situation is out scope of the present work.

Let us stress once more that, in the particular case of formal context induced by graphs, objects coincide with attributes and the relation $\mathcal{R}$ is defined on the same set, as in case of rough-set indiscernibility relation. Hence, FCA tools can be compared and/or combined with RST ones. For instance, from the fact that both $''$ and $\mathbf{u}$ are closure operators on objects, we have that $O \subseteq O''$ and $O \subseteq \mathbf{u}(O)$. So, we can wonder which is the relationship among the two mappings O'' and $\mathbf{u}(O)$. In case of complete bipartite graphs we have that $O'' = \mathbf{u}(O)$ (and also $O' = \mathbf{e}(O)$), as can be seen by Propositions 3.3 and 4.5. Also in case of complete graphs and $A = V(G)$ we have $O'' = \mathbf{u}(O) = O$ (by Propositions 3.2 and 4.4). However, in the general case, nothing can be said as it is shown by the following example.

Example 4.3. Let us consider the following (bipartite and not complete) graph:

If we set $O = \{v_1, v_3\}$, then we get $O'' = \{v_1, v_3, v_4\}$ and $\mathbf{u}(O) = O$. So, $\mathbf{u}(O) \subset O''$. On the other hand, considering the complete graph of Fig. 1 with $O = \{2, 3\}$ and $A = \{1, 2\}$, we have $\mathbf{u}(O) = \{2, 3, 4\}$ and $O'' = O$ leading to $O'' \subset \mathbf{u}(O)$.

5   Conclusion

We laid bare the possibility to investigate graphs using techniques from Formal Concept Analysis and Rough Set Theory. Several results exploring the corresponding on graphs of operators in the two theories have been given. The picture, however, is far from being complete. Indeed, a complete description of the structure of oppositions arising from FCA and RST in the case of graphs, as well as the interaction between FCA and RST operators is still missing. Moreover, as far as RST is concerned, we only explored the approximations defined by the standard indiscernibility relation. A natural extension would be to consider more general rough set models and to explore other concepts such as rough membership, attribute dependencies and reducts.

References

1. Alexe, G., Alexe, S., Crama, Y., Foldes, S., Hammer, P.L., Simeone, B.: Consensus algorithms for the generation of all maximal bicliques. Discrete Appl. Math. **145**, 11–21 (2004)
2. Ciucci, D., Dubois, D., Prade, H.: The structure of oppositions in rough set theory and formal concept analysis - toward a new bridge between the two settings. In: Beierle, C., Meghini, C. (eds.) FoIKS 2014. LNCS, vol. 8367, pp. 154–173. Springer, Heidelberg (2014)
3. Dawande, M., Keskinocak, P., Swaminathan, J.M., Tayur, S.: On bipartite and multipartite clique problems. J. Algorithms **41**, 388–403 (2001)

4. Dubois, D., Dupin de Saint Cyr, F., Prade, H.: A possibility-theoretic view of formal concept analysis. Fundamenta Informaticae **75**, 195–213 (2007)
5. Dubois, D., Prade, H.: From Blanché's hexagonal organization of concepts to formal concept analysis and possibility theory. Log. Univers. **6**, 149–169 (2012)
6. Ganter, B., Wille, R.: Formal Concept Analysis: Mathematical Foundations. Springer, Heidelberg (1999)
7. Gaume, B., Navarro, E., Prade, H.: A parallel between extended formal concept analysis and bipartite graphs analysis. In: Hüllermeier, E., Kruse, R., Hoffmann, F. (eds.) IPMU 2010. LNCS, vol. 6178, pp. 270–280. Springer, Heidelberg (2010)
8. Kuznetsov, S.O., Obiedkov, S.A.: Comparing performance of algorithms for generating concept lattices. J. Exp. Theor. Artif. Intell. **14**, 189–216 (2002)
9. Li, J., Liu, G., Li, H., Wong, L.: Maximal biclique subgraphs and closed pattern pairs of the adjacency matrix: a one-to-one correspondence and mining algorithms. IEEE Trans. Knowl. Data Eng. **19**, 1625–1637 (2007)
10. Pawlak, Z.: Rough Sets. Theoretical Aspects of Reasoning about Data. Kluwer Academic Publisher, The Netherlands (1991)

A Conceptual Approach for Relational IR: Application to Legal Collections

Nada Mimouni[✉], Adeline Nazarenko, and Sylvie Salotti

LIPN (UMR 7030), Université Paris 13 - Sorbonne Paris Cité and CNRS,
93430 Villetaneuse, France
{Nada.Mimouni,Adeline.Nazarenko,Sylvie.Salotti}@lipn.univ-paris13.fr

Abstract. Everyone who works in the legal field is faced with the complexity of documentary sources of law, that are highly interrelated and interdependent of each others. It is essential for legal practitioners to rely on systems that retrieve all the sources related to the legal cases they are working on and not only the most relevant ones. The challenge for legal IR is to achieve exhaustivity and handle this complexity by retrieving documents on the basis of the semantic content and the intertextual relationships. This work proposes an IR approach for legal sources that goes beyond existing systems. It is based on Formal and Relational Concept Analysis to structure, query and browse collections of legal documents.

Keywords: Legal information retrieval · Intertextuality · Relational queries · Exploration · Formal concept analysis · Relational concept analysis

1 Introduction

This paper presents an application of Formal and Relational Concept Analysis (FCA/RCA) to relational Information Retrieval (IR) in order for IR systems to take in account, not only the semantic content of documents, but also their intertextual links. Actually, documents often need to be enhanced with contextual information and IR systems must return to users not only lists of documents but lists of graphs of interlinked documents.

This need is especially challenging in the legal domain where law sources refer to each other by various types of relations (*e.g.* amendment, transposition, jurisprudence), thus forming a large network of highly interlinked documents. At any time, legal experts and citizens consult and query the legal collections. A legal case cannot be analyzed without taking into account all the relevant legal sources and their interrelationships. This appears in legal experts' queries and the way they explore the collections of legal sources navigating through the links of retrieved documents. They call for IR systems able to answer "relational queries", such as *"Which legislative texts cite the article 4 of the Labour Code?"*, *"Which are consolidated versions of the data protection act?"*, *"what are data protection act amendements?"* or *"Is there any local decree developing the law about tax exemption for low energy consumption housings?"*.

© Springer International Publishing Switzerland 2015
J. Baixeries et al. (Eds.): ICFCA 2015, LNAI 9113, pp. 303–318, 2015.
DOI: 10.1007/978-3-319-19545-2_19

These needs are not met by existing legal IR systems such as Legifrance[1] or UKLegislation[2]. Those systems traditionally rely on a logical approach rather than on a vectorial one because exhaustivity matters more than result ranking. However, they do not allow for querying on intertextual criteria. Besides, the hypertext navigation facilities show their limits on large collections, as users easily get lost.

This paper tackles the problem of semantic search in a collection of linked documents. The goal is to handle a new category of queries (hereafter "relational queries") to retrieve documents using both content and cross-references as search criteria. We propose an approach based on Formal and Relational Concept Analysis (FCA [8] and RCA [10]) to model a collection of documents and answer various types of relational queries.

The overview of the proposed search approach is presented first (Sect. 2). Section 3 shows how the collections are modeled while Sects. 4 and 5 present our querying and exploration methods. This approach is compared to related works in Sect. 6 and the conclusion is given in Sect. 7.

2   Overview of the Relational IR Approach

The proposed approach is composed of four main steps (see Fig. 1):

- The document collection is first structured into formal concept lattices based on the semantic descriptors associated with the documents (Semantic content modeling).
- Those lattices are then enriched with intertextuality, *i.e.* relational information, which produces a relational lattice family (Intertextuality modeling).
- Users' queries, possibly combining semantic descriptors and cross-references constraints, are matched against the relational lattice family, which gives graphs of documents as answers (Direct querying).
- The lattice structure can be further exploited to retrieve approximate results as an alternative or in addition to direct ones (Browsing).

3   Modeling Semantic and Intertextual Features

3.1   Legal Collection as an Attributed Graph

This work is part of the Légilocal project, a French project aiming at offering easy access to local legal data for citizens and legal experts. The Légilocal collection is composed of various types of documents (legislative texts, administrative acts and editorial documents) linked to each other by various types of links.

We assume here that any document i has a unique type j, that the documents are semantically annotated (they are associated with semantic descriptors)

---

[1] www.legifrance.gouv.fr/.

[2] www.legislation.gov.uk/.

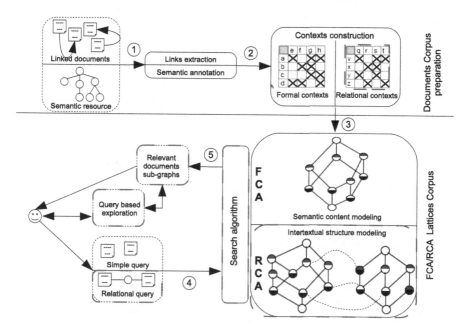

Fig. 1. Overview of the relational IR approach

and the citations have been typed and formalized as labelled and oriented links between documents. In the following, $Type(d_i, t_j)$, $Att(d_i, s_j)$ and $Rel(d_i, r_l, d_{i'})$ respectively indicate that the document d_i is of type t_j, is associated with the descriptor s_j and is the source of a r_l link, which target is $d_{i'}$.

A document collection C is modeled as an oriented, labeled and attributed graph $C = \mathcal{G}(D, R, A)$ where (i) the nodes are documents of D, (ii) the documents are associated with attributes, types of T and semantic descriptors of S ($A = T \cup S$) and (iii) the edges are binary oriented and labeled links, which types belong to R. There is no constraint on the number of nodes, attributes and links in that graph nor on the combination of attributes and links for a given document unit.

Such a collection graph is described by a formula of the following language (see Fig. 2 for an example):

$graph_{coll} \rightarrow pred_c$ ['$\wedge$ ' $pred_c$]*
$pred_c \rightarrow$ 'Type' '('id_{doc}',' id_{type}')' | 'Att' '('id_{doc}',' id_{sem}')' | 'Rel' '('id_{doc}',' id_{rel}',' id_{doc}')'
$id_{doc} \rightarrow$ 'd$_1$' | 'd$_2$' | ... $id_{type} \rightarrow$ 't$_1$' | 't$_2$' | ...
$id_{sem} \rightarrow$'s$_1$' | 's$_2$' | ... $id_{rel} \rightarrow$ 'r$_1$' | 'r$_2$' | ...
where ($\forall$ i, j, k, l) (d$_i \in$ D, s$_j \in$ S, t$_k \in$ T and r$_l \in$ R).

A subset ($Coll_L$) of the Légilocal collection is used for illustration in the following. For sake of readability, it is kept to a minimum. It is composed of two types of documents (municipal orders and legislation) and a single link type (visa legislation). See Table 1 for a detailed description.

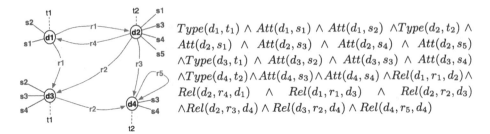

$$Type(d_1, t_1) \wedge Att(d_1, s_1) \wedge Att(d_1, s_2) \wedge Type(d_2, t_2) \wedge$$
$$Att(d_2, s_1) \wedge Att(d_2, s_3) \wedge Att(d_2, s_4) \wedge Att(d_2, s_5)$$
$$\wedge Type(d_3, t_1) \wedge Att(d_3, s_2) \wedge Att(d_3, s_3) \wedge Att(d_3, s_4)$$
$$\wedge Type(d_4, t_2) \wedge Att(d_4, s_3) \wedge Att(d_4, s_4) \wedge Rel(d_1, r_1, d_2) \wedge$$
$$Rel(d_2, r_4, d_1) \quad \wedge \quad Rel(d_1, r_1, d_3) \quad \wedge \quad Rel(d_2, r_2, d_3)$$
$$\wedge Rel(d_2, r_3, d_4) \wedge Rel(d_3, r_2, d_4) \wedge Rel(d_4, r_5, d_4)$$

Fig. 2. Example of graph modeling a collection composed of 4 documents.

Table 1. Vocabulary used to describe the Légilocal example collection

Types	Descriptions		
OrderMun	Municipal order (*arrêté municipal*)		
Legislation	Legislation (*texte législatif*) : code article, law, decree		

Relations	Descriptions		
vl	Visa législation : a municipal order having a visa to a legislative text		

Descriptors	Corresponding terms	Descriptors	Corresponding terms
VTM	*Land motor vehicle*	Qd	*Quad*
4 × 4	*4-wheeled vehicle*	Mtr	*trial bikes*
RegC	*traffic regulation*	IntC	*driving ban*
PEN	*protection of natural areas*	EspV	*green spaces*
CheR	*rural road*	CheF	*forest road*
SigR	*road sign*	TraV	*quiet neighborhood*
PNat	*natural park*	TraP	*public tranquility*
PPM	*Mayor's police authority*	PolC	*traffic police*
PRM	*Mayor's regulatory authority*	PolM	*municipal police*

Identifiers	Referred documents
AC	Order 97-17 of Champigné
A07	Order 2007-31 of Villecresnes
A17	Order 2012-17 of Villecresnes
A48	Order 2012-48 of Villecresnes
A11	Order 2011-22 of Villecresnes
AA	Order of 04/07/97 of municipality of Ance
A94	Order of 24/05/94 of Magny-le-Feule
L362	Article L362-1 of Environmental Code
L21	Article L2122-28 of the General Code of Local Authorities
L1	Article L2213-1 of the General Code of Local Authorities
L2	Article L2213-2 of the General Code of Local Authorities
L4	Article L2213-4 of the General Code of Local Authorities
L12-1	Article L2212-1 of the General Code of Local Authorities
L91	Law 91-2 of January 3rd, 1991
D92	Decree 92-258 of March 20th, 1992
C05	OLIN Circular of September 6th, 2005
C1	Article 131-1 of Municipal Code

3.2 Structuring the Collection

FCA [8] and RCA [16] enable to structure the document collection into a family
of interrelated concept lattices. Structuring such a collection as a lattice accounts
to pre-computing the answers to all the satisfiable elementary queries. The relational lattice family structure extends this approach to relational queries.

Modeling Semantic Content. The semantic content of a collection is modeled
as a formal context, *i.e.* as a binary relation (*object* × *attributes*) between a set

of objects (the documents) and a set of attributes (the semantic descriptors associated to the documents). A formal concept lattice is then built from the formal context.

For legal collections, we build independent formal contexts and lattices for each type of documents, because the document typology plays an important role in legal drafting and reasoning and because it is easier to handle several small lattices than a single large one corresponding to a large document collection. This means that the document types (t1 and t2 on Fig. 2) are not encoded as document attributes as opposed to semantic descriptors (s1, s2, ...).

The formalization of the content of the first document type of $Coll_L$ is given by the formal context $\mathcal{K}_{ord} = (A, S, I)$, where A is the set of orders (*e.g.* AC, A07), S is a set of semantic descriptors of the domain (*e.g.* VTM, Mtr) as described in Table 1 and I is a binary relation between A and S. For a document $a \in A$ and a semantic descriptor $s \in S$, $(a, s) \in I$ holds if the document a is semantically annotated by the semantic descriptor s. Table 2 presents the formal contexts, $\mathcal{K}_{ord}$ and $\mathcal{K}_{leg}$, modeling the semantic content of $Coll_L$ (resp. for municipal orders and legislation).

Table 2. The formal contexts of orders ($\mathcal{K}_{ord}$) and legislation ($\mathcal{K}_{leg}$)

	VTM	Qd	4 × 4	Mtr	RegC	IntC	PEN	EspV	CheR	CheF	SigR	TarV
AC	×				×	×	×		×			
A07	×	×			×	×	×	×	×			
A17	×				×	×				×		
A48					×						×	
A11					×				×			×
AA	×				×	×						
A94	×		×	×	×	×			×			

	VTM	4 × 4	Qd	RegC	IntC	PEN	PNat	CheR	TraP	PolC	PPM	PRM	PolM
L362	×			×	×	×							
L21											×		
L1									×	×			
L2				×	×				×	×			
L4				×	×	×		×	×	×			
L12-1													×
L91	×	×		×	×	×	×	×					
D92	×			×		×							
C05	×		×	×		×							
C1													×

Two concept lattices ($\mathcal{L}(\mathcal{C})$ for orders and $\mathcal{L}(\mathcal{C}')$ for legislation) are derived from $\mathcal{K}_{ord}$ and $\mathcal{K}_{leg}$ respectively (Figs. 3 and 4)[3], thus structuring the collection in the form of concept hierarchies. The formal concepts represent classes of documents (extents) characterized or described by sets of descriptors (intents). For example, the concept 2 in $\mathcal{L}(\mathcal{C})$ represents the set of documents which share the descriptors CheR and RegC, *i.e.* documents AC, A11 and A94. The relation between the concepts 6 and 2 is interpreted as a relation of generalization/specialization between the classes represented by those concepts.

---

[3] Lattices are build using Galicia platform [17].

In an information retrieval perspective [4], the lattice built by FCA gathers all possible combinations of document attributes represented by the intents of concepts, each of which corresponding to a specific elementary query. We propose to extend that approach to relational queries using RCA.

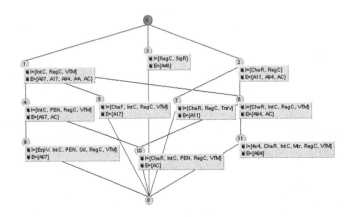

Fig. 3. The concept lattice $\mathcal{L}(\mathcal{C})$

Modeling Intertextual Relationships. RCA is a relational extension of FCA, that allows for extracting formal concepts from sets of data described by intrinsic and relational attributes. The data set is modeled as a relational context family (RCF). Formally, a RCF, $\mathcal{R}$, is a pair $(\mathbf{K}, \mathbf{R})$, where $\mathbf{K}$ is a set of contexts $\mathcal{K}_i = (O_i, A_i, I_i)$, $\mathbf{R}$ is a set of relations $r_k \subseteq O_i \times O_j$ where O_i and O_j are the object sets of the formal contexts $\mathcal{K}_i$ and $\mathcal{K}_j$.

The Légilocal RCF contains the binary contexts of Table 2 and an additional relational context representing the vl relation. Since that relation has the set of municipal orders and the set of legislative texts as domain and range, it is represented as a *municipal orders × legislative texts* relation (Table 3).

The instances of a relation are called links. The links are "scaled" in order to be included as binary attributes in the formal context of the domain of the relation. This mechanism is called *relational scaling*[4]. It builds a relational lattice family (RLF) from the RCF. Formally, given a relation $r \subseteq O_i \times O_j$ and a lattice $\mathcal{L}_j$ on $\mathcal{K}_j = ran(r)$, the **existential scaling** operator $sc_\times^{(r,\mathcal{L}_j)} : \mathbf{K} \to \mathbf{K}$ is defined as:

$$sc_\times^{(r,\mathcal{L}_j)}(\mathcal{K}_i) = (O_i^{(r,\mathcal{L}_j)}, A_i^{(r,\mathcal{L}_j)}, I_i^{(r,\mathcal{L}_j)})$$

where $O_i^{(r,\mathcal{L}_j)} = O_i$, $A_i^{(r,\mathcal{L}_j)} = A_i \cup \{r : c \mid c \in \mathcal{L}_j\}$, and
$I_i^{(r,\mathcal{L}_j)} = I_i \cup \{(o, r : c) \mid o \in O_i, c \in \mathcal{L}_j, r(o) \cap extent(c) \neq \emptyset\}$.

The lattice of municipal orders enriched by scaling on the vl relation, $\mathcal{L}_{vl}(\mathcal{C})$, is given by Fig. 5. The whole collection of documents is now represented as a RLF

---

[4] Various types of scaling can be defined but we rely only on the existential one here.

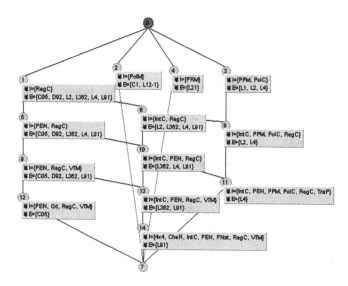

Fig. 4. The concept lattice $\mathcal{L}(\mathcal{C}')$

Table 3. Formal relational context of the VisasL relation

	L362	L21	L1	L2	L4	L12-1	L91	D92	C05	C1
AC			×		×		×	×		
A07	×				×		×	×	×	
A17		×		×						
A48		×		×						
A11						×				
AA				×						
A94										×

composed of the lattices $\mathcal{L}_{vl}(\mathcal{C})$ and $\mathcal{L}(\mathcal{C}')$ related by the vl relation. One can observe that the intents of $\mathcal{L}_{vl}(\mathcal{C})$ have been enriched with relational attributes. For instance, the intent of the concept 14 in $\mathcal{L}_{vl}(\mathcal{C})$ combines two semantic descriptors {CheR, RegC} with two relational attributes $\{vl : c0, vl : c2\}$. The presence of the attribute $vl : c2$ means that each document in the extent $\{A11, A94\}$ has at least one vl relation with a document belonging to the extent of the concept 2 in $\mathcal{L}(\mathcal{C}')$.

4   Querying the Conceptual Relational Structures

4.1   Querying as Graph Matching

Modeling document collections as graphs leads to consider information retrieval as graph querying where queries are themselves formalized as graphs.

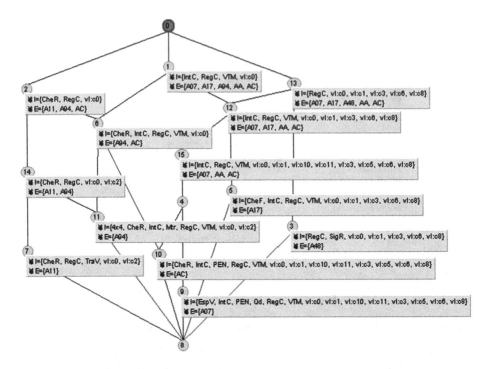

Fig. 5. Lattice $\mathcal{L}_{vl}(\mathcal{C})$ given by the scaling of $\mathcal{L}(\mathcal{C})$ on the vl relation.

A query graph is similar to that of a collection but it possibly contains (i) variables in place of document identifiers, (ii) inequality constraints on these variables, (iii) a focus for restricting the answer to a subset of the query variables. A query graph is a formula of the form [*focus:*] *graph*$_q$ [with *constraint*]. Elementary queries are described without any relational predicate as opposed to relational ones (Fig. 6).

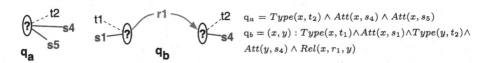

Fig. 6. Examples of query graphs: q_a represents an elementary query and q_b a relational query, which is represented with a focus (x, y) in the formula.

Matching queries and documents amounts to instantiate the query graph onto the graph of the collection. An answer is (i) a set containing all the subgraphs of the collection graph that instantiate the query graph if it has no explicit focus, (ii) a set containing all the tuples of identifiers instantiating the query focus if there is one, (iii) an empty set if the query graph cannot be instantiated.

4.2   Answering Elementary Queries

An elementary query is represented as a query concept $\mathcal{Q}_s$ which extent is a virtual query object, Q_{vo}, that represents the target document of the query, and which intent Q_I is the set of query attributes a_i. Answering that query from the lattice $\mathcal{L}$ based on $\mathcal{K} = (O, A, Inc)$ amounts to insert the concept $\mathcal{Q}_s$ in the lattice using an incremental algorithm [9] and to locate in $\mathcal{L}$ the lowest concept with Q_{vo} in its extent (by design, its intent includes Q_I). The query is satisfiable if that concept contains documents besides Q_{vo}, those documents being answers to the initial query.

4.3   Answering Relational Queries

For relational queries, the matching process is more complex. It is divided into four steps. The graph query is first decomposed into an ordered list of elementary queries that are answered in turn until the last query node is matched against the relational lattice, which gives the set of documents satisfying that node if there is any. If the expected answer is a set of graphs of documents, the answer graphs are then reconstructed using the family of relational contexts. Finally, the graph constraints are applied, which filters out the list of answer graphs.

Query Decomposition. A query graph is first transformed as a query tree, $Q_T = < R, Q_{T1}, Q_{T2}, ..., Q_{Tn} >$, where R is the root node and the Q_{Ti} are the subtrees rooted in the nodes that are domain or range of a relation having R respectively as range or domain[5]. If there is a cycle in the query graph, it is broken: two distinct child nodes are created and an equality constraint between them is added to the query. Any graph can be represented as a tree associated to few equality constraints, since any oriented link can be substituted by its inverse relation.

The query tree is then transformed into a list of elementary queries corresponding to the various tree nodes, which are listed in reverse breadth-first order, *e.g.* with the root node at the back (Fig. 7).

Root Concept Localisation. Once a query graph is transformed into a sequence of elementary queries, those elementary queries are matched in turn against the relational lattice family (see Fig. 8). The node type of each query determines the document lattice on which it must be matched. At the end of this evaluation process, the query corresponding to the root node is matched against the collection lattice family (Step 4 on Fig. 8). If that root query is satisfiable then the initial relational query is also satisfiable. The documents answering the root query are returned as a set of elementary answers to the initial query.

---

[5] If the query graph has a focus, the node corresponding to the first focus variable is chosen as root. Otherwise the root node is chosen arbitrarly.

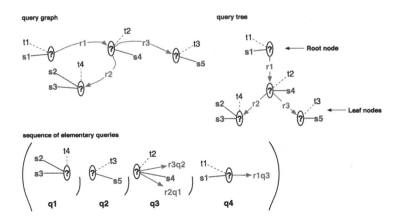

Fig. 7. Example of a graph query transformed as a sequence of elementary queries

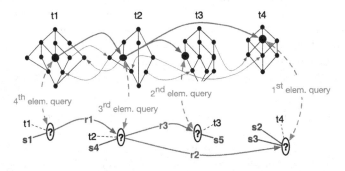

Fig. 8. Iterative query matching process

Answer Completion. Inserting the sequence of elementary queries in the lattice family prouves that the initial graph query is satisfiable and enables to locate a list of formal concepts that play a role in the identification of the elementary answers (set of documents) but it doesn't give the actual graphs of documents that answer the relational query.

Actually, with the existential scaling, a relational attribute $r_x{:}C_y$ in a concept C intent indicates that any document belonging to the extent of C is related to at least one document of C_y by the relation r_x. However, identifying which documents in C_y are actually related to which documents in C requires to go back to the formal and relational contexts underlying the relational lattice family.

The exploration of the formal and relational context family is done in reverse order of localisation. The completion process starts with the documents of the root node (elementary answers). Each of these answers is completed by identifying in the corresponding relational context to which documents the root document is actually related. Step by step, the elementary answers, which are documents, are extended to graphs.

Constraint Checking. The final step simply consists in erasing from the list of answers the documents, graphs of documents or tuples of documents, that violate any query constraint, be it a constraint of the initial query or a constraint generated by the graph to tree transformation.

4.4   Example

Let's consider an elementary query related to the Légilocal sub-collection: "*Which are legislative texts* (Legislation) *concerning traffic regulation* (RegC) *of land motor vehicles* (VTM) *on rural roads* (CheR)?*". To answer that query, a virtual object Q_{leg} described by the attributes RegC, VTM, CheR is created. It is inserted in $\mathcal{L}(\mathcal{C}')$ (Fig. 9). The most specific concept containing Q_{leg} in its extent is the concept 12. The answer is given by the documents associated with Q_{leg} in the extent of the concept 12. There is a single answer: the document L91.

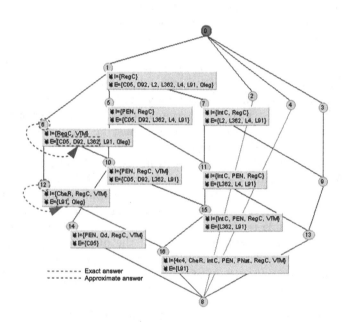

Fig. 9. Simple query on Legislation lattice (Color figure online)

A second example is given by the relational query : "*Which municipal orders* (OrderMun) *concerning rural roads* (CheR) *make a visa* (vl) *to legislative texts* (Legislation) *about traffic regulation* (RegC) *of land motor vehicles* (VTM)?*". The query must be matched against the relational lattice family composed of the lattices of municipal orders and legislation, enriched with the vl relation, *i.e.* the interrelated lattices $\mathcal{L}_{vl}(\mathcal{C})$ and $\mathcal{L}(\mathcal{C}')$ (Fig. 10).

To answer the query, a virtual object Q_{leg} representing the query object of a first elementary query is created with the intent {RegC, VTM}. Its document

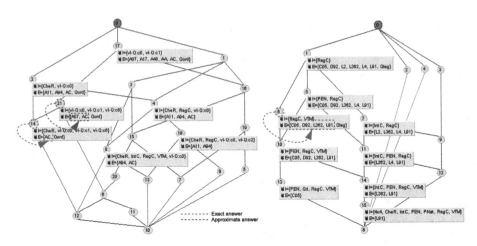

Fig. 10. Relational lattice family composed of the enriched lattice of orders ($\mathcal{L}_{vl}(\mathcal{C})$) and the lattice of legislative texts ($\mathcal{L}(\mathcal{C}')$) (Color figure online)

type being a legislative text, Q_{leg} is inserted in the lattice $\mathcal{L}(\mathcal{C}')$, which gives the concept 6 as answer. In a second step, Q_{ord} is created with the semantic attribute CheR and the relational attribute vl:c6. It is inserted in the lattice of municipal orders enriched with the relation vl: the localized concept is $c14$.

Since the query has no focus, the answer must be completed into a list of graphs, composed of documents belonging to the extents of $c14$ and $c6$ and related by vl. The concept $c14$ contains in its extent the object Q_{ord} and the document AA which share the semantic descriptor CheR and the relation vl to the concept $c6$ of the other lattice ($\mathcal{L}(\mathcal{C}')$). $c6$ contains in its extent Q_{leg} and the documents {C05, D92, L362, L91} which share the semantic descriptors {RegC, VTM}. According to the relational context, two exact answers are given: $\mathcal{G}_1 = $ AA $\rightarrow_{vl}$ D92 and $\mathcal{G}_2 = $ AA $\rightarrow_{vl}$ L91. If the query had a single variable focus, on the order lattice for example, the answer would simply be the set of documents contained in the extent of the concept $c14$, *i.e.* the document AA.

5   Browsing the Conceptual Relational Structures

As explained above, structuring the collection has an advantage: it accounts to pre-computing the answers to all the satisfiable queries, either elementary or relational ones. It can also be exploited to find approximate answers to users' queries, relying on the strategy of [12,19].

5.1   Exploring Facilities

If the user gets too few answers to a query, he/she can broaden his/her search, explore the lattice family and get approximate answers, *i.e.* sets of (graphs of) documents with almost all but not all the searched properties. Reversely, in case of numerous exact answers, the user can introduce additional constraints to narrow the search. In most cases, this query generalization and specialization process does not require any additional computation, as the relevant information is already structured in the relational lattice family.

Exploring variations around an elementary query consists in navigating in the target lattice around the formal concept associated to the initial query. Going upward (resp. downward) allows for broadening (resp. restricting) the initial query, and thus for returning more (resp. less) numerous answers. The lattice structure shows which relaxed or restricted queries are satisfiable and how many documents could be returned in each case. On the 1st diagram of Fig. 11, the query node can be relaxed in its two parent nodes. Reversely, the child nodes correspond to query graph restrictions and subclasses of results.

Exploring a relational lattice family around a relational query is more complex. Some variants can be answered without additional computation:

- If the variant simply consists in suppressing or adding a semantic descriptor to the intent of the root node, it goes back to the previous elementary case (2nd diagram on Fig. 11).
- Suppressing a relational attribute of the root node is also a form of query generalization. This cuts one of the graph branches. The 3rd diagram of Fig. 11 shows in grey the erased part of the graph.
- A complex query can also be restricted by adding a new relation which has a node of the answer graph as domain or co-domain (4th diagram of Fig. 11).

The other cases require additional computation for locating a new query root concept in the relational lattice family.

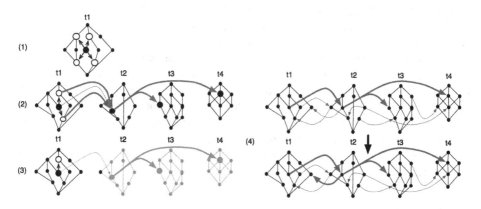

Fig. 11. Query broadening and restricting

5.2   Example

Let's consider the example of Sect. 4.4: *Which are legislative texts (*Legislation*) concerning traffic regulation (*RegC*) of land motor vehicles (*VTM*) on rural roads*

(CheR)?". The answer of the elementary query described by the attributes RegC, VTM, CheR is given by the concept 6 on the lattice of Legislative texts (Fig. 9) but it is possible to relax a semantic constraint. Removing the semantic attribute CheR gives a more general answer containing three documents: {L362, D92, C05} (blue highlighted elements).

The relational lattice family can similarly be explored (Fig. 10). Starting with the relational query "*Which municipal orders (OrderMun) concerning rural roads (CheR) make a visa (vl) to legislative texts (Legislation) about traffic regulation (RegC) of land motor vehicles (VTM)?*", one can relax a semantic or a relational constraint on the lattice of orders. For instance, removing the descriptor CheR gives a new order (A07) that shares relational attributes with AC (blue highlighted elements). A relaxed answer-graph $\mathcal{G}_r$ is given by the document A07 linked to documents {C05, D92, L362, L91}: $\mathcal{G}_{r1}$ = A07 $\rightarrow_{vl}$ C05, $\mathcal{G}_{r2}$ = A07 $\rightarrow_{vl}$ D92, $\mathcal{G}_{r3}$ = A07 $\rightarrow_{vl}$ L362 and $\mathcal{G}_{r4}$ = A07 $\rightarrow_{vl}$ L91.

6   Related Work

The diversity and multiplicity of links between legal documents is considered as the major factor of complexity in this domain [2]. Many works have studied this complexity by analyzing the topology of citation networks of legal collections [7,15,18]. The detailed analysis of the semantics of intertextual links and their use for search purposes has deserved less attention.

General IR approaches focus on the semantic content of documents and return a list of independent documents without considering their context. Graph-based document analysis is used to improve the ranking of retrieved documents (*e.g.* PageRank [3]) but not for retrieving documents.

Lattice theory and particularly FCA has been used as a base for IR models [14]. It has been applied to documentary search [4,5] where objects and attributes of the formal contexts correspond to documents and their keywords. Several studies have investigated the contribution of FCA for retrieval by navigation and browsing data collections and proved its usefulness [6,11,19]. Based on a logical model, it is well suited for search in the legal domain. Relational search based on RCA was firstly introduced in [1] where authors work on the problem of selecting suitable webservices for instantiating an abstract calculation workflow. The use of RCA to handle the multi-relational structure of a given collection of documents was introduced in [13].

7   Conclusion

We have presented a conceptual IR approach able to handle simple and relational queries, which is critical for legal practitioners. It is based on formal and relational concept analysis and it structures a collection of documents into a hierarchy of document classes. Each class is characterized by content descriptors and relations to other classes. The conceptual relational structure allows to represent in a unified way the intrinsic and relational properties of documents.

An advantage of the resulting hierarchical structure is the possibility to explore the collection without extra calculations. It allows for approximate answers if too or few documents are returned and it indicates the number of documents that can be returned by each relaxed or generalized query. This process of query reformulation is essential for logical IR systems that do not return ordered lists of results.

One limitation of the proposed approach is the computational cost for large datasets. However, in real applications, it has been estimated that the maximal complexity is never reached and many solutions have been proposed for complexity reduction such as iceberg lattices which reduce the lattices size by limiting the exploration depth of the concept set. The legal domain characteristic which induces cutting up the set of documents into small collections should help to cope with the complexity issue. We are also considering building the lattice on a subset or results returned by a traditional query.

Acknowledgments. This work has been partially funded by the Légilocal project (French Single Inter-Ministry Fund, FUI-9, 2010–2013) and is supported by Labex EFL (French National Research Agency, ANR-10-LABX-0083).

References

1. Azmeh, Z., Rouane, M-H., Huchard, M., Napoli, A., Valtchev, P.: Querying relational concept lattices. In: Proceedings of the 8th International Conference on Concept Lattices and their Applications (CLA 2011), Nancy, France, pp. 377–392, October 2011

2. Bourcier, D.: Sciences juridiques et complexitë. Un nouveau modéle d'analyse. Droit et Cultures **61**(1), 37–53 (2011)

3. Brin, S., Page, L.: The anatomy of a large-scale hypertextual web search engine. In: Proceedings of the Seventh International Conference on World Wide Web (WWW7), pp. 107–117 (1998)

4. Carpineto, C., Romano, G.: Using concept lattices for text retrieval and mining. In: Ganter, B., Stumme, G., Wille, R. (eds.) Formal Concept Analysis. LNCS (LNAI), vol. 3626, pp. 161–179. Springer, Heidelberg (2005)

5. Codocedo, V., Lykourentzou, I., Napoli, A.: A semantic approach to concept lattice-based information retrieval. Ann. Math. Artif. Intell. **72**(1–2), 169–195 (2014)

6. Ferré, S.: Camelis: a logical information system to organise and browse a collection of documents. Int. J. Gen. Syst. **38**(4), 379–403 (2009)

7. Fowler, J.H., Johnson, T.R., Spriggs, J.F., Jeon, S., Wahlbeck, P.J.: Network analysis and the law: measuring the legal importance of precedents at the u.s. supreme court. Polit. Anal. **15**, 324–346 (2007)

8. Ganter, B., Stumme, G., Wille, R. (eds.): Formal Concept Analysis, Foundations and Applications. LNCS (LNAI), vol. 3626. Springer, Heidelberg (2005)

9. Godin, R., Missaoui, R., Alaoui, H.: Incremental concept formation algorithms based on galois (concept) lattices. Comput. Intell. **11**, 246–267 (1995)

10. Huchard, M., Hacene, M.R., Roume, C., Valtchev, P.: Relational concept discovery in structured datasets. Ann. Math. Artif. Intell. **49**, 39–76 (2007)

11. Kotters, J.: Object configuration browsing in relational databases. In: Jäschke, R. (ed.) ICFCA 2011. LNCS, vol. 6628, pp. 151–166. Springer, Heidelberg (2011)

12. Messai, N., Devignes, M.-D., Napoli, A., Smaïl-Tabbone, M.: Querying a bioinformatic data sources registry with concept lattices. In: Dau, F., Mugnier, M.-L., Stumme, G. (eds.) ICCS 2005. LNCS (LNAI), vol. 3596, pp. 323–336. Springer, Heidelberg (2005)

13. Mimouni, N., Fernàndez, M., Nazarenko, A., Bourcier, D., Salotti, S.: A relational approach for information retrieval on XML legal sources. In: International Conference on Artificial Intelligence and Law, (ICAIL), Rome, Italy, pp. 212–216 (2013)

14. Priss, U.: Faceted knowledge representation. Electron. Trans. Artif. Intell. 4(C), 21–33 (2000)

15. Romain, B., Mazzega, P., Bourcier, D.: A network approach to the french system of legal codes- part i: analysis of a dense network. J. Artif. Intell. Law 19, 333–355 (2011)

16. Rouane, M.H., Huchard, M., Napoli, A., Valtchev, P.: Relational concept analysis: mining concept lattices from multi-relational data. Ann. Math. Artif. Intell. 67(1), 81–108 (2013)

17. Valtchev, P., Grosser, D., Roume, C., Hacene, M.R.: Galicia: an open platform for lattices. In: Using Conceptual Structures (ICCS 2003), pp. 241–254. Shaker Verlag (2003)

18. Winkels, R., de Ruyter, J.: Survival of the fittest: network analysis of dutch supreme court cases. In: Palmirani, M., Pagallo, U., Casanovas, P., Sartor, G. (eds.) AICOL-III 2011. LNCS, vol. 7639, pp. 106–115. Springer, Heidelberg (2012)

19. Wray, T., Eklund, P.W.: Exploring the information space of cultural collections using formal concept analysis. In: Jäschke, R. (ed.) ICFCA 2011. LNCS, vol. 6628, pp. 251–266. Springer, Heidelberg (2011)

Author Index

Adarichev, Vyacheslav   39
Adaricheva, Kira   39
Alibek, Kenneth   39
Amanbekkyzy, Adina   39
Antoni, Lubomir   183

Balcázar, José L.   3
Borchmann, Daniel   219
Brucker, François   169
Buzmakov, Aleksey   153, 200

Cabrera, I.P.   114
Chiaselotti, Giampiero   287
Ciucci, Davide   287
Codocedo, Victor   61
Cordero, P.   114

Dubois, Didier   21

Ferré, Sébastien   271

García-Pardo, F.   114
Gentile, Tommaso   287

Konecny, Jan   81
Krajči, Stanislav   183
Krídlo, Ondrej   183
Kuznetsov, Sergei O.   200

Leeuwenberg, Artuur   153
Lumpe, Lars   145

Melo Mora, Luis Felipe   236
Mimouni, Nada   303

Napoli, Amedeo   61, 153, 200
Nation, J.B.   39
Nazarenko, Adeline   303

Ojeda-Aciego, M.   114
Okimoto, Gordon   39

Peláez-Moreno, Carmen   97
Prade, Henri   21
Préa, Pascal   169

Rudolph, Sebastian   252

Săcărea, Christian   252
Sailanbayev, Alibek   39
Salotti, Sylvie   303
Sarkar, Shuchismita   39
Schmidt, Stefan E.   145
Seidalin, Nazar   39
Soldano, Henry   128

Toussaint, Yannick   153, 236
Troancă, Diana   252

Valverde-Albacete, Francisco J.   97

Printed in the United States
By Bookmasters